New Trends in Probability and Statistics – Volume 3

Multivariate Statistics and Matrices in Statistics

NEW TRENDS IN PROBABILITY AND STATISTICS Volume 3

Multivariate Statistics and Matrices in Statistics

Proceedings of the 5th Tartu Conference,
Tartu – Pühajärve, Estonia, 23-28 May 1994

Editors
E.-M. Tiit, T. Kollo and H. Niemi

TEV
VILNIUS, LITHUANIA

///VSP///
UTRECHT, THE NETHERLANDS
TOKYO, JAPAN

VSP BV
P.O. Box 346
3700 AH Zeist
The Netherlands

TEV Ltd.
Akademijos 4
Vilnius
Lithuania

First published in 1995
ISBN 90-6764-195-2 (VSP)
ISBN 9986-546-03-6 (TEV)

Typeset by TEV Ltd., Vilnius
Printed in Lithuania by Spindulys, Kaunas

CONTENTS

III. TESTS IN MULTIVARIATE STATISTICS

IV. MULTIVARIATE NONPARAMETRIC MODELS AND THEIR APPLICATIONS

V. DISCRIMINATION AND CLASSIFICATION

VI. MATRICES IN STATISTICS

PREFACE

In Tartu, an old university city in Estonia, the tradition of organizing international conferences in multivariate statistics and its applications began in the early seventies. The working language of the conferences was Russian because the majority of the participants came from different regions of the Soviet Union. After Estonia regained independence, the conference tradition continued.

In 1994, the same organizing committee, the academic staff of the Department of Mathematical Statistics, tried to organize a truly international conference on the same topic – multivariate statistics. The organizers expected this conference to be a meeting point for scientists from the West and East, but, unfortunately, only a few guests from the East participated in the conference. In any case, the conference turned out to be a success. About 70 participants from 18 countries travelled to Tartu, where the conference was opened. The following days were spent in the picturesque village of Pühajärve. The atmosphere of the conference was "friendly and stimulating", quoting our most distinguished guest Professor C. R. Rao. After the official conference, a one-day satellite workshop followed on the matrices in statistics, which was organized by Professor G. Styan.

The majority of participants were interested in publishing the proceedings of the conference. Most of the papers were sent to the Editorial Board before the intended deadline. A selection of papers was made based upon referee reports and revisions. This collection of papers reflects the volume and range of the problems considered in the conference.

The papers are clusterized into 6 parts, which are close to, but not identical with, the sections of the conference. The clusterization process has been rather statistical, where the clusters might overlap. According to some traditions in ordering the clusters and the papers in clusters (in the multivariate space) as the first characteristic the scale 'theoretical – practical' has been used, where the ordering number is stochastically increasing with the practical use of the paper. Using the rules of democracy and equality of all our guests, we did not separate the papers of the invited speakers in to special chapters, but improved the existing chapters by including the texts of invited speakers as introductions.

On behalf of the organizing committee of the conference, I would like to express our gratitude to the invited speakers, honourable professors C. R. Rao, K.-T. Fang, Y. Fujikoshi, I. Olkin and G. Styan, who determined the creative atmosphere of the conference.

The local organizing committee is very thankful to professors C. M. Cuadras, O. Frank, H. Neudecker, H. Niemi, Š. Raudys and G. Styan for their participation in the International Organizing Committee and their help during the whole organizing process of the conference.

Many thanks to all participants, especially to the authors of papers, those who gave software demonstrations and poster presentations, and also to the accompanying guests, who helped to create a friendly atmosphere during the conference.

I wish to express the deepest gratitude of the Editorial Board to all our referees who efficiently gave many valuable recommendations to our authors and helped to compile the proceedings.

I hope that this collection of papers preserves and passes onto the readers the stimulating and creative atmosphere of the 5th Tartu Conference and encourages them to participate in the 6th Tartu Conference on Multivariate Statistics.

E.-M. Tiit

Part One

ASYMPTOTIC DISTRIBUTIONS AND EXPANSIONS

Multivar. Statist., pp. 3–15
E.–M. Tiit *et al.* (Eds)

ERROR BOUNDS FOR ASYMPTOTIC EXPANSIONS OF THE DISTRIBUTIONS OF THE CLASSIFICATION STATISTIC *W* AND RELATED STATISTICS

YASUNORI FUJIKOSHI

Department of Mathematics, Hiroshima University,
Higashi-Hiroshima, 724, Japan

ABSTRACT

The classification statistic W is used to classify an observation as coming from one of two multivariate normal populations with common covariance matrix and different means when all these parameters are estimated from a sample from each of the populations. Let Δ be the Mahalanobis distance between the populations. Then it is known that the limiting distribution of $(W - \frac{1}{2}\Delta^2)/\Delta$ as the sample sizes approach infinity is the standard normal distribution if the observation is from the first population. Further, an asymptotic expansion for its distribution has been obtained. The purpose of this paper is to obtain explicit bounds for the errors in these approximations. Similar results are obtained for the distributions of the Studentized classification statistic and other related statistics.

1. INTRODUCTION

One of the important interests in discriminant analysis is to classify an observation as coming from one of two observations, based on a sample from each of the two populations. Suppose that we have a sample $x_1^{(1)}, \dots, x_{N_1}^{(1)}$ from Π_1: $N_p(\mu^{(1)}, \Sigma)$ and sample $x_1^{(2)}, \dots, x_{N_2}^{(2)}$ from Π_2: $N_p(\mu^{(2)}, \Sigma)$ where $\mu^{(1)} \neq \mu^{(2)}$, Σ is positive definite and all the parameters are unknown. Let x be a new observation which comes from one of Π_1 and Π_2. Then the observation x may be classified into Π_1 or Π_2 by using the classification statistic,

$$W = \left(\bar{x}^{(1)} - \bar{x}^{(2)}\right)' S^{-1} \left[x - \frac{1}{2}\left(\bar{x}^{(1)} + \bar{x}^{(2)}\right)\right], \tag{1.1}$$

where $\bar{x}^{(i)}$ are the sample mean vectors,

$$nS = \sum_{i=1}^{2} \sum_{j=1}^{N_i} \left(x_j^{(i)} - \bar{x}^{(i)}\right)\left(x_j^{(i)} - \bar{x}^{(i)}\right)',$$

and $n = N_1 + N_2 - 2$. The distribution of W is needed, to know the probability of misclassification or the probability of correct classification. The distributions of related statistics are also needed for the statistical inference in the discriminant analysis. In general, the exact distributions are hard to obtain, but their asymptotic approximations including asymptotic expansions have been obtained. For a review of these results, see (Siotani, 1982).

The purpose of this paper is to study error bounds for asymptotic approximations of the distributions. Usually the error terms have been known as the ones of the first, second or third order with respect to $(N_1^{-1}, N_2^{-1}, n^{-1})$. However, such error estimators are not sufficient, since the results are obtained under the assumption if N_1 and N_2 are sufficiently large. More precisely, such results give no precise information about how N_1 and N_2 should be large in order to make the actual errors less than a given constant. It is hoped to obtain error estimators under given values of N_1 and N_2. Unfortunately there is little research done on explicit error bounds (for a review, see (Fujikoshi, 1993)). In Section 2 we give a fundamental result in obtaining an error bound. In Section 3 the results are applied to obtain error bounds as well as validities for asymptotic approximations of the distribution of W under given values of N_1 and N_2. Similar results are obtained for the Studentized classification statistic and other related statistics, which are discussed in Section 4.

2. A FUNDAMENTAL RESULT

Explicit error bounds have been studied for the distributions of scale mixtures. For a review, see (Fujikoshi, 1993). In this paper we consider an extension of a scale mixture defined by

$$X = VZ - U, \tag{2.1}$$

where Z and (V, U) are independent variables such that $V > 0$ with probability one. We assume that Z is distributed as the standard normal distribution with the cumulative distribution function Φ and the density function ϕ. Then we can write the distribution function of X as

$$P(X \leqslant x) = F(x) = E_{(V,U)}\left[\Phi\left(V^{-1}(x + U)\right)\right]. \tag{2.2}$$

Now we will make an approximation for $\Phi(v^{-1}(x+u))$ which is useful in the following typical situation: V and U depend on a parameter n, $V = 1 + O_p(n^{-1/2})$, $U = O_p(n^{-1/2})$ and the first few moments of V^2 (or V^{-2}) are more computable than the ones of V (or V^{-1}). The following two types of approximations for $\Phi(v^{-1}x)$ have been introduced by Fujikoshi (1985, 1987) and Shimizu (1987):

$$\Phi_{\delta,k}(x, v) = \Phi(x) - \sum_{j=1}^{k-1} \frac{\delta^j}{2^j j!} a_{\delta j}(x)\phi(x)(v^{2\delta} - 1)^j, \tag{2.3}$$

where $\delta = -1$ or $\delta = 1$. Here we use the notation δ for the expression of two types of approximations. The functions $a_{-1,j}$ and $a_{1,j}$ are defined by the relations

$$(\partial^j/\partial y^j)\Phi(x(\sqrt{y})^\delta)\big|_{y=1} = -(2\delta)^j a_{\delta,j}(x)\phi(x).$$

More precisely,

$$a_{1,j}(x) = H_{2j-1}(x),$$

$$a_{-1,j}(x) = \sum_{i=0}^{j-1}(2i-1)!!\binom{j-1}{i}x^{2j-2i-1},$$

where $(2i-1)!! = (2i-1)\cdot(2i-3)\cdots3\cdot1$, with the convention $(-1)!! = 1$ and $H_j(x)$ is the Hermite polynomial defined by $(\mathrm{d}^j/\mathrm{d}x^j)\phi(x) = (-1)^j H_j(x)\phi(x)$. Now we introduce the following two types of approximations for $\Phi(v^{-1}(x+u))$:

$$\begin{aligned}\Phi_{\delta,k,k}(x,v,u) = \Phi_{\delta,k}(x,v) &+ \sum_{j=1}^{k-1}\frac{1}{j!}(-1)^{j-1}H_{j-1}(x)\phi(x)u^j \\ &- \sum_{j=1}^{k-1}\sum_{i=1}^{k-j-1}\frac{\delta^j}{2^j j! i!}a_{\delta,j,i}(x)\phi(x)(v^{2\delta}-1)^j u^i,\end{aligned} \tag{2.4}$$

where

$$a_{\delta,j,i}(x)\phi(x) = (\partial^i/\partial x^i)\{a_{\delta,j}(x)\phi(x)\}.$$

The error bounds for the approximations $\Phi_{\delta,k,k}$ depend on the quantities

$$\alpha_{\delta,k} = \sup\{|a_{\delta,k}(x)|\phi(x)/(2^k k!)\} \tag{2.5}$$

and

$$\beta_{\delta,k} = \left(\alpha_{\delta,k}^{1/k} + D_{\delta,k}^{1/k}\right)^k, \tag{2.6}$$

where the constants $D_{\delta,k}$ are upper bounds for

$$D_{\delta,k} = \sup_{x,0<y<1}\left|\Phi(xy^{-\delta/2}) - \Phi_{\delta,k}(x,y)\right|.$$

Some numerical values of $\alpha_{\delta,k}$ are given in Table 1. The numerical values when certain upper bounds $D_{\delta,k}$ (see (Shimizu and Fujikoshi, 1994)) are used are given in Table 2. Further, it has been proved (Shimizu and Fujikoshi, 1994) that for $\delta = -1$, we can set $\beta_{-1,k} = 0.5$.

Table 1.
The numerical values of $\alpha_{\delta,k}$

	k			
δ	1	2	3	4
1	0.121	0.069	0.048	0.037
-1	0.121	0.079	0.061	0.050

Table 2.
The numerical values of $\beta_{\delta,k}$

	k			
δ	1	2	3	4
1	0.621	0.940	1.94	3.47
-1	0.621	0.977	1.67	2.98

LEMMA 2.1. *Let* $\Phi(x) = \int_{-\infty}^{x} \phi(u)\,du$ *and* $\phi(u) = (2\pi)^{-1/2}\exp(-\frac{1}{2}x^2)$. *Then, it holds that for any* $x, v > 0$ *and* u,

$$(1)\ \left|\Phi\left(v^{-1}(x+u)\right) - \Phi_{\delta,k,k}(x,v,u)\right| \leqslant \alpha_{\delta,k}\left(v^2 \vee v^{-2} - 1\right)^k + \frac{1}{k!} h_{k-1}|u|^k + \sum_{j=1}^{k-1} \gamma_{\delta,j,k-j}\left|v^{2\delta} - 1\right|^j |u|^{k-j},$$

$$(2)\ \alpha_{\delta,k}\left(v^2 \vee v^{-2} - 1\right)^k \leqslant \beta_{\delta,k}\left|v^{2\delta} - 1\right|^k,$$

where $a \vee b = \max\{a, b\}$,

$$h_k = \sup\{|H_k(x)|\phi(x)\} \tag{2.7}$$

and

$$\gamma_{\delta,j,k-j} = \sup\left\{|a_{\delta,j,k-j}(x)|\phi(x)/\left(2^j j!(k-j)!\right)\right\}. \tag{2.8}$$

Proof. It is known (see, e.g. (Fujikoshi, 1993)) that

$$\left|\Phi\left(v^{-1}(x+u)\right) - \Phi_{\delta,k}(x+u,v)\right| \leqslant \alpha_{\delta,k}\left(v^2 \vee v^{-2} - 1\right)^k.$$

Considering Taylor's expansion for each of the terms of $\Phi_{\delta,k}(x+u,v)$ with respect to u, we have

$$\left|\Phi_{\delta,k}(x+u,v) - \Phi_{\delta,k,k}(x,v,u)\right| \leqslant \frac{1}{k!} h_{k-1}|u|^k + \sum_{j=1}^{k-1} \gamma_{\delta,j,k-j}\left|v^{2\delta} - 1\right|^j |u|^{k-j}.$$

Therefore, the first result follows from the two above-mentioned inequalities. For a proof of the second result, see (Shimizu and Fujikoshi, 1994).

For the case $\delta = 1$, we have

$$a_{1,j,i}(x) = (-1)^j H_{2j-1+i}(x), \qquad \gamma_{1,j,k-j} = \frac{1}{2^j j!(k-j)!} h_{k+j-1}. \tag{2.9}$$

Table 3.
The numerical values of h_k

k	0	1	2	3	4	5	6
h_k	0.399	0.242	0.399	0.550	1.197	2.309	5.984

Some numerical values of h_j are given in Table 3.

Now we consider an approximation for the distribution function of $X = VZ - U$, which is defined by

$$\Phi_{\delta,k,k}(x) = \mathbf{E}_{(V,U)}\{\Phi_{\delta,k,k}(x, V, U)\}. \tag{2.10}$$

The approximation $\Phi_{\delta,k,k}(x)$ is meaningful under the assumption that there exist the $(k-1)$-th moments of $V^{2\delta}$ and U. Using Lemma 2.1, we have the following fundamental result.

THEOREM 2.1. *Suppose that a random variable X can be expressed as $X = VZ - U$, where Z is the standard normal variable, V is positive with probability one, and Z and (V,U) are mutually independent. Further, assume that a given integer $k > 0$,*

$$A(\delta)\colon\ \mathbf{E}\{V^{2\delta j}|U|^{k-j}\} < \infty, \quad j = 0, 1, \ldots, k,$$

where $\delta = 1$ or -1. Then, it holds that

$$\begin{aligned}|\mathbf{P}(X \leqslant x) - \Phi_{\delta,k,k}(x)| &\leqslant \beta_{\delta,k}\mathbf{E}(|V^{2\delta} - 1|^k) + \frac{1}{k!}h_{k-1}\mathbf{E}(|U|^k) \\ &\quad + \sum_{j=1}^{k-1}\gamma_{\delta,j,k-j}\mathbf{E}\{|V^{2\delta} - 1|^j|U|^{k-j}\}.\end{aligned} \tag{2.11}$$

In a special case $U \equiv 0$, we have

$$|\mathbf{P}(X \leqslant x) - \Phi_{\delta,k}(x)| \leqslant \beta_{\delta,k}\mathbf{E}(|V^{2\delta} - 1|^k), \tag{2.12}$$

under the assumption of $A(\delta)$, where $\Phi_{\delta,k}(x) = \mathbf{E}_V\{\Phi_{\delta,k}(x, V)\}$. We note that if we can replace $A(\delta)$ by a strong assumption

$$\tilde{A}(\delta)\colon\ A(\delta)\&\mathbf{E}(V^{-2\delta k}) < \infty,$$

we have the upper bound obtained from the upper bound in (2.8) by replacing

$$\begin{aligned}\beta_{\delta,k}\mathbf{E}\{|V^{2\delta} - 1|^k\} \quad \text{by} \quad &\alpha_{\delta,k}\mathbf{E}\{(V^2 \vee V^{-2} - 1)^k\} \\ \text{or} \quad &\alpha_{\delta,k}[\mathbf{E}\{|V^2 - 1|^k\} + \mathbf{E}\{|V^{-2} - 1|^k\}].\end{aligned}$$

3. THE CLASSIFICATION STATISTIC W

The distribution of W depends on the parameters $\mu^{(1)}$, $\mu^{(2)}$, and Σ through the Mahalanobis squared distance $\Delta^2 = (\mu^{(1)} - \mu^{(2)})'\Sigma^{-1}(\mu^{(1)} - \mu^{(2)})$, which can be estimated by

$$D^2 = \left(\bar{x}^{(1)} - \bar{x}^{(2)}\right)' S^{-1} \left(\bar{x}^{(1)} - \bar{x}^{(2)}\right). \tag{3.1}$$

The limiting distribution of W as $N_1 \to \infty$ and $N_2 \to \infty$ is normal with variance Δ^2 and mean $\frac{1}{2}\Delta^2$ if x is from Π_1 and mean $-\frac{1}{2}\Delta^2$ if x is from Π_2, i.e.,

$$L(W \mid \Pi_1) \longrightarrow N\Big(\frac{1}{2}\Delta^2, \Delta^2\Big), \qquad L(W \mid \Pi_2) \longrightarrow N\Big(-\frac{1}{2}\Delta^2, \Delta^2\Big). \tag{3.2}$$

Okamoto (1963) gave asymptotic expansions of the distributions of $(W - \frac{1}{2}\Delta^2)/\Delta$ for x coming from Π_1 and $(W + \frac{1}{2}\Delta^2)/\Delta$ for x from Π_2, i.e.,

$$P_1(x\colon\ \Delta) = \mathbf{P}\left\{\frac{W - \frac{1}{2}\Delta^2}{\Delta} \leqslant x \mid \Pi_1\right\},$$

$$P_2(x\colon\ \Delta) = \mathbf{P}\left\{\frac{W + \frac{1}{2}\Delta^2}{\Delta} \leqslant x \mid \Pi_2\right\},$$

up to terms of the second order wiht respect to $(N_1^{-1}, N_2^{-1}, n^{-1})$ when $N_1 \to \infty$, $N_2 \to \infty$ and $N_1/N_2 \to$ a positive limit. The result of $P_1(x\colon\ \Delta)$ up to terms of the first order can be written as

$$P_1(x\colon\ \Delta) = \Phi(x) + \phi(x)\left\{\frac{g_1}{N_1} + \frac{g_2}{N_2} + \frac{g_3}{n}\right\} + \mathrm{O}_2, \tag{3.3}$$

where O_j denotes the term of the j-th order with respect to $(N_1^{-1}, N_2^{-1}, n^{-1})$,

$$g_1 = -\frac{1}{2\Delta^2}\{x^3 + (p-3)x - p\Delta\},$$

$$g_2 = -\frac{1}{2\Delta^2}\{x^3 + 2\Delta x^2 + (p - 3 + \Delta^2)x + (p-2)\Delta\},$$

$$g_3 = -\frac{1}{4}\{4x^3 + 4\Delta x^2 + (6p - 6 + \Delta^2)x + (2p-1)\Delta\}.$$

Asymptotic expansion for $P_2(x\colon\ \Delta)$ is obtained from the relation

$$P_2(x\colon\ \Delta) = 1 - P_1^*(-x\colon\ \Delta),$$

where $P_1^*(w\colon\ \Delta)$ is the $P_1(w\colon\ \Delta)$ with N_1 and N_2 interchanged. So, we consider the case of $P_1(x\colon\ \Delta)$.

Our interest is to study error bounds for the approximations (3.2) and (3.3). It is easily seen that

$$P_1(x:\ \Delta) = \mathbf{E}_{(V,U)}\{\Phi(V^{-1}(x+U))\}, \tag{3.4}$$

where

$$\begin{aligned} V &= \frac{\tilde{D}}{\Delta}, \\ U &= \frac{1}{\Delta}\Big\{(\bar{x}^{(1)} - \bar{x}^{(2)})' S^{-1}(\bar{x}^{(1)} - \mu^{(1)}) + \frac{1}{2}(\Delta^2 - D^2)\Big\}, \\ \tilde{D}^2 &= (\bar{x}^{(1)} - \bar{x}^{(2)})' S^{-1}\Sigma S^{-1}(\bar{x}^{(1)} - \bar{x}^{(2)}). \end{aligned} \tag{3.5}$$

Using Theorem 2.1 with $\delta = 1$ and $k = 2$, we obtain an approximation to $P_1(x:\ \Delta)$,

$$\Phi_{1,2,2}(x) = \Phi(x) + \phi(x)\Big\{\mathbf{E}(U) - \frac{1}{2}x\mathbf{E}(V^2 - 1)\Big\} \tag{3.6}$$

the error bound of which is given by

$$B_{1,2,2} = \beta_{1,2}\mathbf{E}\{(V^2-1)^2\} + \frac{1}{2}h_1\mathbf{E}(U^2) + \frac{1}{2}h_2\mathbf{E}\{|(V^2-1)U|\}. \tag{3.7}$$

From (3.6) and (3.7), we obtain a useful upper bound for the standard normal approximation given by

$$|P_1(x:\ \Delta) - \Phi(x)| \leqslant G, \tag{3.8}$$

where

$$\begin{aligned} G = {} & h_0|\mathbf{E}(U)| + \frac{1}{2}h_1|\mathbf{E}\{(V^2-1)\}| \\ & + \beta_{1,2}\mathbf{E}\Big\{(V^2-1)^2\Big\} + \frac{1}{2}h_1\mathbf{E}(U^2) \\ & + \frac{1}{2}h_2\Big[\mathbf{E}\Big\{(V^2-1)^2\Big\}\mathbf{E}(U^2)\Big]^{1/2}. \end{aligned}$$

The expectations that appear in the expression of G can be evaluated by using the following Lemmas.

LEMMA 3.1. *Let nS be distributed as a Wishart distribution $W_p(I_p, n)$. Then*

1) $\mathbf{E}(S^{-1}) = n(n-p-1)^{-1}I_p$, *if* $n-p-1>0$,
2) $\mathbf{E}(S^{-2}) = \frac{(n-1)n^2}{d}I_p$, *if* $n-p-3>0$,
3) *For any symmetric matrices A and B,*

$$\mathbf{E}\begin{bmatrix} \mathrm{tr}AS^{-1}\,\mathrm{tr}BS^{-1} \\ \mathrm{tr}AS^{-1}BS^{-1} \end{bmatrix} = \frac{n^2}{d}\begin{bmatrix} n-p-2 & 2 \\ 1 & n-p-1 \end{bmatrix}\begin{bmatrix} \mathrm{tr}A\mathrm{tr}B \\ \mathrm{tr}AB \end{bmatrix},$$

if $n-p-3>0$, *where* $d = (n-p)(n-p-1)(n-p-3)$.

LEMMA 3.2. *Let D^2 and $\tilde{D}^2$ be the statistics defined by* (3.1) *and* (3.5), *respectively. Then the statistics D^2 and $\tilde{D}^2$ are distributed like*

1) $D^2 = \frac{Nn}{N_1N_2} \cdot \frac{Y_1}{Y_2}$,

2) $\tilde{D}^2 = \frac{Nn^2}{N_1N_2} \cdot \frac{Y_1}{Y_2}\left(1 + \frac{Y_3}{Y_4}\right)$,

where

$$Y_1 \sim \chi_p^2(\tau^2), \quad Y_2 \sim \chi_{n-p+1}^2, \quad Y_3 \sim \chi_{p-1}^2,$$
$$Y_4 \sim \chi_{n-p+2}^2, \quad \tau^2 = (N_1N_2/N)\Delta^2,$$

and Y_j's are mutually independent.

For a proof of (1) and (2) in Lemma 3.1, see (Das Gupta, 1968). The third result can be obtained by modifying the method as in (Fujikoshi, 1987). The first result (1) of Lemma 3.2 is essentially the same as Hotelling's T^2-statistic. The second result (2) can be obtained by the same way as in (Fujikoshi, 1993).

Using Lemmas 3.1 and 3.2, we can evaluate each term in the upper bound G in (3.8) as follows:

$$\mathbf{E}(U) = \frac{n}{2m}\Delta^{-1}\left\{p\left(\frac{1}{N_1} - \frac{1}{N_2}\right) - \frac{1}{n}(p+1)\Delta^2\right\},$$

$$\begin{aligned}\mathbf{E}(U^2) = \frac{n^2}{m(m-2)}\Delta^{-2}\Bigg[&\frac{1}{4}\left(\frac{1}{N_1^2} + \frac{1}{N_2^2}\right)p(p+2) \\ &- \frac{p^2}{2N_1N_2} + \left\{\frac{1}{N_2} - \frac{1}{2n}\left(\frac{1}{N_1} - \frac{1}{N_2}\right)p(p+3)\right\}\Delta^2 \\ &+\frac{1}{4}\left\{\frac{2}{n} + \frac{1}{n^2}(p+1)(p+3)\right\}\Delta^4\Bigg],\end{aligned}$$

$$\begin{aligned}\mathbf{E}(V^2 - 1) = \frac{n^2}{m(m-2)}\Bigg\{&\frac{2}{n}(p+2) - (p+1)(p+3) \\ &+\frac{p-1}{m+1} + p\left(1 + \frac{p-1}{m+1}\right)\tau^{-2}\Bigg\},\end{aligned} \tag{3.9}$$

$$\mathbf{E}\left\{(V^2-1)^2\right\} = \mathbf{E}\left\{\left(\frac{\tilde{D}^2}{\Delta^2}\right)^2\right\} - 1 - 2\mathbf{E}(V^2 - 1),$$

$$\begin{aligned}\mathbf{E}\left\{\left(\frac{\tilde{D}^2}{\Delta^2}\right)^2\right\} = &\frac{n^4\{1 + 2(p+2)\tau^{-2} + p(p+2)\tau^{-4}}{m(m-2)(m-4)(m-6)} \\ &\times\left\{1 + 2\frac{p-1}{m+1} + \frac{p^2-1}{(m+1)(m-1)}\right\},\end{aligned}$$

where $m = n - p - 1$ and $\tau^2 = (N_1N_2/N)\Delta^2$. We note that the inequality (3.8) with the above moments is valid for all N_1 and N_2 so that $n - p - 7 > 0$. Using Theorem 2.1 with $\delta = 1$ and $k = 4$, we obtain an approximation to $P_1(x\colon \Delta)$,

$$|P_1(x\colon \Delta) - \Phi_{1,4,4}(x)| \leqslant B_{1,4,4}, \tag{3.10}$$

where

$$\begin{aligned}\Phi_{1,4,4}(x) = \Phi_{1,2,2}(x) - \phi(x)\Big[&\frac{1}{8}H_3(x)\mathbf{E}\big\{(V^2-1)^2\big\}\\ &-\frac{1}{2}H_2(x)\mathbf{E}\{(V^2-1)U\} + \frac{1}{2}H_1(x)\mathbf{E}(U^2)\\ &+\frac{1}{48}H_5(x)\mathbf{E}\big\{(V^2-1)^3\big\} - \frac{1}{8}H_4(x)\mathbf{E}\big\{(V^2-1)^2U\big\}\\ &+\frac{1}{4}H_3(x)\mathbf{E}\big\{(V^2-1)U^2\big\} - \frac{1}{6}H_2(x)\mathbf{E}(U^3)\Big]\end{aligned}$$

and

$$\begin{aligned}B_{1,4,4} = \beta_{1,4}\mathbf{E}\big\{(V^2-1)^4\big\} &+ \frac{1}{48}h_6\mathbf{E}\big\{|(V^2-1)^3U|\big\}\\ + \frac{1}{16}h_5\mathbf{E}\big\{(V^2-1)^2U^2\big\} &+ \frac{1}{12}h_4\mathbf{E}\big\{|(V^2-1)U^3|\big\} + \frac{1}{24}h_3\mathbf{E}(U^4).\end{aligned}$$

4. RELATED STATISTICS

4.1. *The studentized classification statistic W*

Anderson (1973, 1984) derived asymptotic expansions for the distributions of the Studentized W, i.e., $(W - \frac{1}{2}D^2)/D$ and $(W + \frac{1}{2}D^2)/D$ under Π_1 and Π_2, respectively. The results are expressed as follows:

$$\begin{aligned}\tilde{P}_1(x\colon \Delta) &= \mathbf{P}\left(\frac{W - \frac{1}{2}D^2}{D} \leqslant x \mid \Pi_1\right)\\ &= \Phi(x) - \phi(x)\left\{\frac{1}{N_1}\left(\frac{x}{2} - \frac{p-1}{\Delta}\right)\right.\\ &\quad\left.+\frac{1}{n}\left[\frac{x^3}{4} + \left(p - \frac{3}{4}\right)x\right]\right\} + O_2,\end{aligned} \tag{4.1}$$

and $\mathbf{P}(-(W + \frac{1}{2}D^2)/D \leqslant x \mid \Pi_2)$ is (4.1) with N_1 and N_2 interchanged. In the derivation it is assumed that N_1/N_2 tends to a positive constant as $N_1 \to \infty$ and $N_2 \to \infty$.

Similarly we can write

$$\tilde{P}_1(x\colon \Delta) = \mathbf{E}_{(V,U)}\{\Phi(V^{-1}(x+U))\}, \tag{4.2}$$

where

$$\begin{aligned}V &= \frac{\tilde{D}}{D},\\ U &= \frac{1}{D}(\bar{x}^{(1)} - \bar{x}^{(2)})'S^{-1}(\bar{x}^{(1)} - \mu^{(1)}).\end{aligned} \tag{4.3}$$

Therefore we can apply the results (3.8) and (3.9) with V and U given by (4.3) to the approximations of $\tilde{P}_1(x\colon \Delta)$ and their error bounds. In this case, from Lemma 3.2 we can write

$$V^2 = \frac{\tilde{D}^2}{D^2} = \frac{n}{Y_2}\left(1 + \frac{Y_3}{Y_4}\right), \tag{4.4}$$

which does not depend on Δ. The computations of the moments of V^2 in (4.4) are much simpler than the one in the case of W statistic. On the other hand, the computations of exact moments of U are not simple. For the computation of G, we can use

$$|U| \leqslant \left\{(\bar{x}^{(1)} - \mu^{(1)})' S^{-1} (\bar{x}^{(1)} - \mu^{(1)})\right\}^{1/2} = T \tag{4.5}$$

and hence

$$|\mathbf{E}(U)| \leqslant \left\{\mathbf{E}(T^2)\right\}^{1/2} = \left\{\frac{pn}{N_1(n-p-1)}\right\}^{1/2}. \tag{4.6}$$

Using (4.4), we obtain the following results on the moments of V in (4.3) :

$$\begin{aligned} \mathbf{E}\left(V^2 - 1\right) &= \frac{1}{(m+1)m}\{2pn - p(p+1)\}, \\ \mathbf{E}\left\{\left(V^2 - 1\right)^2\right\} &= \frac{1}{m(m+1)(m-1)(m-2)}\{2n^3 + 4(p-1)(p+2)n^2 \\ &\quad - 2\left(2p^3 + 8p^2 + 6p - 3\right)n + p(p+1)(p+2)(p+3)\}, \end{aligned} \tag{4.7}$$

where $m = n - p - 1$. Using (4.6) and (4.7), we obtain an explicit upper bound (3.8), which depends on the sample sizes (N_1, N_2) only, but its order is $\mathrm{O}_{1/2}$. It is intresting to find a sharp bound for $|\mathbf{E}(U)|$ whose order is O_1. We note that for the studentized classification statistic, the inequalities (3.8) and (3.10) are valid under the conditions $n - p - 3 > 0$ and $n - p - 7 > 0$, respectively. On the other hand, the corresponding conditions for the classification statistic W are $n - p - 7 > 0$ and $n - p - 15 > 0$, respectively.

4.2. *The expectation of* $Q_D = \Phi(-D/2)$

The distribution of $Q_D = \Phi(-D/2)$ plays an important role in estimating the probabilities of misclassification. McLachlan (1973) gave an asymptotic expansion for the expectation of Q_D in the following form:

$$\begin{aligned} \mathbf{E}\{Q_D\} = \Phi\left(-\frac{1}{2}\Delta\right) + \phi\left(-\frac{1}{2}\Delta\right)\left\{\frac{1}{16\Delta}\left(\frac{1}{N_1} + \frac{1}{N_2}\right)[\Delta^2 - 4(p-1)]\right. \\ \left. + \frac{1}{32\Delta n}[\Delta^2 - 4(2p+1)]\right\} + \mathrm{O}_2. \end{aligned} \tag{4.8}$$

Now we will obtain upper bounds for the errors of the approximations. We can write

$$\mathbf{E}\{Q_D\} = \mathbf{E}_V\{\Phi(V^{-1}\xi)\}, \tag{4.9}$$

where $\xi = -\Delta/2$ and $V = \Delta/D$. Therefore, using Theorem 2.1 with $U \equiv O$, we have a general approximation

$$\Phi_{\delta,k}(\xi) - \sum_{j=1}^{k-1} \frac{\delta^j}{2^j j!} a_{\delta,j}(\xi)\phi(\xi)\mathbf{E}\left\{(V^{2\delta}-1)^j\right\}. \tag{4.10}$$

In this case we have to use a general approximation with $\delta = -1$, since the moments of V^{-2} are computable, but those of V^2 are not. Letting $k = 2$, we obtain

$$\begin{aligned}\left|\mathbf{E}\{Q_D\} - \Phi\left(-\frac{1}{2}\Delta\right)\right| &\leqslant \alpha_{-1,1}\left|\mathbf{E}\left(\frac{D^2}{\Delta^2}-1\right)\right| \\ &\quad + \beta_{-1,2}\mathbf{E}\left\{\left(\frac{D^2}{\Delta^2}-1\right)^2\right\}.\end{aligned} \tag{4.11}$$

Further, letting $k = 4$, we obtain

$$\begin{aligned}&\left|\mathbf{E}\{Q_D\} - \left[\Phi\left(-\frac{1}{2}D\right) - \sum_{j=1}^{3} \frac{(-1)^j}{2^j j!} a_{-1,j}\left(-\frac{1}{2}\Delta\right)\phi\left(-\frac{1}{2}\Delta\right)\right.\right. \\ &\left.\left.\times\,\mathbf{E}\left\{\left(\frac{D^2}{\Delta^2}-1\right)^j\right\}\right]\right| \leqslant \alpha_{-1,4}\mathbf{E}\left\{\left(\frac{D^2}{\Delta^2}-1\right)^4\right\}.\end{aligned} \tag{4.12}$$

The moments of $(\frac{D^2}{\Delta^2} - 1)$ in (4.11) and (4.12) are given as follows:

$$\begin{aligned}\mathbf{E}\left(\frac{D^2}{\Delta^2}-1\right) &= \frac{n}{m}\left(\frac{1}{n}(p+1) + p\tau^{-2}\right), \\ \mathbf{E}\left\{\left(\frac{D^2}{\Delta^2}-1\right)^2\right\} &= \frac{n^2}{m(m-2)}\left[\frac{2}{n} + \frac{1}{n^2}(p+1)(p+3)\right. \\ &\quad \left. +2\left\{2+\frac{1}{n}p(p+3)\right\}\tau^{-2} + p(p+2)\tau^{-4}\right], \\ \mathbf{E}\left\{\left(\frac{D^2}{\Delta^2}-1\right)^3\right\} &= \frac{n^3}{m(m-2)(m-4)}\left[\frac{1}{n^2}\left\{2(3p+11)\right.\right. \\ &\quad \left. + \frac{1}{n}(p+1)(p+3)(p+5)\right\} \\ &\quad + \frac{3}{n}\left\{2(3p+10) + \frac{1}{n}p(p+3)(p+5)\right\}\tau^{-2} \\ &\quad + 3(p+2)\left\{4+\frac{1}{n}p(p+5)\right\}\tau^{-4} \\ &\quad \left. + p(p+2)(p+4)\tau^{-6}\right],\end{aligned} \tag{4.13}$$

$$
\begin{aligned}
\mathbf{E}\left\{\left(\frac{D^2}{\Delta^2}-1\right)^4\right\} = {} & \frac{n^4}{m(m-2)(m-4)(m-6)} \\
& \times\left[\frac{1}{n^2}\left\{12+\frac{4}{n}\left(3p^2+28p+61\right)\right.\right. \\
& \left.+\frac{1}{n^2}(p+1)(p+3)(p+5)(p+7)\right\} \\
& +\frac{4}{n}\left\{12+\frac{2}{n}(p+3)(6p+35)\right. \\
& \left.+\frac{1}{n^2}p(p+2)(p+3)(p+7)\right\}\tau^{-2} \\
& +6\left\{8+\frac{2}{n}(p+2)(5p+38)\right. \\
& \left.+\frac{1}{n^2}p(p+2)(p+5)(p+7)\right\}\tau^{-4} \\
& +4(p+2)(p+4)\left\{6+\frac{1}{n}p(p+7)\right\}\tau^{-6} \\
& \left.+p(p+2)(p+4)(p+6)\tau^{-8}\right],
\end{aligned}
$$

where $m = n-p-1$ and $\tau^2 = (N_1N_2/N)/\Delta^2$. We note that the inequality (4.11) and (4.12) are valid under the conditions $n-p-3>0$ and $n-p-7>0$, respectively. Further, the orders of the resultant upper bounds in (4.11) and (4.12) are O_1 and O_2 with respect to $(N_1^{-1}, N_2{-}1, n^{-1})$, respectively.

REFERENCES

Anderson, T. W. (1973). An asymptotic expansion of the distribution of the studentized classification statistic *W*. *Ann. Statist.* **1**, 964–972.

Anderson, T. W. (1984). *An Introduction to Multivariate Statistical Analysis.* 2nd ed. Wiley, New York.

Das Gupta, S. (1968). Some aspects of discrimination function coefficients. *Sankhya, A* **30**, 387–400.

Fujikoshi, Y. (1985). An error bound for an asymptotic expansion of the distribution function of an estimate in a multivariate linear model. *Ann. Statist.* **13**, 827–831.

Fujikoshi, Y. (1987). Error bounds for asymptotic expansions of scale mixtures of distributions. *Hiroshima Math. J.* **17**, 301–324.

Fujikoshi, Y. (1993). Error bounds for approximations to the distributions of the standardized and studentized estimates in a multivariate linear model. *J. Statist. Plann. Inf.* **36**, 165–174.

Fujikoshi, Y. (1993). Error bounds for asymptotic approximations of some distribution functions. In: *Multivariate Analysis: Future Directions*, pp. 181–208, C. R. Rao (Ed.), North-Holland, Amsterdam.

MsLachlan, G. J. (1973). An asymptotic expansion of the expectation of the estimated error rate in discriminant analysis. *Austral. J. Statist.* **15**, 210–214.

Okamoto, M. (1963). An asymptotic expansion for the distribution of the linear discriminant function. *Ann. Math. Statist.* **34**, 1286–1301.

Shimizu, R. (1987). Error bounds for asymptotic expansion of the scale mixtures of the normal distribution. *Ann. Inst. Statist. Math.* **39**, 611–622.

Shimizu, R. and Fujikoshi, Y. (1994). Sharp error bounds for asymptotic expansions of the distribution functions of scale mixtures. Submitted for publication.

Siotan, M. (1982). Large sample approximations and asymptotic exepansions of classification statistic. In: *Handbook of Statistics, Vol. 2*, pp. 47–60, P. R. Krishnaiah and L. N. Kanal (Eds), North-Holland, Amsterdam.

Multivar. Statist., pp. 17–33
E.–M. Tiit *et al.* (Eds)

SOME RESULTS ON THE EMPIRICAL SPACINGS PROCESS AND ITS BOOTSTRAPPED VERSION

S. RAO JAMMALAMADAKA[1] and JING QIN[2]
[1]University of California, Santa Barbara

[2]University of Maryland

ABSTRACT

This paper considers the CLT and SLLN for the empirical spacings process, indexed by functions. A bootstrapped version of this process is shown to work and a strong approximation rate established for this bootstrapped version.

1. INTRODUCTION AND PRELIMINARIES

Let $X_1, X_2, \ldots, X_{n-1}$ be independently and identically distributed (i.i.d.) random variables (r.v.s) with a common continuous distribution function (d.f.) F with support on R^1. An important basic question of interest is whether these observations are from a specified distribution function — the goodness of fit problem. A simple probability integral transformation on these random variables, lets us equate the specified distribution to the uniform distribution on $[0, 1]$. Thus from now on, we shall assume that this reduction has been done and under the null hypothesis of interest, the observations have a $U(0, 1)$ distribution. A broad class of procedures for testing this null hypothesis are based on the spacings, namely

$$D_{i:n} = (X_{i,n} - X_{i-1,n}), \qquad i = 1, 2, \ldots, n, \tag{1.1}$$

where $0 = X_{0,n} \leqslant X_{1,n} \leqslant \cdots \leqslant X_{n-1,n} \leqslant X_{n,n} = 1$ are the order statistics from $U(0, 1)$ distribution. See for instance (Pyke, 1965, 1972; Rao and Sethuraman, 1975; Shorack, 1972; Aly *et al.*, 1984).

If Z_i, $i = 1, 2, \ldots$, is a sequence of i.i.d. exponential random variables with mean 1, i.e., with d.f.

$$F(z) = 1 - e^{-z}, \tag{1.2}$$

then it is well known (see, e.g. (Pyke, 1965)) that

$$\{nD_{i:n}\}_{i=1}^{n} \sim \left\{\frac{Z_i}{\overline{Z}_n}\right\}_{i=1}^{n}, \tag{1.3}$$

where $\overline{Z}_n = \sum_{i=1}^n Z_i/n$ and $\sim$ denotes equivalence in distribution. Indeed there is a probability space $(\Omega, \mathcal{A}, \mathcal{P})$ on which both $D_{i:n}$ and Z_i can be defined, so that $\sim$ in (1.3) can be replaced by $=$ almost surely. Let

$$F_n(t) = n^{-1} \sum_{i=1}^{n} I(Z_i \leqslant t) \tag{1.4}$$

and

$$\widehat{F}_n(t) = n^{-1} \sum_{i=1}^{n} I\left(\frac{Z_i}{\overline{Z}} \leqslant t\right) = F_n(\overline{Z}t), \tag{1.5}$$

denote the empirical d.f. (e.d.f.) of the exponential sequence and the normalized spacings sequence $\{nD_{i:n}\}_{i=1}^n$ respectively. We call

$$\hat{\alpha}_n(t) = \sqrt{n}\big(\widehat{F}_n(t) - F(t)\big), \qquad t \geqslant 0, \tag{1.6}$$

the empirical spacings process. Denote by

$$\nu_n(\cdot) = \sqrt{n}\big(P_n(\cdot) - P(\cdot)\big) \tag{1.7}$$

and

$$\hat{\nu}_n(\cdot) = \sqrt{n}\big(\widehat{P}_n(\cdot) - P(\cdot)\big), \tag{1.8}$$

where $\widehat{P}_n, P_n$, and P are probability measures corresponding to $\widehat{F}_n, F_n$, and F respectively.

Limit theory for empirical processes has grown enormously over the last few decades, as evidenced by the voluminous book by Shorack and Wellner (1986). Recently the study of empirical processes indexed by sets or functions has become very important and the results are quite general in scope. See, for instance (Pollard, 1984; Sheehy and Wellner, 1992), etc. In Section 2, we explore the central limit theorem (CLT) and the strong law of large numbers (SLLN) for the spacings process, indexed by functions. It is easily checked that the usual bootstrap method fails for the spacings process, so in Section 3, we introduce a resampling scheme for which the bootstrap approximation is shown to work. This allows one to get critical values for any spacings test statistics, which are known to be notoriously slow in converging to their limiting distributions. Finally in Section 4, a bootstrap strong approximation rate is derived.

2. SPACINGS PROCESSES INDEXED BY VC FUNCTIONS

For $1 \leqslant s < \infty$ and for some probability measure Q on R^d, we denote by $L^s(R^d, Q)$ the space of measurable real functions g on R^d with $(\int |g|^s \, dQ)^{1/s} < \infty$. In most of what follows, Q will be empirical measure P_n or the theoretical measure P. We denote the $L^s(R^d, P_n)$ (pseudo) norm by

$$\|\cdot\|_{s,n} = \left(\int |\cdot|^s \, \mathrm{d}P_n\right)^{1/s}$$

and we sometimes call this the "empirical norm" whose theoretical counterpart is

$$\|\cdot\|_s = \left(\int |\cdot|^s \, \mathrm{d}P\right)^{1/s}.$$

For $\mathcal{F}$ a class of functions, the envelope H of $\mathcal{F}$ is defined as

$$H = \sup_{f \in \mathcal{F}} |f|.$$

Moreover, for $\mathcal{F} \subseteq L^s(R^d, Q)$, we define the covering number $N_s(\delta, Q, \mathcal{F})$ as the smallest value of k such that there exists $f_1, \ldots, f_k$ in $\mathcal{F}$, for all $f \in \mathcal{F}$

$$\min_{1 \leqslant j \leqslant k} \left(\int |f - f_j|^s \, \mathrm{d}Q\right)^{1/s} < \delta. \tag{2.1}$$

The logarithm of $N_s(\delta, Q, \mathcal{F})$ is called the δ-entropy of $\mathcal{F}$ for the metric

$$\left(\int |\cdot|^s \, \mathrm{d}Q\right)^{1/s}.$$

LEMMA 1. *Suppose $\mathcal{F}$ is a permissible (measurable) class with envelope H, then*

$$\sup_{f \in \mathcal{F}} \left|\int f \, \mathrm{d}(P_n - P)\right| \longrightarrow 0, \tag{2.2}$$

almost surely iff both $H \in L^1(R^d, P)$ and

$$n^{-1} \log N_1(\delta, P_n, \mathcal{F}) \longrightarrow 0, \qquad P. \tag{2.3}$$

Moreover if $H \in L^1(R^d, P)$ and (2.3) holds, then for all $\delta > 0$, the theoretical covering number $N_1(\delta, P, \mathcal{F})$ is finite, i.e.,

$$N_1(\delta, P, \mathcal{F}) < \infty. \tag{2.4}$$

Proof. See (for example (Van de Geer, 1988, Theorem 2.2.1 and 2.2.2)).

THEOREM 1 (Strong law of large numbers). *For a given class of functions $\mathcal{F}$, define*

$$\mathcal{F}_\alpha = \{f(ct) \mid f(t) \in \mathcal{F},\ c \in [1-\alpha, 1+\alpha]\},$$

where $0 < \alpha < 1$. Let H be the envelope for $\mathcal{F}_\alpha$. If

(i) $$n^{-1}\log N_1(\delta, P_n, \mathcal{F}_\alpha) \longrightarrow 0, \quad P, \tag{2.5}$$

(ii) $$\int_0^\infty H(y)\exp\left(\frac{\alpha-1}{1+\alpha}y\right)\,dy < \infty \tag{2.6}$$

and

(iii) $g_f(c) = E\{f(cZ_i)\}$ *are continuous for any* $f \in \mathcal{F}$,

then

$$\sup_{f\in\mathcal{F}}\left|n^{-1}\sum_{i=1}^{n} f\left(\frac{Z_i}{\bar{Z}}\right) - Pf\right| \longrightarrow 0 \qquad \text{a.s.} \tag{2.7}$$

Proof. When $|1/\bar{Z} - 1| < \delta \leqslant \alpha$, we have

$$\begin{aligned}
&\sup_{f\in\mathcal{F}}\left|n^{-1}\sum_{i=1}^{n} f\left(\frac{Z_i}{\bar{Z}}\right) - Pf\right| \\
&\leqslant \sup_{f\in\mathcal{F}}\left|n^{-1}\sum_{i=1}^{n} f\left(\frac{Z_i}{\bar{Z}}\right) - g_f\left(\frac{1}{\bar{Z}}\right)\right| + \sup_{f\in\mathcal{F}}\left|g_f\left(\frac{1}{\bar{Z}}\right) - g_f(1)\right| \\
&\leqslant \sup_{f\in\mathcal{F},\, c\in[1-\delta,1+\delta]}\left|n^{-1}\sum_{i=1}^{n} f(cZ_i) - g_f(c)\right| \\
&\quad + \sup_{f\in\mathcal{F}}\left|g_f\left(\frac{1}{\bar{Z}}\right) - g_f(1)\right| \\
&\leqslant \sup_{f\in\mathcal{F}_\alpha}\left|n^{-1}\sum_{i=1}^{n} f(Z_i) - Pf\right| + \sup_{f\in\mathcal{F}}\left|g_f\left(\frac{1}{\bar{Z}}\right) - g_f(1)\right|.
\end{aligned}$$

Since $\mathcal{F}_\alpha$ satisfies (2.3), so does $\mathcal{F}$. By Lemma 1, (2.4) is true. For any $\delta > 0$, there exist $f_1, f_2, \ldots, f_k \in \mathcal{F}$ such that $k < \infty$ and (2.1) holds true with $Q = P$ and $s = 1$. Note that

$$\begin{aligned}
|g_f(c) - g_f(1)| &= |P[f(cZ) - f(Z)]| \\
&\leqslant \left|P\left[f_j(cZ) - f_j(Z)\right]\right| + \left|P\left[f(cZ) - f_j(cZ)\right]\right| + \left|P\left[f(Z) - f_j(Z)\right]\right|,
\end{aligned}$$

while

$$\begin{aligned}
\left|P\left[f(cZ) - f_j(cZ)\right]\right| &= \left|\int_0^\infty \left[f(cz) - f_j(cz)\right]\,\exp(-z)\,dz\right| \\
&\leqslant c^{-1}\int |f(y) - f_j(y)|\,\exp(-y/c)\,dy
\end{aligned}$$

$$\leqslant (1-\alpha)^{-1}\left\{\int |f(y)-f_j(y)|\,\exp(-y)\,dy\right\}^{1/2}$$
$$\times\left\{\int |f(y)-f_j(y)|\,\exp\,[2y(1/2-1/(1+\alpha))]\,dy\right\}^{1/2}$$
$$\leqslant \frac{2}{1-\alpha}\left\{\int H(y)\exp[y(\alpha-1)/(\alpha+1)]\,dy\right\}^{1/2}$$
$$\times\left\{\int |f(y)-f_j(y)|\,\exp(-y)\,dy\right\}^{1/2}$$
$$\leqslant A^{1/2}\delta^{1/2},$$

where

$$A=\left[\frac{2}{1-\alpha}\right]^2\int H(y)\exp[y(\alpha-1)/(\alpha+1)]\,dy<\infty.$$

So for any $\varepsilon>0$, choose $\delta>0$ such that

$$|P[f(cZ)-f_j(cZ)]|+|P[f(Z)-f_j(Z)]|\leqslant A^{1/2}\delta^{1/2}+\delta<\varepsilon/2.$$

Hence combining the above, when $|1/\bar{Z}-1|<\delta$,

$$\sup_{f\in\mathcal{F}}\left|n^{-1}\sum_{i=1}^{n}f\left(\frac{Z_i}{\bar{Z}}\right)-Pf\right|$$
$$\leqslant \sup_{f\in\mathcal{F}_\alpha}\left|n^{-1}\sum_{i=1}^{n}f(Z_i)-Pf\right|+k\max_{1\leqslant j\leqslant k}\left|g_{f_j}\left(\frac{1}{\bar{Z}}\right)-g_{f_j}(1)\right|+\frac{\varepsilon}{2}.$$

Since k is finite, $g_{f_j}(\cdot)$ is continuous and $|1/\bar{Z}-1|\to 0$ a.s., and note the fact that

$$P\left\{\sup_{f\in\mathcal{F}}\left|n^{-1}\sum_{i=1}^{n}f\left(\frac{Z_i}{\bar{Z}}\right)-Pf\right|I\left(\left|\frac{1}{\bar{Z}}-1\right|>\delta\right)>\varepsilon\right\}$$
$$\leqslant P\left\{\left|\frac{1}{\bar{Z}}-1\right|>\delta\right\}\leqslant\frac{c}{n^2}.$$

By the Borel–Cantelli lemma, the theorem now follows.

COROLLARY 1. *Let* $\mathcal{F} = \{I(-\infty, x]\colon x \in R\}$, *then* $\mathcal{F}_\alpha = \mathcal{F}$ *and*

$$\sup_x \left| n^{-1} \sum_{i=1}^{n} I\left(\frac{Z_i}{\overline{Z}} < x\right) - F(x) \right| \longrightarrow 0, \qquad \text{a.s.}$$

In order to study the central limit theorem, we need Pollard's sparseness condition. Let

$$J(\delta) = J(\delta, P, \mathcal{F}) = \int_0^\delta \left[2 \log (N_2(x, P, \mathcal{F})^2/x)\right]^{1/2} \mathrm{d}x.$$

LEMMA 2 (Pollard, 1982, 1984). *Suppose that* $\mathcal{F} \subseteq L^2(R, P)$ *is permissible. Let the random covering number satisfies the uniformity condition: for each* $\eta > 0$ *and* $\varepsilon > 0$ *there exists a* $\gamma > 0$ *such that*

$$\limsup P\{J(\gamma, P_n, \mathcal{F}) > \eta\} < \varepsilon. \tag{2.8}$$

Then the central limit theorem holds for the class $\mathcal{F}$.

THEOREM 2. *Let* $\mathcal{F} \subseteq L^2(R, P)$ *be a continuous permissible function class such that* $\mathcal{F}_\alpha$ *satisfies* (2.8). *If*

$$\int H^2(y) \exp\left[-y(1-\alpha)/(1+\alpha)\right] \mathrm{d}y < \infty,$$

where H *is the envelope of* $\mathcal{F}_\alpha$, *then we have the representation*

$$\begin{aligned} &\frac{1}{\sqrt{n}} \sum_{i=1}^{n} \left[f\left(\frac{Z_i}{\overline{Z}}\right) - Pf(Z)\right] \\ &= \frac{1}{\sqrt{n}} \sum_{i=1}^{n} [f(Z_i) - Pf(Z)] + \sqrt{n}\left[g_f\left(\frac{1}{\overline{Z}}\right) - g_f(1)\right] + \mathrm{o}_p(1), \end{aligned} \tag{2.9}$$

where $\mathrm{o}_p(1)$ *holds uniformly for* $f \in \mathcal{F}$.

Proof. Since $\mathcal{F}_\alpha$ satisfies (2.1), so does $\mathcal{F}$. For any $\delta > 0$, there exist $f_1, f_2, \ldots, f_k \in \mathcal{F}$, such that $k < \infty$ and (2.1) is true with $Q = P$ and $s = 2$. By Minkowski's inequality

$$\begin{aligned} &\sup_{f\in\mathcal{F}} \left\{\int [f(cZ) - f(Z)]^2 \,\mathrm{d}P\right\}^{1/2} \\ &\leqslant \sup_{f\in\mathcal{F}} \left\{\int [f(cZ) - f_j(cZ)]^2 \,\mathrm{d}P\right\}^{1/2} \\ &\quad + \sup_{f\in\mathcal{F}} \left\{\int [f_j(cZ) - f_j(Z)]^2 \,\mathrm{d}P\right\}^{1/2} \\ &\quad + \sup_{f\in\mathcal{F}} \left\{\int [f_j(Z) - f(Z)]^2 \,\mathrm{d}P\right\}^{1/2}. \end{aligned} \tag{2.10}$$

Note that for $c \in [1-\alpha, 1+\alpha]$,

$$
\begin{aligned}
&\int [f(cZ) - f_j(cZ)]^2 \,\mathrm{d}P \\
&= \frac{1}{c}\int_0^\infty [f(y) - f_j(y)]^2 \exp(-y/c)\,\mathrm{d}y \\
&\leqslant \frac{2}{1-\alpha}\int H(y)|f(y) - f_j(y)| \exp(-y/2) \\
&\quad \times \exp\left[-y\left(\frac{1}{1+\alpha} - \frac{1}{2}\right)\right] \mathrm{d}y \qquad (2.11)\\
&\leqslant \frac{2}{1-\alpha}\left\{\int H^2(y)\exp\left[-y(1-\alpha)/(1+\alpha)\right]\mathrm{d}y\right\}^{1/2} \\
&\quad \times \left\{\int |f_j(y) - f(y)|^2 \exp(-y)\,\mathrm{d}y\right\}^{1/2} \\
&\leqslant (C-2)\delta,
\end{aligned}
$$

where $0 < C - 2 < \infty$. So there exists a $\delta_1(\delta) > 0$ such that if $|c-1| < \delta_1$,

$$
\sup_{f\in\mathcal{F}}\left\{\int [f(cZ - f(Z)]^2\,\mathrm{d}P\right\}^{1/2} \leqslant C\delta
$$

or

$$
\begin{aligned}
[C\delta] &\equiv \left\{(f_1(t), f_2(t)) \mid f_1, f_2 \in \mathcal{F}_\alpha,\ \int (f_1 - f_2)^2\,\mathrm{d}P < C\delta\right\} \\
&\supset \left\{(f(ct), f(t)) \mid f \in \mathcal{F},\ |c-1| < \delta_1(\delta)\right\} \\
&\equiv [\delta_1]
\end{aligned}
$$

and

$$
\begin{aligned}
&P\Big\{\sup_{f\in\mathcal{F}} |\nu_n(f(Z/\overline{Z})) - \nu_n(f(Z))| > \eta\Big\} \\
&\leqslant P\Big\{\sup_{f\in\mathcal{F}} |\nu_n(f(Z/\overline{Z})) - \nu_n(f(Z))| > \eta, |1/\overline{Z} - 1| < \delta_1\Big\} \\
&\quad + P\{|1/\overline{Z} - 1| > \delta_1\} \qquad (2.12)\\
&\leqslant P\Big\{\sup_{[\delta_1]} |\nu_n(f(cZ)) - \nu_n(f(Z))| > \eta\Big\} + P\{|1/\overline{Z} - 1| > \delta_1\} \\
&\leqslant P\Big\{\sup_{[C\delta]} |\nu_n(f_1(Z)) - \nu_n(f_2(Z))| > \eta\Big\} + P\{|1/\overline{Z} - 1| > \delta_1\}.
\end{aligned}
$$

By the stochastically equicontinuous property of $\{\nu_n\}$ (see (Pollard, 1984, p. 139–150)), for n large enough, the right hand side of (2.12) can be arbitrarily small. Hence

$$
|\nu_n(f(Z/\overline{Z})) - \nu_n(f(Z))| = o_p(1) \qquad \text{uniformly in} \quad f \in \mathcal{F},
$$

i.e.,

$$\frac{1}{\sqrt{n}}\sum_{i=1}^{n}[f(Z_i/\overline{Z})-g_f(1/\overline{Z})]-\frac{1}{\sqrt{n}}\sum_{i=1}^{n}[f(Z_i)-Pf(Z)]=o_p(1).$$

COROLLARY 2. *Under the conditions of Theorem 2, if*

$$\lim_{t\to 0}\sup_{f\in\mathcal{F}}\left|\frac{g_f(1+t)-g_f(1)}{t}-g_f'(1)\right|=0,$$

then

$$\frac{1}{\sqrt{n}}\sum_{i=1}^{n}[f(\dot{Z}_i/\overline{Z})-Pf(Z)]\longrightarrow B_f+g_f'(1)\int F^{-1}\,\mathrm{d}B_f,$$

where B_f is a Brownian bridge indexed by $\mathcal{F}$.

This result can be generalized to empirical processes subject to perturbations and scale factors, the more general setting described in (Rao and Sethuraman, 1975).

3. BOOTSTRAPPING FOR THE EMPIRICAL SPACINGS PROCESS

Let $X_1, X_2, \ldots$ be i.i.d. with distribution P on $(\Omega, \mathcal{A}, \mathcal{P})$ and let

$$P_n = n^{-1}\sum_{i=1}^{n} I(X_i \leqslant x)$$

be the empirical measure of the first n X's. The nonparametric bootstrap proceeds by sampling from $P_n^{\omega} = P_n(\cdot,\omega)$. Suppose that $X_{n1}^*, \ldots, X_{nm}^*$ are i.i.d. with the distribution P_n^{ω} on $(\Omega, \mathcal{A})$. Let

$$P_m^* = m^{-1}\sum_{i=1}^{m} I\big(X_{ni}^* \leqslant x\big), \qquad \nu_{n,m}^* = \sqrt{m}\big(P_m^* - P_n\big).$$

Thus $X_{n1}^*, \ldots, X_{nm}^*$ is the "bootstrap sample", P_m^* is the "bootstrap empirical measure" and $\nu_{m,n}^*$ is the "bootstrap empirical process". Under very weak conditions, Sheehy and Wellner (1992) show that bootstrap approximation for empirical processes indexed by functions works. However, for the empirical spacings processes, the usual bootstrap procedure fails, i.e., if we resample $U_1^*, \ldots, U_{m-1}^*$ from $\widetilde{F}_n(t) = n^{-1}\sum_{i=1}^{n} I(U_i \leqslant t)$, where U are uniform distribution random variables. Construct "bootstrap spacings"

$$\delta_{i:m}^* = U_{i:m}^* - U_{i-1:m}^*, \qquad i = 1, 2, \ldots, m,$$

where $0 = U^*_{0:m} \leqslant U^*_{1:m} \leqslant \dots \leqslant U^*_{m:m} = 1$ are the order statistics of $U^*_1, \dots, U^*_m$. Then the "bootstrap spacings process" is

$$\tilde{\nu}^*_m = \sqrt{m}\big[\widetilde{F}^*_m(t) - \widehat{F}_n(t)\big], \qquad \widetilde{F}^*_m(t) = m^{-1}\sum_{i=1}^{m} I\big(\delta^*_{i:m} \leqslant t\big).$$

In the bootstrap sample, the possibility of ties is high and this makes too many bootstrapped spacings to be zero. Thus, it can be checked that $\tilde{\nu}^*_m$ and $\hat{\nu}_n$ have different limit distributions and the usual bootstrapping fails. Hence we suggest the following modified resampling scheme:

Let $Z^*_1, \dots, Z^*_m$ be i.i.d. random variables with the distribution function $\widehat{F}_n(t)$, the empirical spacings processes defined in (1.5) and let

$$F^*_m(t) = m^{-1}\sum_{i=1}^{m} I\big(Z^*_i \leqslant t\big),$$

$$\widehat{F}^*_m(t) = m^{-1}\sum_{i=1}^{m} I\left(\frac{Z^*_i}{\overline{Z}^*} \leqslant t\right) = F^*_m\big(\overline{Z}^* t\big).$$

We define the modified "bootstrap empirical spacing processes" as

$$\nu^*_m = \sqrt{m}\big(F^*_m(t) - \widehat{F}_n(t)\big), \qquad \hat{\nu}^*_m = \sqrt{m}\big(\widehat{F}^*_m(t) - \widehat{F}_n(t)\big). \tag{3.1}$$

We will now prove that this modified Bootstrap empirical spacings process (3.1) and the empirical spacings process have the same asymptotic distribution.

LEMMA 3 (Sheehy and Wellner, 1992). *Suppose that $\mathcal{F}$ is nearly linearly Deviation Measurable for* ($\{P_n\}$) (see (Gine and Zinn, 1984, p. 935) *for a definition*) *and that*

i) *H is $\{P_n\}$ uniformly square integrable,*

ii) *$(\mathcal{F}, H)$ is a sparse,*

iii) $\|P_n - P_0\|_{\mathcal{G}} \to 0$ *as* $n \to \infty$,

where $\mathcal{G} = \mathcal{F} \cup \mathcal{F}^2 \cup \mathcal{F}'^2$ and $\mathcal{F}^2 = \{f^2 \mid f \in \mathcal{F}\}$, $\mathcal{F}'^2 = \{(f-g)^2 \mid f \in \mathcal{F}, g \in \mathcal{F}\}$. Then $\mathcal{F} \in CLT(\{P_n\}_{n \geqslant 0})$ (i.e., central limit theorem holds uniformly with respect to P_n).

LEMMA 4. *Under the conditions of Lemma 3, if $\mathcal{F}$ is a continuous function class, then*

$$\|\widehat{P}_n - P\|_{\mathcal{G}_{\alpha/2}} \longrightarrow 0 \qquad as \quad n \longrightarrow \infty,$$

where $\mathcal{G}_{\alpha/2} = \mathcal{F}_{\alpha/2} \cup \mathcal{F}^2_{\alpha/2} \cup \mathcal{F}'^2_{\alpha/2}$.

Proof. We only need to prove

$$\|\widehat{P}_n - P\|_{\mathcal{F}'^2} = \sup_{f_1, f_2 \in \mathcal{F}_{\alpha/2}} \left|(\widehat{P}_n - P)(f_1 - f_2)^2\right| \longrightarrow 0 \qquad \text{as} \quad n \to \infty,$$

since the other conditions are the same. Let

$$g_{f,h}(c_1, c_2) = P\big[f(c_1 Z) - h(c_2 Z)\big]^2.$$

When $c_1/\overline{Z} \in [1-\alpha, 1+\alpha)$ and $c_2/\overline{Z} \in [1-\alpha, 1+\alpha)$, we have

$$\begin{aligned}
&\sup_{f_1, f_2 \in \mathcal{F}_{\alpha/2}} \left|(\widehat{P}_n - P)(f_1 - f_2)^2\right| \\
&= \sup_{f,h \in \mathcal{F}} \sup_{c_1, c_2 \in [1-\alpha/2, 1+\alpha/2]} \left|(\widehat{P}_n - P)\big(f(c_1 Z) - h(c_2 Z)\big)^2\right| \\
&\leqslant \sup_{f,h \in \mathcal{F}} \sup_{c_1, c_2 \in [1-\alpha/2, 1+\alpha/2]} \Big| n^{-1} \sum_i \big[f\big(c_1 Z_i/\overline{Z}\big) - h\big(c_2/\overline{Z}\big)\big]^2 \\
&\quad - g_{f,h}\big(c_1/\overline{Z}, c_2/\overline{Z}\big)\Big| \qquad\qquad (3.2) \\
&\quad + \sup_{f,h \in \mathcal{F}} \sup_{c_1, c_2 \in [1-\alpha/2, 1+\alpha/2]} \left|g_{f,h}\big(c_1/\overline{Z}, c_2/\overline{Z}\big) - g_{f,h}(c_1, c_2)\right| \\
&\leqslant \sup_{f,h \in \mathcal{F}_\alpha} \Big| n^{-1} \sum_i \big[f(Z_i) - h(Z_i)\big]^2 - g_{f,h}(1,1)\Big| \\
&\quad + \sup_{f,h \in \mathcal{F}} \sup_{c_1, c_2 \in [1-\alpha/2, 1+\alpha/2]} \left|g_{f,h}\big(c_1/\overline{Z}, c_2/\overline{Z}\big) - g_{f,h}(c_1, c_2)\right|.
\end{aligned}$$

The first term of (3.2) goes to zero by the Theorem 12 in (Pollard, 1982). We need to prove that the second term of (3.2) also goes to zero. For $a \in [1-\alpha/2, 1+\alpha/2]$, consider

$$\begin{aligned}
&\left|g_{f,h}(c_1 a, c_2 a) - g_{f,h}(c_1, c_2)\right| \\
&= \Big|P\big[f(c_1 aZ) - h(c_2 aZ)\big]^2 - P\big[f_i(ac_1 Z) - f_j(ac_2 Z)\big]^2 \\
&\quad + P\big[f_i(ac_1 Z) - f_j(ac_2 Z)\big]^2 - P\big[f(c_1 Z) - h(c_2 Z)\big]^2 \\
&\quad + P\big[f_i(c_1 Z) - f_j(c_2 Z)\big]^2 - P\big[f_i(c_1 Z) - f_j(c_2 Z)\big]^2\Big| \\
&\leqslant P\Big\{\big[\left|f(c_1 aZ) - f_i(c_1 aZ)\right| + \left|h(c_2 aZ) - f_j(c_2 aZ)\right|\big] 4H(Z)\Big\} \\
&\quad + P\Big\{\big[\left|f(c_1 Z) - f_i(c_1 Z)\right| + \left|h(c_2 Z) - f_j(c_2 Z)\right|\big] 4H(Z)\Big\} \\
&\quad + \Big|P\big[f_i(ac_1 Z) - f_j(ac_2 Z)\big]^2 - P\big[f_i(c_1 Z) - f_j(c_2 Z)\big]^2\Big|.
\end{aligned}$$

Rest of the proof is similar to that of Theorem 2.2.

THEOREM 3. *Under the same conditions as in Theorem 2, for almost all sample sequences* $X_1, X_2, \ldots, X_n$

$$\sqrt{m}\left[m^{-1}\sum_{i=1}^{m} f\left(\frac{Z_i^*}{\overline{Z}^*}\right) - n^{-1}\sum_{i=1}^{n} f\left(\frac{Z_i}{\overline{Z}}\right)\right]$$
$$= \sqrt{m}\left[m^{-1}\sum_{i=1}^{m} f(Z_i^*) - n^{-1}\sum_{i=1}^{n} f(Z_i/\overline{Z})\right] \tag{3.3}$$
$$+ \frac{\sqrt{m}}{n}\left[\sum_{i=1}^{m} f\left(\frac{Z_i}{\overline{Z}\,\overline{Z}^*}\right) - f\left(\frac{Z_i}{\overline{Z}}\right)\right] + o_p(1), \quad \text{a.s.}$$

where $o_p(1)$ *is uniform for all* $f \in \mathcal{F}$.

Proof. We show that $\mathcal{F}_{\alpha/2} \in CLT(\{P_n\}_{n\geqslant 0})$ by using Lemma 3.

i) Let $H_1(t) = \sup_{f\in\mathcal{F}_{\alpha/2}} |f(t)|$ be an envelope for $\mathcal{F}_{\alpha/2}$. Since $H(t)$ is

$$\sup_{c\in(1-\alpha/2, 1+\alpha/2)} |H_1(ct)| = \sup_{c\in(1-\alpha/2,1+\alpha/2)} \sup_{f\in\mathcal{F}_{\alpha/2}} |f(ct)| \leqslant H(t),$$

then $H_1(t)$ is a.s. $\{\widehat{P}_n\}$ uniformly square integrable. In fact when $|1/\overline{Z} - 1| < \alpha/2$,

$$\begin{aligned}\widehat{P}_n H_1^2 I(H_1 \geqslant \lambda) &= n^{-1}\sum H_1^2(Z_i/\overline{Z}) I\big(H_1(Z_i/\overline{Z}) > \lambda\big)\\ &\leqslant n^{-1}\sum H^2(Z_i) I(H(Z_i) > \lambda) \longrightarrow P\big(H^2 I(H \geqslant \lambda)\big) \qquad \text{a.s.}\end{aligned}$$

iii) is correct for $\mathcal{G}_{\alpha/2}$ by Lemma 4. Then by Lemma 3, $\mathcal{F}_{\alpha/2} \in CLT(\{\widehat{P}_n\})$. For any $\delta > 0$, there exists a $\delta_1(\delta) > 0$, $\delta_1 < \alpha/4$ such that

$$\sup_{f\in\mathcal{F},\ |c-1|<\delta_1} P\big(f(cZ) - f(Z)\big)^2 < \delta/2. \tag{3.4}$$

Note that

$$\begin{aligned}&P^*\Big\{\sup_{f\in\mathcal{F}} \big|\nu_m^*(f(Z/\overline{Z}^*)) - \nu_m^*(f(Z))\big| > \eta\Big\}\\ &\leqslant P^*\Big\{\sup_{f\in\mathcal{F}} \big|\nu_m^*\big(f(Z/\overline{Z}^*)\big) - \nu_m^*(f(Z))\big| > \eta,\ \big|1/\overline{Z}^* - 1\big| < \delta_1\Big\}\\ &\quad + P^*\big\{\big|1/\overline{Z}^* - 1\big| > \delta_1\big\}\\ &\leqslant P^*\Big\{\sup_{f\in\mathcal{F},\ c\in[1-\alpha/4, 1+\alpha/4]} \big|\nu_m^*(f(cZ)) - \nu_m^*(f(Z))\big| > \eta\Big\}\\ &\quad + P^*\big\{\big|1/\overline{Z}^* - 1\big| > \delta_1\big\}.\end{aligned}$$

By Lemma 4

$$\sup_{f\in\mathcal{F},\ c\in[1-\alpha/4,1+\alpha/4]} \left|\widehat{P}_n\big(f(cZ)-f(Z)\big)\right|^2 \longrightarrow \sup_{f\in\mathcal{F},\ c\in[1-\alpha/4,1+\alpha/4]} \left|P\big(f(cZ)-f(Z)\big)\right|^2 < \delta/2,$$

i.e., the equicontinuity holds for the process. By the equivalent property between $\mathcal{F}_{\alpha/2} \in CLT(\{P_n\}_{n\geqslant 0})$ and the asymptotic equicontinuity of $\mathcal{F}_{\alpha/2}$ uniformly in P_n (see (Sheehy and Wellner, 1992)), the right hand side of the above can be made arbitrarily small for almost samples. Hence, we have

$$\nu_m^*\big(f\big(Z/\overline{Z}^*\big)\big) - \nu_m^*\big(f(Z)\big) = o_p(1) \quad \text{a.s. uniformly for} \quad f\in\mathcal{F}.$$

and we have the desired results.

It is easy to see the bootstrap version of spacings processes indexed by functions has the same asymptotical distribution with the spacings processes.

4. BOOTSTRAP STRONG APPROXIMATION RATE

This section is based on the results of Aly *et al.* (1984) where they discuss the k-spacings processes. For convenience we restrict attention to bootstrapping the simple spacings process (with $k = 1$).

LEMMA 5 (Aly *et al.*, 1984). *Define a sequence of Gaussian processes* $\{\widehat{B}(t)\colon\ 0\leqslant t<\infty\}$ *on* $(\Omega,\mathcal{A},\mathcal{P})$ *such that*

$$\sup_{0\leqslant t<\infty} \left|\hat{\nu}_n(t)-\widehat{B}(F(t))\right| = O\big(n^{-1/4}(\log n)^{3/4}\big) \qquad \text{a.s.}$$

and

$$P\Big\{\sup_{0\leqslant t<\infty} \left|\hat{\nu}_n(t)-\widehat{B}\big(F(t)\big)\right| > An^{-1/4}(\log n)^{3/4}\Big\} \leqslant Cn^{-\varepsilon} \qquad \textit{for any} \quad \varepsilon>0,$$

where

$$\hat{\nu}_n(t) = \sqrt{n}\big(\widehat{F}_n(t)-F(t)\big), \qquad \widehat{B}(t) = B(t) + WF^{-1}(t)f\big(F^{-1}(t)\big) \qquad (4.1)$$

with $W=\int_0^1 F^{-1}\,\mathrm{d}B \sim N(0,\sigma^2)$ *and* $B(t)$ *is a standard Brownian bridge.*

LEMMA 6 (Dvoretzky–Kiefer–Wolfowitz inequality).

$$P\Big(\sup_t\left|\widehat{F}_n(t)-F(t)\right| > d\Big) \leqslant c_1 \exp\big(-c_2nd^2\big), \qquad d>0, \qquad (4.2)$$

where c_1, c_2 *are positive constants.*

Proof. We have

$$P\Big(\sup_t |\widehat{F}_n(t) - F(t)| > d\Big)$$
$$\leqslant P\Big(\sup_t |F_n(\overline{Z}t) - F(\overline{Z}t)| > d/2\Big) + P\Big(\sup_t |F(\overline{Z}t) - F(t)| > d/2\Big),$$

but $F(\overline{Z}t) - F(t) = (\overline{Z} - 1)tf(\eta t)$, where $f(\cdot)$ is the density function of exponential (1), η lies between 1 and $\overline{Z}$. Note the fact

$$\sup_{|\eta-1|<d<1,\ -\infty<t<\infty} |tf(\eta t)| = A < \infty.$$

$$P\Big(\sup_t |F(\overline{Z}t) - F(t)| > d/2\Big)$$
$$\leqslant P(|\overline{Z} - 1| > d/2) + P\Big(\sup_t |\overline{Z} - 1|\,tf(\eta t)| > d/2, |\overline{Z} - 1| \leqslant d/2\Big)$$
$$\leqslant P(|\overline{Z} - 1| > d/2) + P(A|\overline{Z} - 1| > d/2).$$

For exponential(1), the moment condition holds for Bernstein's inequality (see (Shorack and Wellner, 1986, p. 855), we have

$$P(|\overline{Z} - 1| > d/2) \leqslant \exp\big(-nc_1d^2\big). \tag{4.3}$$

So by the D–K–W inequality for empirical distribution function

$$P\Big(\sup_t |F_n(\overline{Z}t) - F(\overline{Z}t)| > d/2\Big) \leqslant P\Big(\sup_t |F_n(t) - F(t)| > d/2\Big)$$
$$\leqslant c_2 \exp\big(-cnd^2\big).$$

Combining the above inequalities we have the result.

LEMMA 7. *Assume that* $0 < \lim_{n\to\infty} \inf(m/n) \leqslant \lim_{n\to\infty} \sup(m/n) < \infty$. *Then*

$$P\left\{\left|\sqrt{m}(\overline{Z}^* - 1) - \int_0^\infty t\,\mathrm{d}B(F(t))\right| > An^{-1/2}(\log n)^2\right\} \leqslant Cn^{-\varepsilon} \tag{4.4}$$

and

$$P\{|\overline{Z}^* - 1| > An^{-1/2}(\log n)^{1/2}\} = \mathrm{O}(n^{-\varepsilon}). \tag{4.5}$$

Proof. Let $\eta_1, \eta_2, \ldots, \eta_m$ be i.i.d. $U(0,1)$, then $F_m^*(t) = U_m(\widehat{F}_n(t))$, where

$$U_m(t) = \frac{1}{m}\sum_{i=1}^m I(U_i < t), \qquad \alpha_m(t) = \frac{1}{\sqrt{m}}\sum_{i=1}^m \big[I(U_i < t) - t\big].$$

Note that

$$\overline{Z}^* - 1 = \int_0^\infty t\,\mathrm{d}\big(F_m^*(t) - \widehat{F}_n(t)\big) = -\int_0^\infty \big[F_m^*(t) - \widehat{F}_n(t)\big]\,\mathrm{d}t$$

$$= -\int_0^\infty \big[U_m\big(\widehat{F}_n(t)\big) - \widehat{F}_n(t)\big]\,\mathrm{d}t = -\frac{1}{\sqrt{m}}\int_0^\infty \alpha_m\big(\widehat{F}_n(t)\big)\,\mathrm{d}t$$

and

$$\left|\sqrt{m}(\overline{Z}^* - 1) + \int_0^\infty B\big(F(t)\big)\,\mathrm{d}t\right|$$

$$= \left|\int_0^\infty \big(\alpha_m\big(\widehat{F}_n(t)\big) - B(F(t)\big)\,\mathrm{d}t\right|$$

$$= \left|\int_0^{Z_{(n)}/\overline{Z}} \big(\alpha_m\big(\widehat{F}_n(t)\big) - B(F(t)\big)\,\mathrm{d}t\right| + \left|\int_{Z_{(n)}/\overline{Z}}^\infty \big(0 - B(F(t)\big)\,\mathrm{d}t\right|$$

$$\leqslant \frac{Z_{(n)}}{\overline{Z}} \sup_{0\leqslant t<\infty} \left|\alpha_m\big(\widehat{F}_n(t)\big) - B\big(F(t)\big)\right| + \int_{Z_{(n)}/\overline{Z}}^\infty \left|B\big(F(t)\big)\right|\mathrm{d}t.$$

Observe that

$$P\{Z_{(n)} > c\log n\} = 1 - P^n\{Z < c\log n\} = 1 - \{1 - n^{-c}\}^n = \mathrm{O}(n^{-\varepsilon})$$

when $c > 1$. (In fact $P(\lim_{n\to\infty} Z_{(n)}/\log n = 1) = 1$, see (Galambos, 1978, p. 224). Rest of the proof is similar to that in (Aly *et al.*, 1984, Lemma 3.1 and 3.2.).

LEMMA 8. *Under the same conditions of Lemma* 7,

$$P\left\{\left|\sqrt{m}\big(\widehat{F}_n\big(\overline{Z}^*t\big) - \widehat{F}_n(t)\big) - tf(t)\int_0^\infty t\,\mathrm{d}B(F(t))\right| > n^{-1/4}(\log n)^{3/4}\right\} = \mathrm{O}(n^{-\varepsilon}). \tag{4.6}$$

Proof. We have

$$\begin{aligned}
&\sqrt{m}\big[\widehat{F}_n\big(\overline{Z}^*t\big) - \widehat{F}_n(t)\big]\\
&= \sqrt{m}\big[\widehat{F}_n\big(\overline{Z}^*t\big) - F\big(\overline{Z}^*t\big)\big] - \sqrt{m}\big[\widehat{F}_n(t) - F(t)\big] + \sqrt{m}\big[F\big(\overline{Z}^*t\big) - F(t)\big]\\
&= \sqrt{m/n}\big[\hat{\nu}_n\big(\overline{Z}^*t\big) - \hat{\nu}_n(t)\big] + \sqrt{m}\big[F\big(\overline{Z}^*t\big) - F(t)\big]\\
&= \sqrt{m/n}\big\{\big[\hat{\nu}_n\big(\overline{Z}^*t\big) - \widehat{B}\big(F\big(\overline{Z}^*t\big)\big)\big] - \big[\hat{\nu}_n(t) - \widehat{B}\big(F(t)\big)\big]\big\}\\
&\quad + \sqrt{m/n}\big[\widehat{B}\big(F\big(\overline{Z}^*t\big)\big) - \widehat{B}\big(F(t)\big)\big] + \sqrt{m}\big(F\big(\overline{Z}^*t\big) - F(t)\big).
\end{aligned}$$

By one-term Taylor expansion, we have

$$|F(\overline{Z}^*t) - F(t)| = f(t\eta)t|\overline{Z}^* - 1|,$$

where η lies between 1 and $\overline{Z}^*$. Let $A_n(\delta) = \{\omega\colon |\overline{Z}^* - 1| < \delta\}$, n large enough to make $An^{-1/2}(\log n)^{1/2} < \delta$, then $P\{A_n^c(\delta)\} \leqslant Bn^{-\varepsilon}$ and on $A_n(\delta)$, we have

$$tf(t\eta) \leqslant \sup_{1-\delta<\eta<1+\delta, 0\leqslant t<\infty} tf(t\eta) = M < \infty.$$

Then by using Lemma 7,

$$\begin{aligned} &P\Big\{\sup_{0\leqslant t<\infty} |F(\overline{Z}^*t) - F(t)| > Mn^{-1/2}(\log n)^{1/2}\Big\} \\ &\leqslant P\{A_n^c(\delta)\} \\ &\quad + P\Big\{A_n(\delta)\bigcap\Big\{\sup_{0\leqslant t<\infty} |F(\overline{Z}^*t) - F(t)| > Mn^{-1/2}(\log n)^{1/2}\Big\}\Big\} \\ &\leqslant Cn^{-\varepsilon} + P\Big\{|\overline{Z}^* - 1| > An^{-1/2}(\log n)^{1/2}\Big\} \\ &\leqslant 2Cn^{-\varepsilon} \quad \text{for } n \text{ large enough.} \end{aligned}$$

Note that from (4.1)

$$\widehat{B}(F(\overline{Z}^*t)) - \widehat{B}(F(t)) = B(F(\overline{Z}^*t)) - B(F(t)) + Wt\{\overline{Z}^*f(\overline{Z}^*t) - f(t)\}.$$

Just like the proof in (Aly *et al.*, 1984, Lemma 3.2) it is easy to prove

$$P\Big\{|\widehat{B}(F(\overline{Z}^*t)) - \widehat{B}(F(t))| > n^{-1/4}(\log n)^{3/4}\Big\} \leqslant Cn^{-\varepsilon},$$

$$P\Big\{\big|[\hat{\nu}_n(\overline{Z}^*t) - \widehat{B}(F(\overline{Z}^*t))] - [\hat{\nu}_n(t) - \widehat{B}(F(t))]\big| > n^{-1/4}(\log n)^{3/4}\Big\} \leqslant Cn^{-\varepsilon}.$$

Finally,

$$\sqrt{m}\{F(\overline{Z}^*t) - F(t)\} = \sqrt{m}f(t)t(\overline{Z}^* - 1) + \sqrt{m}f'(t\eta)t^2(\overline{Z}^* - 1)^2,$$

it is easy to get the result.

THEOREM 4. *Assume that* $0 < \lim_{n\to\infty}\inf(m/n) \leqslant \lim_{n\to\infty}\sup(m/n) < \infty$. *Then we can define a sequence of Brownian motions* $\{\widehat{B}(t),\ 0 \leqslant t \leqslant 1\}$ *such that*

$$\begin{aligned} &P\Big\{\sup_{0\leqslant t<\infty} |\sqrt{m}(\widehat{F}_m^*(t) - \widehat{F}_n(t)) - \widehat{B}(F(t))| > An^{-1/4}(\log n)^{3/4}\Big\} \\ &\leqslant Cn^{-\varepsilon}, \end{aligned} \tag{4.7}$$

for any $\varepsilon > 0$.

Proof.

$$\sqrt{m}(\widehat{F}^*_m(t) - \widehat{F}_n(t)) = \sqrt{m}(F^*_m(\overline{Z}^*t) - F^*_m(t)) + \sqrt{m}(F^*_m(t) - \widehat{F}_n(t)).$$

$$\begin{aligned}
&\sqrt{m}(\widehat{F}^*_m(t) - \widehat{F}_n(t)) \\
&= \sqrt{m}\{U_m(\widehat{F}_n(\overline{Z}^*t)) - U_m(\widehat{F}_n(t))\} + \sqrt{m}\{U_m(\widehat{F}_n(t)) - \widehat{F}_n(t)\} \\
&= \{\alpha_m(\widehat{F}_n(\overline{Z}^*t)) - \alpha_m(\widehat{F}_n(t))\} + \sqrt{m}\{\widehat{F}_n(\overline{Z}^*t) - \widehat{F}_n(t)\} \\
&\quad + \alpha_m(\widehat{F}_n(t)).
\end{aligned} \tag{4.8}$$

Using the Komlós, Major, and Tusnády's construction, we can define a sequence of Brownian bridges $B(t)$ which are independent of $(Z_1, \ldots, Z_n)$ such that

$$P\Big\{\sup_{0\leqslant t\leqslant 1} |\alpha_m(t) - B(t)| > A_{11}n^{-1/2}(\log n)^{1/2}\Big\} \leqslant B_{12}n^{-\varepsilon} \tag{4.9}$$

for any $\varepsilon > 0$ (see (Csörgö *et al.*, 1986, p. 158)). By Theorem 2c of Burke *et al.* (1981) (for example (Csörgö and Révész, 1981, Lemma 1.1.1)), we have

$$P\Big\{\sup_{0<s\leqslant 1-h_n}\ \sup_{0\leqslant t\leqslant h_n} |B(s) - B(s+t)| > A_{13}n^{-1/4}(\log n)^{3/4}\Big\} \leqslant B_{13}n^{-\varepsilon} \tag{4.10}$$

with $h_n = A_{12}n^{-1/2}$. By Lemma 6, we have

$$P\Big\{\sup_t |\widehat{F}_n(t) - F(t)| > A_{12}n^{-1/2}(\log n)^{1/2}\Big\} \leqslant B_{12}n^{-\varepsilon}.$$

Note the fact that

$$\begin{aligned}
&\sup_t |\widehat{F}_n(\overline{Z}^*t) - \widehat{F}_n(t)| \\
&\leqslant \sup_t |\widehat{F}_n(\overline{Z}^*t) - F(\overline{Z}^*t)| + \sup_t |\widehat{F}_n(t) - F(t)| + \sup_t |F(\overline{Z}^*t) - F(t)| \\
&\leqslant 2\sup_t |\widehat{F}_n(t) - F(t)| + \sup_t |F(\overline{Z}^*t) - F(t)|.
\end{aligned}$$

Combining the above inequality with Lemma 8, we have

$$P\Big\{\sup_t |\alpha_m(\widehat{F}_n(\overline{Z}^*t)) - \alpha_m(\widehat{F}_n(t))| > An^{-1/4}(\log n)^{3/4}\Big\} \leqslant Cn^{-\varepsilon}$$

and

$$P\Big\{\sup_{0\leqslant t<\infty} |\alpha_m(\widehat{F}_n(t)) - B(F(t))| > A_1n^{-1/4}(\log n)^{3/4}\Big\} \leqslant Cn^{-\varepsilon} \tag{4.11}$$

for any $\varepsilon > 0$ and some constants. From this, the desired result follows.

REFERENCES

Aly, E. E. A. A., Beirlant, J. and Horváth, L. (1984). Strong and weak approximations of k-spacings processes. *Z. Wahrsch. verw. Geb* **66**, 461–484.

Burke, M. D., Csörgö, S. and Horváth, L. (1981). Strong approximations of some biometric estimates under random censorship. *Z. Wahrsch. verw. Geb* **56**, 87–112.

Csörgö, M. and Révész, P. (1981). *Strong Approximations in Probability and Statistics*. Academic Press/ Akademiai Kiado, New York, Budapest.

Csörgö, M., Csörgö, S. and Horváth, L. (1986). *An Asymptotic Theory for Empirical Reliability and Concentration Processes*. Springer, Berlin.

Galambos, J. (1978). *The Asymptotic Theory of Extreme Order Statistics*. Wiley, New York.

Geer, Van de. (1988). Regression analysis and empirical processes. *CWI Tract* **45**, Amsterdam.

Gine, E. and Zinn, J. (1984). Some limit theorems for empirical processes. *Ann. Probab.* **12**, 929–989.

Pyke, R. (1965). Spacings. *J. Roy. Statist. Soc. Ser. B* **27**, 395–449.

Pyke, R. (1972). Spacings revisited. In: *Proc. Sixth Berkeley Symposium Math. Statist. Probab., Vol. 1*, pp. 417–427,

Pollard, D. (1982). A central limit theorem for empirical processes. *J. Austr. Math. Soc. Series A* **33**, 235–248.

Pollard, D. (1984). *Convergence of Stochastic Processes*. Springer, Berlin.

Rao, J. S. and Sethuraman, J. (1975). Weak convergence of empirical distribution functions of random variables subject to perturbations and scale factors. *Ann. Statist.* **3**, 299–313.

Sheehy, A. and Wellner, J. A. (1992). Uniform Donsker classes of functions. *Ann. Probab.* **20**, 1983–2030.

Shorack, G. (1972). Convergence of quantile and spacings processes with applications. *Ann. Math. Statist.* **43**, 1400–1411.

Shorack, G. and Wellner, J. A. (1986). *Empirical Processes with Applications to Statistics*. Wiley, New York.

Multivar. Statist., pp. 35–41
E.-M. Tiit *et al.* (Eds)

ON THE CORNISH–FISHER EXPANSION IN FINITE POPULATION

I. TRAAT

Department of Mathematical Statistics, University of Tartu,
Liivi 2, EE2400 Tartu, Estonia

ABSTRACT

Cornish-Fisher expansions of a quantile of the studentized sample mean and of the ratio estimator are given up to the term $n^{-1/2}$ when sampling from a finite population.

1. INTRODUCTION

Finite population parameters like mean, total and proportion, are of special interest in Survey Sampling practice. It is very important to give the estimates of these parameters together with their confidence intervals. The exact distributions of the corresponding estimators are usually unknown, but for large sample sizes the ordinary normal approximation works quite well. There are still many situations, e.g., in domain estimation, where more exact approximations are needed. Often the unknown distribution of an estimator is approximated by the Edgeworth expansion. For example, Bai and Rao (1992) received the Edgeworth expansion for the ratio of two sample means in i.i.d. case allowing one of the variables to be counting variable. Babu and Singh (1985) treated the finite population case and proved the validity of the one-term Edgeworth expansion for the multivariate sample mean and for some of its functions, from which the special cases for studentized sample mean and ratio estimator follow. Edgeworth expansion helps to calculate the probability that a random variable belongs to a given interval. Cornish–Fisher expansion, in opposite, enables us to construct the interval corresponding to the given probability level. In the current paper two different estimators — studentized sample mean and studentized ratio estimator — are considered. Cornish–Fisher expansions of the quantiles of these estimators are formally given up to the order $n^{-1/2}$. References to the results guaranteeing approximation validity of order $o(n^{-1/2})$ for the corresponding probability are made. In the limit process $n \to \infty$, $N \to \infty$, we assume that $n/N \leqslant 1 - \delta$, for some $\delta > 0$.

2. POPULATION CHARACTERISTICS AND ESTIMATORS

Let us have a two-dimensional vector characterizing finite population of N objects:

$$\begin{pmatrix} y_i \\ x_i \end{pmatrix}, \qquad i = 1, \dots, N.$$

We suppose that distribution function F_N assigns a probability mass $1/N$ to each object in the population and we denote the expectation operator according to this distribution by E_N. We use traditional capital letters for the finite population means

$$\overline{Y} = E_N\, y = \frac{1}{N} \sum_{i=1}^{N} y_i,$$

$$\overline{X} = E_N\, x,$$

and the notation M_k for the higher central moments of y-variable:

$$M_k = E_N(y - E_N\, y)^k = \frac{1}{N} \sum_{i=1}^{N} \left(y_i - \overline{Y}\right)^k.$$

Consider the sample of size n

$$\begin{pmatrix} y_i \\ x_i \end{pmatrix}, \qquad i = 1, \dots, n,$$

drawn without replacement. We are interested in an interval estimate of the population mean $\overline{Y}$. Results for the mean are extendable for the population total through the relationship

$$Y = N\overline{Y},$$

and for the population proportion if to consider the following y-variable:

$$y_i = \begin{cases} 1, & \text{the object has a certain attribute,} \\ 0, & \text{does not have.} \end{cases}$$

We observe two different estimators of $\overline{Y}$ – sample mean $\bar{y}_1$ and ratio estimator $\bar{y}_2$:

$$\bar{y}_1 = \frac{1}{n} \sum_{i=1}^{n} y_i,$$

$$\bar{y}_2 = \frac{\bar{y}}{\bar{x}} \overline{X},$$

where $\bar{x}$ is the sample mean of an auxillary variable. Corresponding studentized statistics are

$$T_{n,j} = \frac{\bar{y}_j - \overline{Y}}{[\widehat{D}\bar{y}_j]^{1/2}}, \qquad j = 1, 2,$$

where $\widehat{D}\bar{y}_j$ is the sample estimate of the variance of $\bar{y}_j$. To find the Cornish–Fisher expansion of the α-quantile $z_{\alpha,j}$ of the statistic $T_{n,j}$, first, cumulants of $T_{n,j}$ are needed.

3. CUMULANTS OF STUDENTIZED MEAN AND RATIO ESTIMATOR

In the case of $\bar{y}_1$ the ordinary sample estimate of its variance is

$$\widehat{D}\bar{y}_1 = \frac{1-f}{n} s^2,$$

where

$$s^2 = \frac{1}{n}\sum_{i=1}^{n}(y_i - \bar{y})^2$$

and the multiplier $1-f$ with $f = n/N$ is a finite population correction. Expanding $T_{n,1}$ with respect to $(\bar{y},\ s^2)$ at the point $(\overline{Y},\ \sigma_N^2)$, where

$$\sigma_N^2 = M_2,$$

we get

$$T_{n,1} = \frac{\sqrt{n}}{\sqrt{1-f}\,\sigma_N}(\bar{y} - \overline{Y}) - \frac{\sqrt{n}}{2\sqrt{1-f}\,\sigma_N^3}(\bar{y} - \overline{Y})(s^2 - \sigma_N^2) + \cdots.$$

It is rather complicated to find expectations of $T_{n,j}$ over the sampling distribution, because there exists dependence between sample elements in the finite population case. For our case some of the necessary moments of $\bar{y}$ and s^2 are given, e.g., in (Kendall and Stuart, 1966, p. 414–420). One also needs to notice that $\bar{y}$ and s^2 are k-statistics, and to use the relationships between cumulants and moments, knowing that r-order cumulants of k-statistics are of order $O(n^{-(r-1)})$. Population size N comes into the expressions through the ratio n/N, which is bounded in the observed limit process. Finally, we have got

$$E\,T_{n,1} = -\frac{\sqrt{1-f}}{2\sqrt{n}}\frac{M_3}{\sigma_N^3} + \mathrm{o}\left(\frac{1}{\sqrt{n}}\right), \tag{3.1}$$

$$E(T_{n,1} - ET_{n,1})^3 = \frac{1}{\sqrt{n}}\left[\frac{1-2f}{\sqrt{1-f}} - 3\sqrt{1-f}\right]\frac{M_3}{\sigma_N^3} + \mathrm{o}\left(\frac{1}{\sqrt{n}}\right). \tag{3.2}$$

To study the ratio estimator $\bar{y}_2$, usually a new variable z is defined

$$z_i = y_i - Rx_i, \qquad i = 1, \ldots, N,$$

where

$$R = \frac{\overline{Y}}{\overline{X}}.$$

Then

$$E_N z = 0, \qquad \bar{z} = \bar{y} - R\bar{x}, \qquad E\bar{z} = 0$$

and the studentized ratio estimator takes the form

$$T_{n,2} = \frac{\bar{z}}{\bar{x}} \frac{\overline{X}}{[(1-f)s^2/n]^{1/2}},$$

where

$$s^2 = \frac{1}{n}\sum_{i=1}^{n}\left(y_i - \frac{\bar{y}}{\bar{x}}x_i\right)^2.$$

After expanding $T_{n,2}$ with respect to $(\bar{x}, \bar{z}, s^2)$ at the point $(\overline{X}, 0, \omega_N^2)$,where

$$\omega_N^2 = \frac{1}{N}\sum_{i=1}^{N}(y_i - Rx_i)^2 = E_N\, z^2,$$

we have

$$T_{n,2} = \frac{\sqrt{n}}{\sqrt{1-f}\,\omega_N}\left[\bar{z} - \frac{1}{\overline{X}}(\bar{x} - \overline{X})\bar{z} - \frac{1}{2\omega_N^2}\left(s^2 - \omega_N^2\right)\bar{z}\right] + \cdots.$$

Taking expectations from here we get the necessary cumulants of $T_{n,2}$:

$$ET_{n,2} = -\frac{\sqrt{1-f}}{2\sqrt{n}}\frac{E_N\, z^3}{\omega_N^3} + o\left(\frac{1}{\sqrt{n}}\right), \tag{3.3}$$

$$E(T_{n,2} - ET_{n,2})^3 = \frac{1}{\sqrt{n}}\left[\frac{1-2f}{\sqrt{1-f}} - 3\sqrt{1-f}\right]\frac{E_N\, z^3}{\omega_N^3} + o\left(\frac{1}{\sqrt{n}}\right). \tag{3.4}$$

The main term of the variance of $T_{n,j}$, $j = 1, 2$, is equal to 1:

$$E(T_{n,j} - ET_{n,j})^2 = 1 + o\left(\frac{1}{\sqrt{n}}\right). \tag{3.5}$$

4. CORNISH–FISHER EXPANSION

Cornish and Fisher (1937) proposed an approximation of the quantile of one distribution through the quantile of the normal distribution. The approximation is given in the form of a power series according to $n^{-1/2}$, including Hermite polynomials and expressions of cumulants of the distribution under interest. Without specifying the expressions of cumulants, the first terms of Cornish–Fisher expansion have been given explicitly in many papers, see (Fisher and Cornish, 1960; Withers, 1984; Lee and Lee, 1992). To present the Cornish–Fisher expansion for some specified statistic, one needs to know the cumulants of that statistic. They are found in the form of power series according to the powers of sample size and usually it is possible to present explicitly only one-two first terms of the expansion. In the i.i.d. case these expressions have been given for several statistics, mainly in the papers dealing with Edgeworth expansions. In the present paper we consider the finite population case. Then, together with n, also population size N comes into expressions. But, when $N \to \infty$, the finite population case tends to the i.i.d. case, and thus, the ordering of the expansions, like Edgeworth and Cornish–Fisher, according the powers of $n^{-1/2}$ is still meaningful.

Lee and Lee (1992) introduce the simple technique to derive the Cornish–Fisher expansion. They look for the relationship

$$z = \xi + \delta, \tag{4.1}$$

so that

$$F(z) = \Phi(\xi), \tag{4.2}$$

where $\Phi(\cdot)$ is the standard normal and $F(\cdot)$ is some other distribution function. The small quantity δ is supposed to be a power series of $n^{-1/2}$

$$\delta = \delta_1 + \delta_2 + \cdots \tag{4.3}$$

with δ_k of order $n^{-k/2}$. Applying the Taylor expansion to $F(\xi + \delta)$ and then the Edgeworth expansion to $F(\xi)$ they get the equation without unknown F

$$\Phi(\xi) = \exp\left\{\delta D + \sum_{k=1}^{\infty} \lambda_k(-D)^k/k!\right\}\Phi(\xi), \tag{4.4}$$

where D is the derivative operator, λ_k, $k \neq 2$, is the k-order cumulant of the distribution F (λ_2 is the second cumulant minus 1). Lee and Lee (1992) consider the adjusted distribution F with $\lambda_1 = \lambda_2 = 0$ and with λ_k of order $n^{1-k/2}$, $k \geqslant 3$. Expanding the exponent in (4.4) and using then the partition of an integer method they write down the general recursive expression for δ_k through λ_k.

In our case, the cumulants do not depend on n is a way as expected by Lee and Lee (1992), and thus, instead of their final result it is better to use the equation (4.4) directly for our specified statistic $T_{n,j}$. Corresponding cumulants are expressed through (3.1)–(3.5). Denoting the main terms of these cumulants by $\lambda_{1,j}$, $\lambda_{3,j}$ and the main term of δ by $\delta_{1,j}$

$$ET_{n,j} = \lambda_{1,j} + \mathrm{o}\left(\frac{1}{\sqrt{n}}\right),$$

$$E(T_{n,j} - ET_{n,j})^3 = \lambda_{3,j} + \mathrm{o}\left(\frac{1}{\sqrt{n}}\right),$$

we obtain from (4.4)

$$\Phi(\xi) = \exp\left(\delta_{1,j}D - \lambda_{1,j}D - \lambda_{3,j}D^3/6 + \cdots\right)\Phi(\xi).$$

After expanding we have

$$\Phi(\xi) = \big(1 + \delta_{1,j}D - \lambda_{1,j}D - \lambda_{3,j}D^3/6 + \cdots\big)\Phi(\xi).$$

Using derivatives, defined by

$$D\Phi(\xi) = \phi(\xi),$$
$$D^3\Phi(\xi) = \big(\xi^2 - 1\big)\phi(\xi),$$

where $\phi(\xi)$ is the density function of the standard normal distribution, we have

$$\delta_{1,j} = \lambda_{1,j} + \lambda_{3,j}\big(\xi^2 - 1\big)/6.$$

If the equation (4.2) equals to α then (4.1) presents the relationship between α-quantile z_α of F and α-quantile ξ_α of the standard normal distribution. For the α-quantile of $T_{n,j}$ we have then obtained the following Cornish–Fisher expansion:

$$z_{\alpha,j} = \xi_\alpha + \lambda_{1,j} + \lambda_{3,j}\big(\xi_\alpha^2 - 1\big)/6, \tag{4.5}$$

where $\lambda_{1,j}, \lambda_{3,j}$ are given by (3.1)–(3.4).

Babu and Singh (1985) have presented the Edgeworth expansion for the studentized one-dimensional functions of multivariate sample mean, including treated here statistics $T_{n,j}$:

$$P(T_{n,j} < x) = \Phi(x) - \big(\lambda_{1,j} + \lambda_{3,j}\big(x^2 - 1\big)/6\big)\phi(x) + \mathrm{o}\big(n^{-1/2}\big).$$

They have proved that this one-term expansion is valid if in the limit process $n \to \infty$, $N \to \infty$ the absolute moments of order $6 + \delta$ of F_N are bounded for some $\delta > 0$, $n/N \leqslant 1 - \delta$ and F_N converges weakly to a strongly nonlattice distribution.

To say something about the accuracy of the quantile expansion, notice, that

$$P(T_{n,j} < z_{\alpha,j}) = P\big(T_{n,j} - \lambda_{1,j} - \lambda_{3,j}\big(\xi_\alpha^2 - 1\big)/6 < \xi_\alpha\big).$$

Since the random variable

$$T_{n,j} - \lambda_{1,j} - \lambda_{3,j}\big(\xi_\alpha^2 - 1\big)/6$$

has the same cumulants as $T_{n,j}$, except the first cumulant (mean), which main term equals to

$$-\lambda_{3,j}\big(\xi_\alpha^2 - 1\big)/6,$$

the $n^{-1/2}$-term in its Edgeworth expansion vanishes and we get

$$P(T_{n,j} < z_{\alpha,j}) = \Phi(\xi_\alpha) + \mathrm{o}\big(n^{-1/2}\big) = \alpha + \mathrm{o}\big(n^{-1/2}\big).$$

As a conclusion, we may state the following.

THEOREM 1. *A one-term Cornish–Fisher expansion in the finite population case for the studentized sample mean $T_{n,1}$ is of the form*

$$z_{\alpha,1} = \xi_\alpha - \frac{\sqrt{1-f}}{6\sqrt{n}} \frac{M_3}{\sigma_N^3} \left[3\xi_\alpha^2 - (\xi_\alpha^2 - 1)\frac{1-2f}{1-f} \right]$$

and for the studentized ratio estimator $T_{n,2}$

$$z_{\alpha,2} = \xi_\alpha - \frac{\sqrt{1-f}}{6\sqrt{n}} \frac{E_N z^3}{\omega_N^3} \left[3\xi_\alpha^2 - (\xi_\alpha^2 - 1)\frac{1-2f}{1-f} \right].$$

If the absolute moments of order $6+\delta$ of F_N are bounded for some $\delta > 0$ in the limit process $n \to \infty$, $N \to \infty$, $n/N \leqslant 1-\delta$ and F_N converges weakly to a strongly nonlattice distribution, then the following is valid for the obtained quantile expansions:

$$P(T_{n,j} < z_{\alpha,j}) = \alpha + o(n^{-1/2}), \qquad j = 1, 2.$$

REFERENCES

Babu, G. J. and Singh, K. (1985). Edgeworth expansion for sampling without replacement from finite populations. *J. Multivariate Anal.* **17**, 261–278.

Bai, Z. D. and Rao, C. R. (1992). A note on Edgeworth expansions for ratio of sample means. *Sankhiā, Series A* **54**, 309–322.

Cornish, E. A. and Fisher, R. A. (1937). Moments and cumulants in the specification of distributions. *Rev. Internat. Statist. Inst.* **5**, 307–320.

Fisher, R. A. and Cornish, E. A. (1960). The percentile points of distributions having known cumulants. *Technometrics* **2**, 209–225.

Kendall, M. G. and Stuart, A. (1966). *The Advanced Theory of Statistics.* Nauka, Moscow.

Lee, Y. S. and Lee, M. C. (1992). On the derivation and computation of the Cornish–Fisher expansion. *Austral. J. Statist.* **34**, 443–450.

Withers, C. R. (1984). Asymptotic expansions for distributions and quantiles with power series cumulants. *J. R. Statist. Soc. B.* **46**, 389–396.

Part Two

GENERAL LINEAR REGRESSION MODELS

Multivar. Statist., pp. 45–56
E.–M. Tiit *et al.* (Eds)

ON PROPORTIONALITY OF REGRESSION COEFFICIENTS IN MISSPECIFIED GENERAL LINEAR REGRESSION MODELS

ULRICH PÖTTER

Max Planck Institute for Human Development and Education,
Lentzeallee 94, D-14195 Berlin, Germany

ABSTRACT

It is shown that estimated regression coefficients in misspecified general linear regression models (the conditional distribution of the variable of interest given the regressors is a function of a linear combination of the regressors) are approximately proportional to the "true" regression coefficients. The constant of proportionality is computed explicitly and a second order approximation is given. The connexion of this result with similar but stronger findings under the additional assumptions of a convex loss function and normally distributed regressors (Li and Duan, 1989) is clarified.

1. INTRODUCTION

Over recent years a large amount of empirical evidence – both obtained from real data sets and simulations – has been accumulated for a peculiar stability property of regression models. Using different and possibly misspecified models results in estimates of regression parameters that are roughly proportional. This is true also if certain kinds of incomplete data are ignored in the analysis. It has long been known that Probit and Logit models often give similar results (Chambers and Cox, 1967). D'Agostino *et al.*(1990) show that a pooled logistic regression model and two variants of Cox' proportional hazards model give nearly identical results for their data sets and offer some reason for the identity. Based on simulations a similar comparison is made by Ingram and Kleinman (1989). Doksum and Gasko (1990) discuss the general relationship between models in binary regression and survival analysis. Addison and Portugal (1987) compare several parametric regressions for a data set on the duration of unemployment and find proportional regression coefficients in all models. Bergström and Edin (1992) give similar results also, using data on unemployment duration. In addition they show that grouping of the dependent variable will attenuate the regression coefficients by a constant factor. Petersen (1991) and Petersen and Koput (1992) treat the effects of grouping in censored regression.

In contrast to the abundance and generality of the empirical findings, theoretical explanations are rare and cover only special cases. Attenuation effects were shown by Bretagnolle and Huber-Carol (1985) for Cox' proportional hazards model if covariates are left out. Ruud (1986) reviews earlier results on proportionality of regression parameters in "limited dependent variable" models with normal regressors. Results for misspecified correlated Logits were obtained by Neuhaus *et al.* (1992). Consistency results for linear location-scale models were given by Gould and Lawless (1988). In the case of least squares estimates in a grouped normal regression Stewart (1983) shows that proportionality is asymptotically exact for jointly normal regressors. Chung and Goldberger (1984) generalize his findings to arbitrary information loss under a linearity condition on the reverse regression. More recently, Goldenshluger and Polyak (1993) showed consistency of the least squares estimator in linear models posing very weak conditions on the error term (the models allow for autoregression and other forms of dependence as well as nearly arbitrary distributions) when the regressors are i.i.d. with a symmetric joint distribution, with expectation 0 and finite fourth moment.

However, there are two general theoretical approaches towards an explanation that seem to cover most of the empirical findings. The first approach places restrictions on the joint distribution of the covariates. It originated from observations on the behaviour of ordinary least squares estimators in nonlinear situations when the regressors are normal (Brillinger, 1983). These results were later generalized for models of binary outcomes and other "limited dependent variable" models (see (Ruud, 1986)). Li and Duan (1989) offered the most general proof to date for the class of general linear regression models. These include models for binary outcomes, transformation models, censored regression and generalized linear models.

The second approach is based on a Taylor approximation argument that requires no conditions on the distribution of the regressors. Solomon (1984, 1986) was apparently the first to use it for the comparison of a proportional hazards model when an accelerated life model is true and for an accelerated life model when a proportional hazards model is true. Struthers and Kalbfleisch (1986) gave more general conditions for the behaviour of the partial likelihood estimator (similar to the ones given later by Hjort (1992)), but also used a Taylor approximation argument to get the proportionality result in the partial likelihood case. In a discussion of a paper by Cox and Reid (1987), Skinner (1987) hinted at the possibility of extending the argument of Solomon to the general linear regression case. Later he expanded on the interpretability of proportional regression coefficients (Skinner, 1989). In that paper he used the same arguments as Li and Duan for the general linear regression case but did not give a proof via the alternative Taylor approximation.

In this paper I will make Skinner's remarks explicit. The constant of proportionality will be computed and the second order approximation will be given. Then I try to elucidate the relation between Skinner's and Li and Duan's result.

2. NOTATION

The conditional distribution of an outcome variable Y given the values of covariates x describes a *general linear regression model* if the distribution of Y depends on x only through a linear combination of the x. The conditional density (or probability) of Y may be written as

$$f(y \mid \alpha_0 + x\alpha), \tag{1}$$

where $f(\cdot \mid \cdot)$ belongs to some class $\mathcal{F}$ of regular conditional densities, x is a $(1 \times p)$ row vector, α_0 is a scalar, and α is the $(p \times 1)$ vector of regression coefficients. It will be assumed that the parameter α is identified at least up to a scalar multiple:

$$\begin{aligned} &f(y \mid \alpha_0 + x\alpha) = g(y \mid \alpha_0^* + x\alpha^*) \qquad \forall x, y; \quad f, g \in \mathcal{F} \\ &\implies \quad \alpha^* = \gamma\alpha, \qquad \gamma \in \mathbb{R}\backslash\{0\}. \end{aligned} \tag{2}$$

The class of general regression models retains the interpretability of relative magnitudes of regression coefficients, a point stressed by Skinner (1989): Let e_i be the i-th unit vector, δ a real scalar. Then

$$\begin{aligned} f(y \mid (x + \delta e_i)\alpha) &= f(y \mid x\alpha + \delta\alpha_i) = f\left(y \mid x\alpha + \delta\,\frac{\alpha_i}{\alpha_j}\,\alpha_j\right) \\ &= f\left(y \mid \left(x + \delta\,\frac{\alpha_i}{\alpha_j}\,e_j\right)\alpha\right), \end{aligned}$$

so that α_i/α_j, $\alpha_j \neq 0$ is the amount of change in x_j required to achieve an effect on the density equivalent to a unit change in x_i. By the identifiability assumption (2) these "equivalent effects" ratios (Skinner, 1989) do not depend on the chosen representation, f.

It is sometimes useful to describe the model given by (1) in terms of random variables instead of densities. Let us write

$$Y = T(h(x\alpha, \varepsilon)), \qquad \varepsilon \sim G(\cdot), \tag{3}$$

where ε is a real valued random variable with distribution function $G(\cdot)$ independent of x, $h(\cdot, \cdot)$ is a strictly monotonic function in its second argument and $T(\cdot)$ is a (not necessarily strictly) monotonic function. A transformation from (1) to the representation (3) may be accomplished by choosing $T(h(\cdot, \cdot))$ to be the generalized inverse of the conditional distribution function in (1) and $G(\cdot)$ to be the uniform distribution on $(0, 1)$. If (3) holds, (1) follows because the independence of ε and x implies that the conditional distribution of Y given $X = x$ only depends on $x\alpha$. Note that the representation (3) is not unique: The functions $T(\cdot)$ and $h(\cdot, \cdot)$ can always be merged into a single $h^*(\cdot, \cdot)$ and

$G(\cdot)$ may be chosen as the uniform distribution by accommodating the function $h^*(\cdot,\cdot)$ accordingly. However, (3) allows an interpretation in terms of a latent variable $Y^* = h(x\alpha, \varepsilon)$ that is observed only after some form of transformation or information loss represented by the function $Y = T(Y^*)$. Most of the "limited dependent variable" models are motivated by such a construction. Moreover, if a continuous covariate x_i with non-zero regression coefficient α_i is present, and $h(\cdot,\cdot)$ is also strictly monotonic in its first argument, then a conceptually simple estimator consistent for $\alpha/\|\alpha\|$ is easily constructed: take the normalized value $\hat{\alpha}$ that maximizes the rank correlation between $x\alpha$ and Y (see (Han, 1987) for details). Obviously this works even if $T(\cdot)$ and $h(\cdot,\cdot)$ are unknown. Note that the existence of consistent estimators implies identifiability as in (2) for that class of models (see, e.g. (Deistler and Seifert, 1978)).

To complete the description of the stochastic setup let $(Y_i, X_i)_{i=1\ldots n}$ be a sequence of i.i.d. vectors where the conditional density of Y given $X = x$ is given by (1), and the marginal distribution of X is nondegenerate with $E(X'X) = \Sigma > 0$ and – without loss of generality – $E(X) = 0$. Sometimes further assumptions on the existence of moments, the differentiability of certain functions and the interchange of differentiation and integration will be needed. These will be clear from the context.

Only maximum likelihood type estimators will be considered explicitly. Denote by

$$\sum_i \ell(y_i, x_i; \beta) = \sum_i \ln g(y_i \mid x_i\beta) \tag{4}$$

the log-likelihood function under an assumed model $g(\cdot \mid \cdot)$ in the class of general regression models but not necessarily identical to the data generating model (1). The first order conditions for a maximum of the log-likelihood function are given by setting the pseudo score functions to 0:

$$\sum_i U(y_i, x_i; \beta) = \sum_i \frac{\partial}{\partial \beta} \ell(y_i, x_i; \beta) = \sum_i x_i' g^*(y_i, x_i\beta) = 0, \tag{5}$$

where $g^*(\cdot,\cdot)$ denotes the derivative of $\ln g(\cdot \mid \cdot)$ with respect to its second argument. The formulation allows for nuisance parameters as long as the solution of (5) is unaffected by them. The sequence of estimators $\hat{\beta}_n$, $n \to \infty$ solving (5) will be Fisher consistent for the solution of

$$\begin{aligned} E\big(U(Y,X;\beta)\big) &= E\big(X'g^*(Y,X\beta)\big) \\ &= E_X\big(X'E_{Y|X}(g^*(Y,x\beta) \mid X = x)\big) = 0, \end{aligned} \tag{6}$$

provided such a solution exists. This generally implies strong consistency. Under some further smoothness assumptions on $\ell(\cdot)$, asymptotic normality of the estimators follows. See (Li and Duan, 1989, p. 1029) and the literature on the behaviour of maximum likelihood type estimators in misspecified models (see, e.g. (Huber, 1967; White, 1982, 1983; Gourieroux *et al.*, 1984; Fahrmeir,

1990)). Hjort (1992) gave similar results for censored data using the counting process framework.

Since most of the arguments below use only the basic structure $E(X'g^*(\cdot,\cdot))$ or similar structures for the expected value of the log-likelihood $\ell(\cdot)$ under the true model, the results will also pertain to any estimators of analogous form. This includes, e.g., Cox' partial likelihood in the proportional hazards case, and any M-estimator derived from maximizing a criterion function by setting its derivative with respect to β to 0.

3. TWO SPECIAL CASES: LINEAR MODELS AND LEAST SQUARES UNDER INFORMATION LOSS

3.1. *Linear models*

The "true" model is given by a location-scale model with a fixed density $f_0(\cdot)$. That is

$$f(y \mid \alpha_0 + x\alpha) = \sigma^{-1} f_0\left(\frac{y - \alpha_0 - x\alpha}{\sigma}\right), \qquad Y = \alpha_0 + x\alpha + \sigma\varepsilon, \quad \varepsilon \sim F_0.$$

Estimation is carried out by assuming another location-scale model with density $g_0(\cdot)$:

$$g(y \mid \beta_0 + x\beta) = \tau^{-1} g_0\left(\frac{y - \beta_0 - x\beta}{\tau}\right). \tag{7}$$

If maximum likelihood estimation based on the model (7) is used, the expected score function $U(\cdot)$ for β becomes

$$E\big(U(Y, X; \beta)\big) = -\frac{1}{\tau} E_X\left(X' E_{Y|X}\left[g_0^*\left(\frac{Y - \beta_0 - x\beta}{\tau}\right) |X = x\right]\right) = 0.$$

Conditional on $X = x$

$$Y - \beta_0 - x\beta = x(\alpha - \beta) + (\alpha_0 - \beta_0) + \sigma\varepsilon. \tag{8}$$

Setting $\beta = \alpha$ makes the inner expectation independent of X so that the expectation with respect to X will be 0 (remember $E(X') = 0$). Therefore $\beta = \alpha$ is a solution of the expected score function which implies consistency of the regression coefficients despite misspecification of the disturbance distribution. This is the result of Gould and Lawless (1988): If a location-scale model is true, then using any other location-scale model for maximum likelihood estimation will lead to consistent estimators of the regression slopes, without any further restrictions on the covariates. Note that the nuisance parameters τ and α_0 do not affect the solution of the expected score equation.

3.2. *Least squares under transformations and information loss*

Assume that a general regression model holds for a latent variable Y^* with fixed distribution of the disturbance ε. Let $Y = T(Y^*)$ be observed, where $T(\cdot)$ is some arbitrary function. Examples include binary regression models with $T(z) = I[z \geqslant 0]$ (where $I[\cdot]$ is the indicator function), grouped data situations, and transformation of the dependent variable as in Box–Cox transforms. The least squares normal equations with expectation

$$E\big(X'(Y - X\beta - \beta_0)\big) = 0$$

are used as estimating equations. The solution for the slopes is

$$\beta = \Sigma^{-1}E(X'Y).$$

Now assume that the conditional expectation of X given Y^* is linear in Y^*:

$$E(X \mid Y^* = y^*) = \delta\big(y^* - E(Y^*)\big). \tag{9}$$

This places some restrictions on the distribution of the regressors X. E.g., the condition is implied by joint normality of the regressors together with the regression structure given by (1).

Observe that δ is the least squares projection of X on Y^*, $\delta = E(X'Y^*)/\operatorname{var}(Y^*)$. Then

$$E(X'Y) = E_{Y^*}\big(YE_{X|Y^*}(X \mid Y^*)\big) = E_{Y^*}\big(Y\delta(Y^* - E(Y^*))\big) = \delta\operatorname{cov}(Y, Y^*).$$

By the preceding example, the regression slopes in the underlying model are given by $\alpha = \Sigma^{-1}E(X, Y^*)$. Therefore

$$\begin{aligned}\beta &= \Sigma^{-1}E(X'Y) = \Sigma^{-1}\delta\operatorname{cov}(Y, Y^*)\\ &= \Sigma^{-1}\big(E(X'Y^*)/\operatorname{var}(Y^*)\big)\operatorname{cov}(Y, Y^*)\\ &= \Sigma^{-1}\big(\Sigma\alpha/\operatorname{var}(Y^*)\big)\operatorname{cov}(Y, Y^*)\\ &= \alpha\big(\operatorname{cov}(Y, Y^*)/\operatorname{var}(Y^*)\big).\end{aligned} \tag{10}$$

The least squares estimator in the model under information loss is proportional to the underlying vector of regression slopes and the proportionality factor is given by the least squares regression slope of Y on Y^*.

A special case of the result was given in (Stewart, 1983) for grouped normal regression. The present version is due to (Chung and Goldberger, 1984). They also point out that it is sufficient to treat the regression slopes of the least squares regression of Y^* on X as the "true" parameter α without assuming an explicit model for the conditional distribution of Y^* given X. Thus, assuming only the existence of moments, a regression result is compared with the one that would have been obtained if there had been no information loss. This operationalistic view may prove to be very helpful for judging empirical findings.

4. SKINNER'S APPROACH

Returning to the general situation, assume that data are generated by a model of the form (1) for some $f(\cdot)$, α_0, α. Score functions from an assumed model $g(\cdot,\cdot)$ of the form (5) are used for estimation. These estimators will converge to the solution of the expected score equation (6). If a solution exists and is unique, Skinner (1987) proposed to use a Taylor approximation to $\beta(\alpha)$ in terms of α around $\alpha = 0$.

In fact, if $\alpha = 0$, setting $\beta = 0$ makes $g^*(\cdot,\cdot)$ independent of X so that the outer expectation in (6) – the covariance between X and $E_{Y|X}(g^*(Y, x\beta) \mid X = x)$ – vanishes. So a solution of (6) exists for $\alpha = 0$.

Taking derivatives with respect to β in the expected score equation (6) at $\alpha = \beta = 0$ – assuming the interchange of derivative and expectation is justified – results in

$$\frac{\partial}{\partial\beta} E\big(X'g^*(Y, X\beta)\big)\big|_{\alpha=\beta=0} = E\big(X'Xg_2^*(Y,0)\big),$$

where $g_2^*(y,\eta)$ is the derivative of $g^*(y,\eta)$ with respect to η. Since $-g_2^*(y,\eta)$ is the "observed" information for η in the assumed model, it will be positive unless the model is unidentified. Therefore its expectation with respect to Y given X will also be positive so that the derivative reduces to

$$\frac{\partial}{\partial\beta} E\big(U(Y, X; \beta)\big) = \kappa\Sigma \tag{11}$$

for some scalar $\kappa < 0$.

Taken together, the implicit function theorem may be invoked and the function $\beta(\alpha)$ may be approximated by

$$\beta(\alpha) \simeq \beta(0) + D\beta(0)\alpha = D\beta(0)\alpha$$

around $\alpha = 0$, where D is the differential operator. It remains to compute $D\beta(0)$ explicitly. Set

$$k(x\alpha, x\beta) = E_{Y|X}\big(g^*(Y, x\beta) \mid X = x\big).$$

Then

$$\begin{aligned} D\beta(0) &= \frac{\partial\beta(\alpha)}{\partial\alpha}\Big|_{\alpha=0} \\ &= -\left[\frac{\partial}{\partial\beta} E_X\big(X'k(X\alpha, X\beta)\big)\big|_{\alpha=\beta=0}\right]^{-1} \\ &\quad \times \left[\frac{\partial}{\partial\alpha} E_X\big(X'k(X\alpha, X\beta)\big)\big|_{\alpha=\beta=0}\right] \\ &= -\big[k_2(0,0)\Sigma\big]^{-1}\big[k_1(0,0)\Sigma\big] = -\frac{k_1(0,0)}{k_2(0,0)} I = \gamma I, \end{aligned} \tag{12}$$

where $k_1(\mu,\eta)$ ($k_2(\mu,\eta)$) are the partial derivatives of $k(\mu,\eta)$ with respect to μ (resp. η) and interchanges of integral and derivative are once again assumed to be valid. Since the terms for the covariances of X cancel, $D\beta(0)$ reduces to a scalar times the identity matrix. Thus the slopes of the regression under the assumed model are (to first order) proportional to the regression slopes under the data generating model.

It is instructive to compare this result to the Example 3.2. When the linearity condition (9) does not hold, the conclusion of the example may not be applicable. Nevertheless, it is possible, with the help of the preceding formula to give an approximation to the least squares estimator under information loss. Specifically, let $T(z) = I[z \geqslant 0]$. That is, least squares method is applied to binary data generated from $T(x\alpha + \alpha_0 + \varepsilon)$, $\varepsilon \sim F(\cdot)$. Then:

$$\frac{\partial}{\partial\beta} E_X\big(X' E_{Y|X}\big(T(x\alpha + \alpha_0 + \varepsilon) - x\beta - \beta_0\big)\big) = -E_X(X'X) = -\Sigma.$$

Moreover,

$$\begin{aligned}\frac{\partial}{\partial\alpha} E_X(X' E_{Y|X}\big(T(x\alpha + \alpha_0 + \varepsilon) - x\beta - \beta_0\big)\big)\Big|_{\alpha=0} \\ = \frac{\partial}{\partial\alpha} E_X\big(X'\big(1 - F(-X\alpha - \alpha_0)\big)\big)\Big|_{\alpha=0} = f(-\alpha_0)\Sigma.\end{aligned} \tag{13}$$

Therefore, the constant of proportionality in this case is given by $\gamma = f(-\alpha_0)$.

If (X,ε) is jointly normal, this coincides with the solution (10). Further explicit expressions for the constant of proportionality in special cases are given in (Solomon, 1984; Galler and Pötter, 1992).

Higher orders of approximation may be obtained from the implicit function theorem together with a further application of the chain rule along the lines of (12). To second order, the Taylor approximation is given by

$$\beta(\alpha) \simeq D\beta(0)\alpha + \frac{1}{2}\big(\alpha' D_l^2\beta(0)\alpha\big)_{l=1,\dots,p}$$

with $D_l^2\beta(0)$ the Hessian of $\beta_l(\alpha)\,|_{\alpha=0}$. Now

$$\begin{aligned}D^2\beta(0) = -\big[E\big(X'Xg_2^*(Y,0)\big)\big]^{-1}\big(D_1^2 E_X\big(X'k(X\alpha, X\beta)\big) \\ + 2D\beta(0)D_{1,2}^2 E_X\big(X'k(X\alpha, X\beta)\big) \\ + \big(D\beta(0)\big)^2 D_2^2 E_X\big(X'k(X\alpha, X\beta)\big)\big),\end{aligned}$$

where D_1^2 is the second order differential operator with respect to α, D_2^2 the one with respect to β and $D_{1,2}^2$ the one with respect to α and β. At $\alpha = \beta = 0$ all terms except the first (inverse) term evaluate to $E(X_iX_jX_l)_{i,j,l\in\{1,\dots,p\}}$ times a constant. Therefore, the second order term vanishes if all third order moments vanish. This happens, e.g., if the joint distribution of X is symmetric with respect to 0, and the third moments exist. Note that no assumption of absolute continuity of the distribution of the covariates needs to be imposed. Some explicit results on the second order terms were given by Solomon (1984) in the special case of the partial likelihood under accelerated failure models.

5. THE APPROACH BASED ON NORMAL REGRESSORS

The second general approach to explain the stability of regression ratios starts with placing restrictions on the distribution of regressors. Let the expected log-likelihood $E(\ell(Y;\beta_0 + X\beta))$ be convex in $X\beta$. If $E_{X|X\alpha}(X\beta \mid x\alpha)$ is linear in $x\alpha$ for all $\beta \in \mathbb{R}^p$, then the maximizer of the likelihood $\hat{\beta}$ converges to $\gamma\alpha$, where γ is a scalar (Li and Duan, 1989, Theorem 2.1). Since α is unknown in applications, the linearity condition will normally be required for all values of α. In this case the proportionality of regression coefficients holds for all values of α, not only approximately for small α. But the restriction on the marginal distribution of the regressors is rather severe: It is, e.g., implied by elliptical distributions, but discrete covariates or non-elliptical distributions are excluded.

The proof of Li and Duan's result is very simple:

$$
\begin{aligned}
E(\ell(Y;b_0 + Xb)) &= E_{X\alpha,\varepsilon}\big(E_{X|X\alpha,\varepsilon}\big(\ell\big(T(h(X\alpha,\varepsilon)); b_0 + Xb\big) \mid X\alpha,\varepsilon\big)\big)\\
&\leqslant E_{X\alpha,\varepsilon}\big(\ell\big(T(h(X\alpha,\varepsilon)); b_0 + E_{X|X\alpha,\varepsilon}(Xb \mid X\alpha,\varepsilon)\big)\big)\\
&\quad \text{(by Jensen's inequality)}\\
&= E_{X\alpha,\varepsilon}\big(\ell\big(T(h(X\alpha,\varepsilon)); b_0 + c + \gamma X\alpha\big)\big)\\
&\quad \text{(by linearity).}
\end{aligned}
$$

Therefore the expected log-likelihood at some b is always smaller than the expected log-likelihood at $\gamma\alpha$, and the result follows.

The convexity condition can be dispensed with at the price of strengthening the distributional assumptions. If X is jointly normal, then the conclusion of the theorem holds without the convexity condition. Li and Duan (1989, Theorem 2.2) prove this by a direct appeal to properties of the multivariate normal distribution. However, it is possible to prove proportionality of regression results using Stein's lemma, thus providing some insight into a possible connexion with the previous results. Recall Stein's lemma: if X is multivariate normal with covariance I (for simplicity) and $g(\cdot)$ a real, differentiable function with $E(\|Dg(X)\|) < \infty$, then $E(X'g(X)) = E(Dg(X))$ (see, e.g. (Ibragimov and Has'minskii, 1981, p. 25)). Now suppose the regressors are multivariate normal. Then

$$
0 = E\big(U(X\beta, Y)\big) = E_X\left(X'k(X\alpha, X\beta)\right) = (\alpha,\beta)E_X\begin{pmatrix} k_1(X\alpha, X\beta)\\ k_2(X\alpha, X\beta)\end{pmatrix}
$$

by Stein's lemma. Therefore,

$$
\beta = -\frac{E_X(k_1(X\alpha, X\beta))}{E_X(k_2(X\alpha, X\beta))}\,\alpha. \tag{14}
$$

This proves proportionality without using the convexity of the log-likelihood, but requires normality. Note that the constant of proportionality agrees with the previous result (12) for $\alpha = \beta = 0$, so that a Taylor argument would lead to the result of Skinner, though using much stronger assumptions. Note also that the proof starts with the expected score function, not with the expected log-likelihood. These similarities together with the second order result (13) prompt for generalizations either via generalizations of Stein's lemma or via Skinner's approach. But trying higher orders of the approximation seems to be less promising: although the accumulating moment conditions may lead to approximations to the normal distribution, it would require increasingly stronger differentiability assumptions for the score function, thus excluding many interesting estimators. Moreover, such assumptions are absent in the approach based on normal regressors.

Another strategy might be to apply generalizations of Stein's lemma to non-normal distributions, perhaps on the lines of Cacoullos and Papathanasiou (1992). They give identities similar to Stein's that hold for a larger class of densities, so that a standard integration by parts argument works. Unfortunately, they also show that the class of densities where this works also has to fulfill a moment condition: The conditional expectation of X_i given all other variables is linear in these variables. This is rather close to the conditions imposed by Li and Duan.

6. CONCLUSIONS

Under a wide variety of circumstances regression models seem to possess a remarkable property of stability: the relative magnitude of the regression coefficients are stable across different models used for estimation and under many forms of information loss. Despite of the many empirical findings a solid theoretical understanding of this property is still missing. I have concentrated here on one specific approach towards an explanation: the use of Taylor approximations via expected estimating equations. This approach has been implicit in the work of Solomon (1984), Skinner (1987) and others. I developed a more explicit version, computed the constant of proportionality and gave a second order approximation. This, and a reanalysis of the other general approach via normality assumptions showed some similarities between them. But still neither could be reduced to the other. Moreover, as the examples in Section 3 demonstrate, neither approach is able to incorporate sufficiently these special results: in the first example the constant of proportionality equals one throughout the parameter space without any conditions on the regressors, in the second linearity of the inverse regression is needed. This is implied by the assumption of normal errors together with the linear form of the models, but not vice versa. Moreover, it was possible to compute the constant of proportionality having a nice interpretation, while the computation from the approximation formula is generally awkward.

While it seems relatively straightforward to extend both approaches to include stochastic forms of data loss (censoring, selection etc.) and to give results for multivariate response models, general results on the set of models and situations in which a given model leads to consistent estimators (up to scale) seem not to be easily derived. The work of Lazrieva and Toronjadze (1991) gives some answers in this direction, though they deal with the partial likelihood scheme which lends itself easily to such questions.

REFERENCES

Addison, J. T. and Portugal, P. (1987). On the distributional shape of unemployment duration. *Rev. Ec. Statist.* **69**, 520–526.

Bergström, R. and Edin, P.-A. (1992). Time aggregation and the distributional shape of unemployment duration. *J. Appl. Econometrics* **7**, 5–30.

Bretagnolle, J. and Huber-Carol, C. (1985). Sous-estimation des contrastes due à l'oublie de variables pertinentes dans le modèle de Cox pour les durées de survie avec censure. *C. R. Acad. Sci. Paris, Sér. I* **300**, 359–362.

Brillinger, D. R. (1983). A generalized linear model with "Gaussian" regressor variables. In: *A Festschrift for E.L. Lehmann*, pp. 97–114, P. J. Bickel *et al.* (Eds), Wadsworth, Belmont, CA.

Cacoullos, T. and Papathanasiou, V. (1992). Lower variance bounds and a new proof of the central limit theorem. *J. Multivariate Anal.* **43**, 173–184.

Chambers, E. A. and Cox, D. R. (1967). Discrimination between alternative binary response models. *Biometrika* **54**, 573–578.

Chung, C.-F. and Goldberger, A. S. (1984). Proportional projections in limited dependent variable models. *Econometrica* **52**, 531–534.

Cox, D. R. and Reid, N. (1987). Parameter orthogonality and approximate conditional inference. *J. Roy. Statist. Soc., B* **49**, 1–39.

D'Agostino, R. B., Lee, M.-L., Belanger, A. J., Cupples, L. A., Anderson, K. and Kannel, W. B. (1990). Relation of pooled logistic regression to time dependent Cox regression analysis: The Framingham Heart Study. *Statist. Med.* **9**, 1501–1515.

Deistler, M. and Seifert, H. G. (1978). Identifiability and consistent estimability in econometric models. *Econometrica* **46**, 969–980.

Doksum, K. A. and Gasko, M. (1990). On a correspondence between models in binary regression analysis and in survival analysis. *Intern. Statist. Rev.* **58**, 243–252.

Fahrmeir, L. (1990). Maximum likelihood estimation in misspecified generalized linear models. *Statistics* **21**, 487–502.

Galler, H.-P. and Pötter, U. (1992). Zur Robustheit von Schätzmodellen für Ereignisdaten. In: *Herausforderungen an den Wohlfahrtsstaat im strukturellen Wandel*, pp. 379–405, R. Hujer *et al.* (Eds), Campus, Frankfurt.

Goldenshluger, A. V. and Polyak, B. T. (1993). Estimation of regression parameters with arbitrary noise. *Math. Methods Statist.* **2**, 18–29.

Gould, A. and Lawless, J. F. (1988). Consistency and efficiency of regression coefficient estimates in location-scale models. *Biometrika* **75**, 535–540.

Gourieroux, C., Montfort, A. and Trognon, A. (1984). Pseudo maximum likelihood methods: Theory. *Econometrica* **52**, 681–700.

Han, A. K. (1987). Non-parametric analysis of a generalized regression model. *J. Econometrics* **35**, 303–316.

Hjort, N. L. (1992). On inference in parametric survival data models. *Intern. Statist. Rev.* **60**, 355–387.

Huber, P. J. (1967). The behavior of maximum likelihood estimates under nonstandard conditions. In: *Proceedings of the Fifth Berkeley Symposium on Mathematical Statistics and Probability*, pp. 221–233, University of California Press, Berkeley.

Ibragimov, I. A. and Has'minskii, R. Z. (1981). *Statistical Estimation. Asymptotic Theory.* Springer, Berlin.

Ingram, D. D. and Kleinman, J. C. (1989). Empirical comparisons of proportional hazards and logistic regression models. *Statist. Med.* **8**, 525–538.

Lazrieva, N. L. and Toronjadze, T. A. (1991). On stable M-estimators in the partial likelihood scheme. In: *New Trends in Probability and Statistics*, Vol. 1, pp. 567–596, V. V. Sazonov and T. Shervashidze (Eds), VSP, Utrecht/Mokslas, Vilnius.

Li, K.-C. and Duan, N. (1989). Regression analysis under link violation. *Ann. Statist.* **17**, 1009–1052.

Lin, D. Y. and Wei, L. J. (1989). The robust inference for the Cox proportional hazards model. *J. Amer. Statist. Assoc.* **84**, 1074–1078.

Neuhaus, J. M., Hauck, W. W. and Kalbfleisch, J. D. (1992). The effects of mixture distribution misspecification when fitting mixed-effects logistic models. *Biometrika* **79**, 755–762.

Nishii, R. (1988). Maximum likelihood principle and model selection when the true model is unspecified. *J. Multivariate Anal.* **27**, 392–403.

Petersen, T. (1991). Time aggregation bias in continuous-time hazard-rate models. In: *Sociological Methodology*, pp. 261–290, Marsden, P. V. (Ed.), Blackwell, Oxford.

Petersen, T. and Koput, K. W. (1992). Time-aggregation bias in hazard-rate models with covariates. *Soc. Meth. Res.* **21**, 25–51.

Ruud, P. A. (1986). Consistent estimation of limited dependent variable models despite misspecification of distribution. *J. Econometrics* **32**, 157–187.

Skinner, C. J. (1987). Comment on "Parameter orthogonality and approximate conditional inference" by Cox and Reid. *J. Roy. Statist. Soc. B* **49**, 24.

Skinner, C. J. (1989). GLM's and coefficient ratios. In: *GLIM 89 Proceedings, Trento*, Springer.

Solomon, P. J. (1984). Effect of misspecification of regression models in the analysis of survival data. *Biometrika* **71**, 291–298.

Solomon, P. J. (1986). Effect of misspecification of regression models in the analysis of survival data. Correction. *Biometrika* **73**, 245.

Stewart, M. B. (1983). On least squares estimation when the dependent variable is grouped. *Rev. Ec. Studies* **50**, 737–753.

Struthers, C. A. and Kalbfleisch, J. D. (1986). Misspecified proportional hazard models. *Biometrika* **73**, 363–369.

White, H. (1982). Maximum likelihood estimation of misspecified models. *Econometrica* **50**, 1–25.

White, H. (1983). Maximum likelihood estimation of misspecified models. Correction. *Econometrica* **51**, 513.

Multivar. Statist., pp. 57–66
E.–M. Tiit *et al.* (Eds)

MORE ON PARTITIONED POSSIBLY RESTRICTED LINEAR REGRESSION

H. J. WERNER[1,*] and C. YAPAR[2]
[1]Institute for Econometrics and Operations Research,
Econometrics Unit, University of Bonn, D-53113 Bonn, Germany,

[2]Karadeniz Technical University, Faculty of Arts and Sciences,
Dept. of Stat. and Comp. Sci., 61080 Trabzon, Turkey

ABSTRACT

This paper deals with the *general* partitioned linear regression model where the regressor matrix $X = (X_1 \quad X_2)$ may be deficient in column rank, the dispersion matrix V is possibly singular, $\beta^t = (\beta_1^t \quad \beta_2^t)$ — being partitioned according to X — is the vector of unknown regression coefficients, and β_2 is possibly subject to consistent linear equality or inequality restrictions. In particular, we are interested in the set of *generalized least squares* (GLS) *selections* for β_2. Inspired by Aigner and Balestra (1988), as well as by Nurhonen and Puntanen (1992), we also consider a specific reduced model and describe a scenario under which the set of GLS selections for β_2 under the reduced model equals the set of GLS selections for β_2 under the original full model. The results obtained in (Nurhonen and Puntanen, 1992; Aigner and Balestra, 1988) for the unrestricted *standard* (full rank) regression model are reobtained as special cases.

1. INTRODUCTION

Let $\mathbf{R}^n$, $\mathbf{R}^{n,m}$ and $\mathcal{P}^{n,n}$ denote the set of n-dimensional real column vectors, the set of $n \times m$ real matrices and the set of real non-negative definite (nnd) and symmetric $n \times n$ matrices, respectively. Given $A \in \mathbf{R}^{n,m}$, let A^t, $A^\dagger$, $\mathcal{R}(A)$, $\mathcal{N}(A)$ and $\operatorname{rank}(A)$ denote the transpose, the Moore–Penrose inverse, the range space (or column space), the null space and the rank, respectively, of A. In addition, let A^- denote an arbitrary *g-inverse* of A satisfying $AA^-A = A$; denote the set of all g-inverses of A by $\{A^-\}$. Further, let I and 0, respectively, stand for the identity and zero matrix of whatever size is appropriate to the context.

Consider the Gauss–Markov model

$$\mathcal{L}(\mathcal{R}) := (y,\ X_1\beta_1 + X_2\beta_2,\ V \mid \beta \in \mathcal{R})$$

*Financial support by Deutsche Forschungsgemeinschaft, Sonderforschungsbereich 303 at the University of Bonn, is gratefully acknowledged.

in which y is an observable random vector with expectation $X_1\beta_1 + X_2\beta_2$ and dispersion V; the vector $\beta := (\beta_1^t \quad \beta_2^t)^t$ of unknown regression coefficients satisfies the a priori restriction $\beta \in \mathcal{R}$; $\mathcal{R}$ is of one of the following cases:

- $\mathcal{R} = \mathcal{R}_e := \{\beta \mid R\beta = r\}$ (*e*quality constrained case),
- $\mathcal{R} = \mathcal{R}_i := \{\beta \mid R\beta \geqslant r\}$ (*i*nequality constrained case),
- $\mathcal{R} = \mathcal{R}_u := \mathbf{R}^m$, $m := m_1 + m_2$ (*u*nconstrained case);

and the model matrices X_1 $(n \times m_1)$, X_2 $(n \times m_2)$, V, R $(p \times m)$, $r \in \mathbf{R}^p$, are fixed and known. There are no rank assumptions on $X := (X_1 \quad X_2)$ and V. However, for $R = (R_1 \quad R_2)$ being partitioned according to $X = (X_1 \quad X_2)$ we assume throughout that $R_1 = 0$ and $\operatorname{rank}(R) = p$. In addition, we require that the conditions

$$\mathcal{R}(X) \subseteq \mathcal{R}(V) \tag{1.1}$$

and

$$\mathcal{R}(X_1) \cap \mathcal{R}(X_2) = \{0\} \tag{1.2}$$

are always satisfied.

For each $a \in \{e, i, u\}$: let $\mathcal{L}_a$ briefly denote the model $\mathcal{L}(\mathcal{R}_a)$. Note that, under the assumption (1.1),

$$y \in \mathcal{R}(V) \qquad \text{(almost surely)}, \tag{1.3}$$

irrespective of under which of the three models y is observed. Due to invariance (cf. Theorem 2.3 in (Werner and Yapar, 1995)), an arbitrary g-inverse V^- of V can be used to define a norm

$$\|x\|_{V^-} := (x^t V^- x)^{1/2}$$

on $\mathcal{R}(V)$. The mathematical programming problem

$$\text{minimize} \quad \|y - Xb\|_{V^-}^2 \quad \text{subject to} \quad b \in \mathcal{R}_a \tag{1.4}$$

is hence well defined for each realization $y \in \mathcal{R}(V)$. Any optimal solution to this convex-quadratic optimization program, that is, any vector from

$$\arg\min_{b \in \mathcal{R}_a} \|y - Xb\|_{V^-}^2 \tag{1.5}$$

is called a *GLS solution* (for β) under model $\mathcal{L}_a$. Although (1.4) possesses an optimal solution for each $y \in \mathcal{R}(V)$, (1.5) need not be a singleton. In which case there do exist many different functions f with $f(y)$ representing a GLS solution for each $y \in \mathcal{R}(V)$. As in (Werner and Yapar, 1995) we call any such function a *GLS selection* for β under $\mathcal{L}_a$; notice that it seems reasonable to reserve the term *'GLS estimator'* for exactly those situations where there does exist only one GLS selection on $\mathcal{R}(V)$. The set of all GLS selections for β under model $\mathcal{L}_a$ will be denoted by $\tilde{\beta}(\mathcal{L}_a)$.

Next, let $M_1 := I - X_1 X_1^\dagger$. Observe that M_1 coincides with the *orthogonal projector* (with respect to the standard inner product) onto $\mathcal{N}(X_1^t)$, that is, $M_1 = P_{\mathcal{N}(X_1^t)}$; cf. Section 2 in (Werner and Yapar, 1995). Hence, in particular, $M_1 X_1 = 0$, and for each $a \in \{e, i, u\}$ it is clear that model

$$\mathcal{M}_a := (M_1 y,\ M_1 X_2 \beta_2,\ M_1 V M_1 \mid \beta \in \mathcal{R}_a)$$

can be obtained from model $\mathcal{L}_a$ by premultiplying y by M_1. Besides these correctly *transformed* models $\mathcal{M}_a$, $a \in \{e, i, u\}$, the *reduced* models

$$\mathcal{N}_a := (M_1 y,\ M_1 X_2 \beta_2,\ V \mid \beta \in \mathcal{R}_a), \qquad a \in \{e, i, u\},$$

are also of interest to us in this paper. Unlike $\mathcal{M}_a$ in $\mathcal{N}_a$ the dispersion of $M_1 y$ is defined to be V. For each $a \in \{e, i, u\}$: let $\tilde{\beta}_2(\mathcal{M}_a)$ and $\tilde{\beta}_2(\mathcal{N}_a)$ stand for the set of all GLS selections for β_2 under $\mathcal{M}_a$ and under $\mathcal{N}_a$, respectively.

For convenience, we further introduce

$$J_1 := \begin{pmatrix} I \\ 0 \end{pmatrix} \quad \text{and} \quad J_2 := \begin{pmatrix} 0 \\ I \end{pmatrix}$$

such that $X_1 = X J_1$ and $X_2 = X J_2$. For each $a \in \{e, i, u\}$, we define $\tilde{\beta}_2(\mathcal{L}_a) := J_2^t \tilde{\beta}(\mathcal{L}_a)$; that is, $\tilde{\beta}_2(\mathcal{L}_a)$ denotes the set of all GLS selections for β_2 under the full model $\mathcal{L}_a$.

This paper is organized as follows. The main results are established in Section 3. There we prove that for each $a \in \{e, i, u\}$ we have $\tilde{\beta}_2(\mathcal{L}_a) = \tilde{\beta}_2(\mathcal{M}_a)$. In addition, we show that $\tilde{\beta}_2(\mathcal{L}_a) = \tilde{\beta}_2(\mathcal{N}_a)$ holds if $V\mathcal{R}(X_1) \subseteq \mathcal{R}(X_1)$ or, equivalently, $V\mathcal{N}(X_1^t) \subseteq \mathcal{N}(X_1^t)$ is satisfied. Recently (see (Werner and Yapar, 1995)) we have derived some interesting representations relating the GLS selections of an equality or inequality constrained model to the GLS selections of the associated unconstrained model. In Section 2, some of these results, playing a key role in Section 3, are restated in the framework of our particular models.

2. PRELIMINARY RESULTS

For the sake of clarity we begin this section with quoting the following known result; see Theorem 2.4 and Theorem 2.10 (ii) in (Werner and Yapar, 1995).

THEOREM 2.1. *For given $A \in \mathbf{R}^{n,m}$ and $W \in \mathcal{P}^{n,n}$, let $\mathcal{G}(A \mid W)$ denote the set of all those matrices $G \in \mathbf{R}^{m,n}$ which satisfy*

$$G \in \{A^-\} \quad \text{and} \quad W\mathcal{N}(A^t) \subseteq \mathcal{N}(AG). \tag{2.1}$$

Then $\mathcal{G}(A \mid W)$ is nonempty. Moreover, if A is of full column rank, then GWG^t is nnd and invariant for any choice of $G \in \mathcal{G}(A \mid W)$.

Since $\mathcal{R}(X) \subseteq \mathcal{R}(V)$ [see (1.1)], Theorem 2.3 in (Werner and Yapar, 1995) tells us that

$$\Gamma_{\mathcal{L}} := X^t V^- X \tag{2.2}$$

is nnd as well as invariant for any choice of V^-. Consider the restrictor matrix $R = (0 \quad R_2)$ from Section 1. Because $\Gamma_{\mathcal{L}} \in \mathcal{P}^{m,m}$, and since R is assumed to be of full row rank, it follows from Theorem 2.1 that $(R^Q)^t \Gamma_{\mathcal{L}} R^Q$ is nnd and invariant for any choice of $(R^Q)^t \in \mathcal{G}(R^t \mid \Gamma_{\mathcal{L}})$; call this unique matrix $\Omega_{\mathcal{L}}$. Likewise it can be seen that

$$\Gamma_{\mathcal{M}} := (M_1 X_2)^t (M_1 V M_1)^- (M_1 X_2) \in \mathcal{P}^{m_2,m_2} \tag{2.3}$$

and

$$\Gamma_{\mathcal{N}} := (M_1 X_2)^t V^- (M_1 X_2) \in \mathcal{P}^{m_2,m_2}, \tag{2.4}$$

are, respectively, invariant for any choice of $(M_1 V M_1)^-$ and V^-. In other words, $\Gamma_{\mathcal{M}}$ and $\Gamma_{\mathcal{N}}$ are also well and uniquely defined nnd matrices. Since R_2 is of full row rank and as $\Gamma_{\mathcal{M}}$ is nnd, it again follows from Theorem 2.1 that $(R_2^Q)^t \Gamma_{\mathcal{M}} R_2^Q$ is nnd and invariant for any choice of $(R_2^Q)^t \in \mathcal{G}(R_2^t \mid \Gamma_{\mathcal{M}})$; call this matrix $\Omega_{\mathcal{M}}$. And finally, on similar lines we obtain that $(R_2^Q)^t \Gamma_{\mathcal{N}} R_2^Q$, henceforth denoted by $\Omega_{\mathcal{N}}$, is nnd and invariant for any choice of $(R_2^Q)^t \in \mathcal{G}(R_2^t \mid \Gamma_{\mathcal{N}})$.

Next, let $\mathcal{A} \in \{\mathcal{L}, \mathcal{M}, \mathcal{N}\}$. Choose any selection $\hat{\beta}_2 \in \tilde{\beta}_2(\mathcal{A}_u)$, and define $\hat{\mu}_{\mathcal{A}} := R_2 \hat{\beta}_2 - r$. For each realization $y \in \mathcal{R}(V)$ (that is, for each realization $\hat{\mu}_{\mathcal{A}}$) it then follows from [(Werner and Yapar, 1995); see the lines around (3.18)] that

$$\widetilde{M}_{\mathcal{A}}(y) := \arg\min_{\mu \geqslant 0} \|\hat{\mu}_{\mathcal{A}} - \mu\|^2_{\Omega_{\mathcal{A}}} \tag{2.5}$$

is nonempty. Let $\widetilde{M}_{\mathcal{A}}$ be pointwise defined according to (2.5). It is pertinent to mention here that $\widetilde{M}_{\mathcal{A}}$ is invariant for any choice of $\hat{\beta}_2$. This happens although $\hat{\mu}_{\mathcal{A}}$, in general, does not share this property with $\widetilde{M}_{\mathcal{A}}$; for more details, we refer to (Werner and Yapar, 1995).

As announced in Section 1 we next state, for each $a \in \{e, i, u\}$, representations for $\tilde{\beta}(\mathcal{L}_a)$, $\tilde{\beta}_2(\mathcal{M}_a)$ and $\tilde{\beta}_2(\mathcal{N}_a)$. All these representations are readily obtained as special cases from Theorem 3.2, Theorem 3.3 and Theorem 3.4 in (Werner and Yapar, 1995); observe that (1.2) is equivalent to $\mathcal{N}(M_1 X_2) = \mathcal{N}(X_2)$.

THEOREM 2.2. *Consider $\mathcal{L}_i$, $\mathcal{L}_e$ and $\mathcal{L}_u$. Let $\hat{\beta}_{\mathcal{L}}$ be an arbitrary but fixed GLS selection for β under model $\mathcal{L}_u$, and let $\Gamma_{\mathcal{L}}$ be defined as above. If $X^{\mathcal{L}}$ and $(R^Q)^t$ denote arbitrary but fixed matrices from $\mathcal{G}(X \mid V)$ and $\mathcal{G}(R^t \mid \Gamma_{\mathcal{L}})$, respectively, then we have*

$$\begin{aligned}\tilde{\beta}(\mathcal{L}_i) = \{&R^Q(r + \tilde{\mu}) + (I - R^Q R)\hat{\beta}_{\mathcal{L}} \\ &+ P_{\mathcal{N}(X)\cap\mathcal{N}(R)} w \mid \tilde{\mu} \in \widetilde{M}_{\mathcal{L}},\ w \text{ arbitrary}\},\end{aligned} \tag{2.6}$$

$$\tilde{\beta}(\mathcal{L}_e) = \{R^Q r + (I - R^Q R)\hat{\beta}_{\mathcal{L}} + P_{\mathcal{N}(X)\cap\mathcal{N}(R)} w \mid w \text{ arbitrary}\}, \tag{2.7}$$

$$\tilde{\beta}(\mathcal{L}_u) = \{X^{\mathcal{L}} y + P_{\mathcal{N}(X)} w \mid w \text{ arbitrary}\}, \tag{2.8}$$

where $\widetilde{M}_{\mathcal{L}}$ is pointwise defined according to (2.5), and where $P_{\mathcal{N}(X)}$, for instance, denotes as usual the orthogonal projector onto $\mathcal{N}(X)$ [along $\mathcal{R}(X^t)$].

THEOREM 2.3. *Consider $\mathcal{M}_i$, $\mathcal{M}_e$ and $\mathcal{M}_u$. Let $\hat{\beta}_{2,\mathcal{M}}$ be an arbitrary but fixed GLS selection for β_2 under model $\mathcal{M}_u$, and let $\Gamma_{\mathcal{M}}$ be defined as above. If $(M_1X_2)^{\mathcal{M}}$ and $(R_2^Q)^t$ denote arbitrary but fixed matrices from $\mathcal{G}(M_1X_2 \mid M_1VM_1)$ and $\mathcal{G}(R_2^t \mid \Gamma_{\mathcal{M}})$, respectively, then we have*

$$\tilde{\beta}_2(\mathcal{M}_i) = \{R_2^Q(r+\tilde{\mu}) + (I - R_2^Q R_2)\hat{\beta}_{2,\mathcal{M}} + P_{\mathcal{N}(X_2)\cap\mathcal{N}(R_2)}w_2 \mid \tilde{\mu} \in \widetilde{M}_{\mathcal{M}},\ w_2 \text{ arbitrary}\}, \tag{2.9}$$

$$\tilde{\beta}_2(\mathcal{M}_e) = \{R_2^Q r + (I - R_2^Q R_2)\hat{\beta}_{2,\mathcal{M}} + P_{\mathcal{N}(X_2)\cap\mathcal{N}(R_2)}w_2 \mid w_2 \text{ arbitrary}\}, \tag{2.10}$$

$$\tilde{\beta}_2(\mathcal{M}_u) = \{(M_1X_2)^{\mathcal{M}}M_1y + P_{\mathcal{N}(X_2)}w_2 \mid w_2 \text{ arbitrary}\}, \tag{2.11}$$

where $\widetilde{M}_{\mathcal{M}}$ is pointwise defined according to (2.5).

THEOREM 2.4. *Consider $\mathcal{N}_i$, $\mathcal{N}_e$ and $\mathcal{N}_u$. Let $\hat{\beta}_{2,\mathcal{N}}$ be an arbitrary but fixed GLS selection for β_2 under $\mathcal{N}_u$, and let $\Gamma_{\mathcal{N}}$ be defined as above. If $(M_1X_2)^{\mathcal{N}}$ and $(R_2^Q)^t$ denote arbitrary but fixed matrices from $\mathcal{G}(M_1X_2 \mid V)$ and $\mathcal{G}(R_2^t \mid \Gamma_{\mathcal{N}})$, respectively, then we have*

$$\tilde{\beta}_2(\mathcal{N}_i) = \{R_2^Q(r+\tilde{\mu}) + (I - R_2^Q R_2)\hat{\beta}_{2,\mathcal{N}} + P_{\mathcal{N}(X_2)\cap\mathcal{N}(R_2)}w_2 \mid \tilde{\mu} \in \widetilde{M}_{\mathcal{N}},\ w_2 \text{ arbitrary}\}, \tag{2.12}$$

$$\tilde{\beta}_2(\mathcal{N}_e) = \{R_2^Q r + (I - R_2^Q R_2)\hat{\beta}_{2,\mathcal{N}} + P_{\mathcal{N}(X_2)\cap\mathcal{N}(R_2)}w_2 \mid w_2 \text{ arbitrary}\}, \tag{2.13}$$

$$\tilde{\beta}_2(\mathcal{N}_u) = \{(M_1X_2)^{\mathcal{N}}M_1y + P_{\mathcal{N}(X_2)}w_2 \mid w_2 \text{ arbitrary}\}, \tag{2.14}$$

where $\widetilde{M}_{\mathcal{N}}$ is pointwise defined according to (2.5).

3. MAIN RESULTS

We formulate our main theorem as follows.

THEOREM 3.1. *For the linear models introduced in Section 1 we have:*

(i) *$\tilde{\beta}_2(\mathcal{L}_a) = \tilde{\beta}_2(\mathcal{M}_a)$ for each $a \in \{i, e, u\}$.*

(ii) *If $V\mathcal{R}(X_1) \subseteq \mathcal{R}(X_1)$, then $\tilde{\beta}_2(\mathcal{L}_a) = \tilde{\beta}_2(\mathcal{N}_a)$ for each $a \in \{i, e, u\}$.*

For proving this theorem we need some further auxiliary results.

THEOREM 3.2. *Let (1.2) be satisfied for the block partitioned matrix $X = (X_1 \quad X_2) \in \mathbf{R}^{n,m}$, and let $V \in \mathcal{P}^{n,n}$. Moreover, let M_1, J_1 and J_2 be all defined as before. If $X^{\mathcal{L}} \in \mathcal{G}(X \mid V)$, then $J_2^t X^{\mathcal{L}} \in \mathcal{G}(M_1X_2 \mid M_1VM_1)$. In addition, $X_2J_2^tX^{\mathcal{L}}X_1 = 0$ or, equivalently, $J_2^tX^{\mathcal{L}}(I - M_1)z \in \mathcal{N}(X_2)$ for each $z \in \mathbf{R}^n$.*

Proof. Let $X^{\mathcal{L}} \in \mathcal{G}(X \mid V)$. For $i = 1, 2$ put $X_i^{\mathcal{L}} := J_i^t X^{\mathcal{L}}$. Then

$$X^{\mathcal{L}} = \begin{pmatrix} X_1^{\mathcal{L}} \\ X_2^{\mathcal{L}} \end{pmatrix}.$$

By definition of $\mathcal{G}(X \mid V)$, $XX^{\mathcal{L}}X = X$ and $V\mathcal{N}(X^t) \subseteq \mathcal{N}(XX^{\mathcal{L}})$. In view of (1.2), clearly $XX^{\mathcal{L}}X_1 = X_1$ and $XX^{\mathcal{L}}X_2 = X_2$ iff

$$X_i X_i^{\mathcal{L}} X_j = \begin{cases} X_i & \text{if } i = j, \\ 0 & \text{if } i \neq j, \end{cases} \tag{3.1}$$

for all $i, j \in \{1, 2\}$. Therefore $X_2X_2^{\mathcal{L}}M_1 = X_2X_2^{\mathcal{L}}$ or, equivalently, $X_2X_2^{\mathcal{L}}(I - M_1) = 0$. This in turn implies $M_1X_2X_2^{\mathcal{L}}M_1X_2 = M_1X_2X_2^{\mathcal{L}}X_2 = M_1X_2$ so that we also have $X_2^{\mathcal{L}} \in \{(M_1X_2)^-\}$. It thus remains to show that $M_1VM_1\mathcal{N}(X_2^tM_1) \subseteq \mathcal{N}(M_1X_2X_2^{\mathcal{L}})$. Of course, $VM_1\mathcal{N}(X_2^tM_1) = V\mathcal{N}(X^t)$; for recall that M_1 represents the orthogonal projector onto $\mathcal{N}(X_1^t)$ along $\mathcal{R}(X_1)$. From (1.2) it further follows that

$$\mathcal{N}(XX^{\mathcal{L}}) = \mathcal{N}(X_1X_1^{\mathcal{L}}) \cap \mathcal{N}(X_2X_2^{\mathcal{L}}). \tag{3.2}$$

But then

$$M_1X_2X_2^{\mathcal{L}}(M_1VM_1)\mathcal{N}(X_2^tM_1) = M_1X_2X_2^{\mathcal{L}}V\mathcal{N}(X^t) = \{0\},$$

since $V\mathcal{N}(X^t) \subseteq \mathcal{N}(XX^{\mathcal{L}})$ and since $X_2X_2^{\mathcal{L}}M_1 = X_2X_2^{\mathcal{L}}$. This completes the proof.

THEOREM 3.3. *Let $V \in \mathcal{P}^{n,n}$, and let* (1.1) *and* (1.2) *be satisfied for the block partitioned matrix $X = (X_1 \quad X_2) \in \mathbf{R}^{n,m}$. Further, let M_1, J_1, J_2, $\Gamma_{\mathcal{L}}$ and $\Gamma_{\mathcal{M}}$ be all defined as before. For convenience, put $\Gamma_{ij} := J_i^t\Gamma_{\mathcal{L}}J_j$, $i, j \in \{1, 2\}$. As usual, let $[\Gamma_{\mathcal{L}}/\Gamma_{11}]$ denote the (generalized) Schur complement of Γ_{11} in $\Gamma_{\mathcal{L}}$, that is, let $[\Gamma_{\mathcal{L}}/\Gamma_{11}] = \Gamma_{22} - \Gamma_{21}\Gamma_{11}^-\Gamma_{12}$. Then $[\Gamma_{\mathcal{L}}/\Gamma_{11}] = \Gamma_{\mathcal{M}}$. Moreover, $J_2^t\Gamma_{\mathcal{L}}^-J_2 \in \{[\Gamma_{\mathcal{L}}/\Gamma_{11}]^-\}$, irrespective of $\Gamma_{\mathcal{L}}^-$. Hence, in particular, $J_2^t\Gamma_{\mathcal{L}}^-J_2 \in \{\Gamma_{\mathcal{M}}^-\}$.*

Proof. First notice that (1.1) holds iff for $i = 1, 2$ we have $\mathcal{R}(X_i) \subseteq \mathcal{R}(V)$ or, equivalently, $\mathcal{N}(V) \subseteq \mathcal{N}(X_i^t)$. Next observe that V^-V is a projector along $\mathcal{N}(V)$; cf. Section 2 in (Werner and Yapar, 1995). Consequently, $X_i^tV^-V = X_i^t$, for $i = 1, 2$. Since $\mathcal{R}(X_1) \subseteq \mathcal{R}(V)$, (2.6) in (Werner and Yapar, 1995) gives us

$$\mathcal{R}(V) = \mathcal{R}(X_1) \oplus V\mathcal{N}(X_1^t)$$

with $\oplus$ indicating direct sum. In view of $\mathcal{R}(X_2) \subseteq \mathcal{R}(V)$, it should now be clear that there (uniquely) exist two matrices, say Z_1 and Z_2, such that $\mathcal{R}(Z_1) \subseteq \mathcal{R}(X_1)$, $\mathcal{R}(Z_2) \subseteq \mathcal{R}(VM_1) = V\mathcal{N}(X_1^t)$ and $X_2 = Z_1 + Z_2$. Since $P_1 := X_1(X_1^tV^-X_1)^-X_1^tV^-$ is a (generally oblique) projector onto $\mathcal{R}(X_1)$ with

$V\mathcal{N}(X_1^t) \subseteq \mathcal{N}(P_1)$ (compare Theorem 2.5 in (Werner and Yapar, 1995)), on the one hand $(I - P_1)X_2 = Z_2$. On the other hand we also get $VM_1(M_1VM_1)^- \times M_1X_2 = Z_2$ because $VM_1(M_1VM_1)^-M_1$ is a projector onto $\mathcal{R}(VM_1)$ satisfying $VM_1(M_1VM_1)^-M_1X_1 = 0$. Consequently $[I - X_1(X_1^tV^-X_1)^-X_1^tV^-]X_2 = VM_1(M_1VM_1)^-M_1X_2$. Combining all these observations results in

$$\begin{aligned}[\Gamma_\mathcal{L}/\Gamma_{11}] &= \Gamma_{22} - \Gamma_{21}\Gamma_{11}^-\Gamma_{12} \\ &= X_2^tV^-X_2 - X_2^tV^-X_1(X_1V^-X_1)^-X_1^tV^-X_2 \\ &= X_2^tV^-\left[I - X_1(X_1^tV^-X_1)^-X_1^tV^-\right]X_2 \\ &= X_2^tV^-VM_1(M_1VM_1)^-M_1X_2 \\ &= X_2^tM_1(M_1VM_1)^-M_1X_2, \\ &= \Gamma_\mathcal{M}.\end{aligned}$$

As is well known (cf. (Pringle and Rayner, 1971, p. 46)), $J_2^t\Gamma_\mathcal{L}^-J_2 \in \{[\Gamma_\mathcal{L}/\Gamma_{11}]^-\}$. Therefore $J_2^t\Gamma_\mathcal{L}^-J_2 \in \{\Gamma_\mathcal{M}^-\}$, and the proof is complete.

THEOREM 3.4. *Consider the models $\mathcal{L}_i$ and $\mathcal{M}_i$, and let M_1, J_1, J_2, $\Gamma_\mathcal{L}$, $\Gamma_\mathcal{M}$, $\Omega_\mathcal{L}$ and $\Omega_\mathcal{M}$ be all defined as before. For each matrix $(R^Q)^t \in \mathcal{G}(R^t \mid \Gamma_\mathcal{L})$ we then have*

$$(R^Q)^t J_2 \in \mathcal{G}(R_2^t \mid \Gamma_\mathcal{M}) \tag{3.3}$$

and

$$(R^Q)^t\Gamma_\mathcal{L}R^Q = (R^Q)^t J_2\Gamma_\mathcal{M}J_2^tR^Q. \tag{3.4}$$

Therefore $\Omega_\mathcal{L} = \Omega_\mathcal{M}$.

Proof. Let $(R^Q)^t \in \mathcal{G}(R^t \mid \Gamma_\mathcal{L})$. Then, by definition of $\mathcal{G}(R^t \mid \Gamma_\mathcal{L})$,

$$R^Q \in \{R^-\} \quad \text{and} \quad (R^QR)^t\Gamma_\mathcal{L}\mathcal{N}(R) = \{0\}. \tag{3.5a-b}$$

Since R^t is of full column rank, (3.5b) happens iff

$$(R^Q)^t\Gamma_\mathcal{L}\mathcal{N}(R) = \{0\}. \tag{3.6}$$

As $R = (0 \quad R_2)$, $\mathcal{N}(R) = \mathbf{R}^{m_1} \times \mathcal{N}(R_2)$ where $\times$ indicates a cartesian product. But now it is evident that (3.6) is equivalent to

$$(R^Q)^t\Gamma_\mathcal{L}J_1 = 0, \qquad (R^Q)^t\Gamma_\mathcal{L}J_2\mathcal{N}(R_2) = \{0\}. \tag{3.7a-b}$$

For convenience we put $\Gamma_{ij} := J_i^t\Gamma_\mathcal{L}J_j$ and $R_i^Q := J_i^tR^Q$ $(i, j = 1, 2)$. Then

$$\Gamma_\mathcal{L} = \begin{pmatrix} \Gamma_{11} & \Gamma_{12} \\ \Gamma_{21} & \Gamma_{22} \end{pmatrix} \quad \text{and} \quad R^Q = \begin{pmatrix} R_1^Q \\ R_2^Q \end{pmatrix}.$$

By means of this notation, (3.7) can be rewritten as

$$\begin{aligned}&(R_1^Q)^t\Gamma_{11}+(R_2^Q)^t\Gamma_{21}=0,\\&\left[(R_1^Q)^t\Gamma_{12}+(R_2^Q)^t\Gamma_{22}\right]\mathcal{N}(R_2)=\{0\}.\end{aligned}\tag{3.8a-b}$$

Since $\Gamma_{\mathcal{L}} := X^tV^-X$ is nnd, $\mathcal{R}(\Gamma_{12}) \subseteq \mathcal{R}(\Gamma_{11})$ (cf. (Pringle and Rayner, 1971, p. 71)) or, equivalently, $\Gamma_{11}\Gamma_{11}^-\Gamma_{12} = \Gamma_{12}$. Postmultiplication of (3.8a) by $\Gamma_{11}^-\Gamma_{12}$ therefore yields

$$(R_1^Q)^t\Gamma_{12}=-(R_2^Q)^t\Gamma_{21}\Gamma_{11}^-\Gamma_{12}.\tag{3.9}$$

Hence $(R_1^Q)^t\Gamma_{12}+(R_2^Q)^t\Gamma_{22} = (R_2^Q)^t[\Gamma_{\mathcal{L}}/\Gamma_{11}] = (R^Q)^tJ_2\Gamma_{\mathcal{M}}$ (observe Theorem 3.3), that is, we have

$$(R^Q)^t\Gamma_{\mathcal{L}}J_2=(R^Q)^tJ_2\Gamma_{\mathcal{M}}.\tag{3.10}$$

(3.7a) and (3.10) can be written as

$$(R^Q)^t\Gamma_{\mathcal{L}}=\begin{pmatrix}0 & (R^Q)^tJ_2\Gamma_{\mathcal{M}}\end{pmatrix},\tag{3.11}$$

whence we get

$$(R^Q)^t\Gamma_{\mathcal{L}}R^Q=(R^Q)^tJ_2\Gamma_{\mathcal{M}}J_2^tR^Q;$$

this is (3.4). In view of (3.11) and $\{0\} \times \mathcal{N}(R_2) \subseteq \mathcal{N}(R)$, we further get

$$(R^Q)^tJ_2\Gamma_{\mathcal{M}}\mathcal{N}(R_2)=\{0\}$$

directly from (3.6). Since $R = (0 \quad R_2)$, $R_2^Q \in \{R_2^-\}$ follows from (3.5a). But now it is clear that (3.3) holds. From (3.3) and (3.4) we also get $\Omega_{\mathcal{L}} = \Omega_{\mathcal{M}}$; notice the lines directly following Theorem 2.1. This completes the proof.

THEOREM 3.5. *Consider the linear models introduced in Section 1, and let*

$$V\mathcal{R}(X_1)\subseteq\mathcal{R}(X_1)\quad\text{or, equivalently,}\quad V\mathcal{N}(X_1^t)\subseteq\mathcal{N}(X_1^t)$$

be satisfied. Moreover, let M_1, J_1, J_2, $\Gamma_{\mathcal{M}}$, $\Gamma_{\mathcal{N}}$, $\Omega_{\mathcal{M}}$ and $\Omega_{\mathcal{N}}$ be all defined as before. For each $X^{\mathcal{L}} \in \mathcal{G}(X \mid V)$, we then have $J_2^tX^{\mathcal{L}} \in \mathcal{G}(M_1X_2 \mid V)$. In addition, we get $\Gamma_{\mathcal{N}} = \Gamma_{\mathcal{M}}$, and therefore $\mathcal{G}(R_2^t \mid \Gamma_{\mathcal{N}}) = \mathcal{G}(R_2^t \mid \Gamma_{\mathcal{M}})$ as well as $\Omega_{\mathcal{N}} = \Omega_{\mathcal{M}}$.

Proof. Because $M_1 = P_{\mathcal{N}(X_1^t)}$ and since $V\mathcal{N}(X_1^t) \subseteq \mathcal{N}(X_1^t)$, clearly $M_1VM_1 = VM_1$. Since $(M_1VM_1)^t = M_1VM_1$, therefore $M_1VM_1 = VM_1 = M_1V$. Consequently

$$(M_1VM_1)V^-(M_1VM_1)=M_1VV^-VM_1=M_1VM_1,$$

that is, $\{V^-\} \subseteq \{(M_1VM_1)^-\}$. But now trivially $\Gamma_\mathcal{M} = \Gamma_\mathcal{N}$ and $\mathcal{G}(R_2^t \mid \Gamma_\mathcal{M}) = \mathcal{G}(R_2^t \mid \Gamma_\mathcal{N})$. Next, let $X^\mathcal{L} \in \mathcal{G}(X \mid V)$ be arbitrary but fixed. Put $X_i^\mathcal{L} := J_i^t X^\mathcal{L}$ for $i = 1, 2$. Then

$$X^\mathcal{L} = \begin{pmatrix} X_1^\mathcal{L} \\ X_2^\mathcal{L} \end{pmatrix}.$$

By Theorem 3.2, $X_2^\mathcal{L} \in \mathcal{G}(M_1X_2 \mid M_1VM_1)$. Hence, in particular, $X_2^\mathcal{L} \in \{(M_1X_2)^-\}$. So it remains to show that $M_1X_2X_2^\mathcal{L}V\mathcal{N}(X_2^tM_1) = \{0\}$ is also satisfied. Clearly,

$$V\mathcal{N}(X_2^tM_1) = V\left[\mathcal{R}(X_1) \oplus \mathcal{N}(X^t)\right] = V\mathcal{R}(X_1) + V\mathcal{N}(X^t).$$

Since $V\mathcal{R}(X_1) \subseteq \mathcal{R}(X_1)$, it follows in view of (3.1) that $V\mathcal{R}(X_1) \subseteq \mathcal{N}(M_1X_2X_2^\mathcal{L})$. In the proof of Theorem 3.2 we have already seen that $V\mathcal{N}(X^t) \subseteq \mathcal{N}(M_1X_2X_2^\mathcal{L})$. This completes the proof.

Now we are in the position to prove Theorem 3.1

Proof of Theorem 3.1 (i). First, we wish to establish $\tilde{\beta}_2(\mathcal{L}_u) = \tilde{\beta}_2(\mathcal{M}_u)$, that is, $J_2^t\tilde{\beta}(\mathcal{L}_u) = \tilde{\beta}_2(\mathcal{M}_u)$ where J_2 is defined as before. To that end, consider representation (2.8) for $\tilde{\beta}(\mathcal{L}_u)$ with an arbitrary but fixed $X^\mathcal{L} \in \mathcal{G}(X \mid V)$. Observe that it suffices to show that $J_2^t\tilde{\beta}(\mathcal{L}_u)$ is of the form (2.11). As (1.2) is equivalent to $\mathcal{N}(X) = \mathcal{N}(X_1) \times \mathcal{N}(X_2)$, clearly

$$P_{\mathcal{N}(X)} = \begin{pmatrix} P_{\mathcal{N}(X_1)} & 0 \\ 0 & P_{\mathcal{N}(X_2)} \end{pmatrix}.$$

If we now partition $w^t = (\, w_1^t \quad w_2^t \,)$ appropriately, then $J_2^tP_{\mathcal{N}(X)}w = P_{\mathcal{N}(X_2)}w_2$. From Theorem 3.2 we further know that $J_2^tX^\mathcal{L} \in \mathcal{G}(M_1X_2 \mid M_1VM_1)$ and that (for each possible realization y) $J_2^tX^\mathcal{L}y$ can be written alternatively in the form $J_2^tX^\mathcal{L}y = J_2^tX^\mathcal{L}M_1y + z$ for some suitably chosen $z \in \mathcal{N}(X_2)$. With these observations in mind, the claimed result is obvious.

Next, we prove $\tilde{\beta}_2(\mathcal{L}_e) = \tilde{\beta}_2(\mathcal{M}_e)$. Consider representation (2.7) for $\tilde{\beta}(\mathcal{L}_e)$ with an arbitrary but fixed $(R^Q)^t \in \mathcal{G}(R^t \mid \Gamma_\mathcal{L})$. In view of $R = (\,0 \quad R_2\,)$ and (1.2), clearly

$$\mathcal{N}(X) \cap \mathcal{N}(R) = \mathcal{N}(X_1) \times \left[\mathcal{N}(X_2) \cap \mathcal{N}(R_2)\right].$$

As above we therefore get $J_2^tP_{\mathcal{N}(X)\cap\mathcal{N}(R)}w = P_{\mathcal{N}(X_2)\cap\mathcal{N}(R_2)}w_2$, where $w_2 = J_2^tw$. Theorem 3.4 tells us that $(J_2^t(R^Q))^t \in \mathcal{G}(R_2^t \mid \Gamma_\mathcal{M})$. In addition, we get $J_2^t(I - R^QR)\hat{\beta}_\mathcal{L} = (I - J_2^tR^QR_2)J_2^t\hat{\beta}_\mathcal{L}$ as $R = (\,0 \quad R_2\,)$. From the previous step we know that $J_2^t\hat{\beta}_\mathcal{L} \in \tilde{\beta}_2(\mathcal{M}_u)$. In view of all these observations it is now clear that $J_2^t\tilde{\beta}(\mathcal{L}_e)$ is a representation for $\tilde{\beta}(\mathcal{M}_e)$; see (2.10).

In order to prove $\tilde{\beta}_2(\mathcal{L}_i) = \tilde{\beta}_2(\mathcal{M}_i)$, consider representation (2.6) for $\tilde{\beta}(\mathcal{L}_i)$. In the light of (2.6), (2.7), (2.9), (2.10) and the foregoing step it suffices to show that $\widetilde{M}_\mathcal{L} = \widetilde{M}_\mathcal{M}$. But by Theorem 3.4, $\Omega_\mathcal{L} = \Omega_\mathcal{M}$. Hence $\widetilde{M}_\mathcal{L} = \widetilde{M}_\mathcal{M}$ as $\tilde{\beta}_2(\mathcal{L}_u) = \tilde{\beta}_2(\mathcal{M}_u)$.

Proof of Theorem 3.1 (ii). Let $V\mathcal{R}(X_1) \subseteq \mathcal{R}(X_1)$. Then, by Theorem 3.5, $J_2^t X^{\mathcal{L}} \in \mathcal{G}(M_1X_2 \mid V)$ for each $X^{\mathcal{L}} \in \mathcal{G}(X \mid V)$. By the same theorem, $\mathcal{G}(R_2^t \mid \Gamma_{\mathcal{N}}) = \mathcal{G}(R_2^t \mid \Gamma_{\mathcal{M}})$. Therefore, in view of (3.3), $(R^Q)^t J_2 \in \mathcal{G}(R_2^t \mid \Gamma_{\mathcal{N}})$ for each $(R^Q)^t \in \mathcal{G}(R^t \mid \Gamma_{\mathcal{L}})$. As in the proof of part (i), we therefore get $\tilde{\beta}_2(\mathcal{L}_u) := J_2^t\tilde{\beta}(\mathcal{L}_u) = \tilde{\beta}_2(\mathcal{N}_u)$ and $\tilde{\beta}_2(\mathcal{L}_e) := J_2^t\tilde{\beta}(\mathcal{L}_e) = \tilde{\beta}_2(\mathcal{N}_e)$ by means of Theorem 2.2 and Theorem 2.4. By Theorem 3.5, $\Omega_{\mathcal{M}} = \Omega_{\mathcal{N}}$. Since $\tilde{\beta}_2(\mathcal{L}_u) = \tilde{\beta}_2(\mathcal{M}_u) = \tilde{\beta}_2(\mathcal{N}_u)$, therefore $\widetilde{M}_{\mathcal{M}} = \widetilde{M}_{\mathcal{N}}$. But then $\tilde{\beta}_2(\mathcal{L}_i) := J_2^t\tilde{\beta}(\mathcal{L}_i) = \tilde{\beta}_2(\mathcal{N}_i)$, and the proof is complete.

We conclude this paper with mentioning that if X and V are both of full column rank then all the GLS selections considered in this paper are unique, that is, the corresponding GLS estimators do exist. Notice that in such a situation the GLS estimator for β_2 under each of our models is the BLUE for β_2 in that model. The unrestricted full rank case (that is, the case where we have no a priori restrictions and where X and V are both of full column rank) has already been investigated recently in (Nurhonen and Puntanen, 1992; Aigner and Balestra, 1988, p. 970).

REFERENCES

Aigner, D. J. and Balestra, P. (1988). Optimal experimental design for error components models. *Econometrica* **56**, 955–971.

Nurhonen, M. and Puntanen, S. (1992). A property of partitioned generalized regression. *Comm. Statist. Theory Methods* **21**(6), 1579–1583.

Pringle, R. M. and Rayner, A. A. (1971). *Generalized Inverse Matrices with Applications to Statistics*. Griffin, London.

Werner, H. J. and Yapar, C. (1995). On inequality constrained generalized least squares selections in the general possibly singular Gauss–Markov model: a projector theoretical approach. *Linear Algebra Appl.* (to appear).

Multivar. Statist., pp. 67–73
E.–M. Tiit *et al.* (Eds)

SIMPLER TESTS OF LINEAR INEQUALITY CONSTRAINTS IN THE STANDARD LINEAR MODEL

RICHARD WILLIAM FAREBROTHER

Department of Economic Studies, Victoria University of Manchester, Manchester M13 9PL, United Kingdom

ABSTRACT

This paper is concerned with the problem of testing H_E: $R\beta = r$ against H_I: $R\beta \geqslant r$ and with the problem of testing H_I: $R\beta \geqslant r$ against H_U: $\beta \in R^k$ in the context of the standard linear model $y = X\beta + \varepsilon$. We develop two "finite-induced" variants of the conventional likelihood-based tests. In certain circumstances, these finite-induced tests may be further simplified to yield directed tests which are closely related to the (King and Smith, 1986) test. Unfortunately these directed tests seem to lack power and we recommend the use of the finite-induced tests or (Kodde and Palm's, 1986) bounds procedure when the full likelihood-based procedure is impractical.

1. INTRODUCTION

Consider the model

$$y = X\beta + \varepsilon, \qquad \varepsilon \sim N(0, \sigma^2 I_n) \tag{1.1}$$

where y is an $n \times 1$ matrix of observations on the dependent variable, X is an $n \times k$ matrix of observations on the k regressors, β is a $k \times 1$ matrix of parameters, and ε is an $n \times 1$ matrix of disturbances, and where X has full column rank k.

In a series of recent papers (Gouriéroux, Holly and Monfort, 1982), henceforth GHM (Farebrother, 1986, 1987, 1988; Goldberger, 1986; Hillier, 1986; King and Smith, 1986; Kodde and Palm, 1986; Rogers, 1986a, 1986b; Wolak, 1987, 1988, 1989) were directly or indirectly concerned with the problem of testing H_E: $R\beta = r$ against H_I: $R\beta \geqslant r$ or H_I: $R\beta \geqslant r$ against H_U: $\beta \in R^k$ in the context of model (1.1) where R is a given $m \times k$ matrix of rank m and r is a given $m \times 1$ matrix.

Unfortunately the Likelihood Ratio, Kuhn–Tucker and Wald tests proposed by these authors are difficult to use if m is at all large, say $m \geqslant 5$. In this paper we shall therefore use the concept of duality discussed by Wolak (1988) to develop two "finite-induced" variants of these conventional statistics. In certain circumstances these finite-induced variants may be further simplified and the simplified test of H_E: $R\beta = r$ against H_I: $R\beta \geqslant r$ takes a form which is closely related to that of the test proposed by King and Smith (1986).

2. LIKELIHOOD-BASED TESTS

2.1. *Estimation*

It seems reasonable to base our tests of H_E: $R\beta = r$ against H_I: $R\beta \geqslant r$ and H_I: $R\beta \geqslant r$ against H_U: $\beta \in R^k$ on the equality constrained, inequality constrained, and uncostrained least squares estimators $\check{\beta}$, $\tilde{\beta}$ and $\hat{\beta}$ of β in model (1.1). These estimators are defined by the values of b which minimise the sum of the squared residuals

$$S(b) = (y - Xb)'(y - Xb) \tag{2.1}$$

subject to $Rb = r$, $Rb \geqslant r$, and $b \in R^k$, respectively.

Let $V = (X'X)^{-1}$, $A = RVR'$, and $B = (RVR')^{-1}$ then it is well known that the unconstrained least squares estimator $\hat{\beta}$ is given by

$$\hat{\beta} = \hat{\beta} + VR'\hat{\lambda} \tag{2.2}$$

where $\hat{\beta} = (X'X)^{-1}X'y$ and $\hat{\lambda} = 0$; the equality constrained least squares estimator $\check{\beta}$ by

$$\check{\beta} = \hat{\beta} + VR'\check{\lambda}$$

where $\check{\lambda} = -B\hat{\delta}$ and $\hat{\delta} = R\hat{\beta} - r$; and the inequality constrained least squares estimator $\tilde{\beta}$ by the unique solution to equations (2.4) and (2.5)

$$\tilde{\beta} = \hat{\beta} + VR'\tilde{\lambda}, \tag{2.4}$$

$$\tilde{\delta} \geqslant 0, \quad \tilde{\lambda} \geqslant 0 \quad \text{and} \quad \tilde{\delta}'\tilde{\lambda} = 0 \tag{2.5}$$

where $\tilde{\delta} = R\tilde{\beta} - r$ and $\tilde{\lambda}$ are defined implicitly.

Finally we note that

$$\tilde{\delta} = \hat{\delta} + A\tilde{\lambda} \tag{2.6}$$

and

$$\tilde{\lambda} = \check{\lambda} + B\tilde{\delta} \tag{2.7}$$

may be derived from equation (2.4) and that $\check{\delta} = R\check{\beta} - r = 0$.

2.2. *Test statistics*

Now $\hat{\delta} = R\hat{\beta} - r$ is normally distributed with mean $\delta = R\beta - r$ and variance $\sigma^2 A = \sigma^2 B^{-1}$ independently of $\hat{\sigma}^2 = S(\hat{\beta})/(n-k)$ so that it seems reasonable to base our test of H_I: $\delta \geqslant 0$ against H_U: $\delta \in R^m$ on the statistic Q_W/σ^2 if σ^2 is known, and on $Q_W/\hat{\sigma}^2$ if σ^2 is not known, where

$$\begin{aligned} Q_W &= (\tilde{\delta} - \hat{\delta})'B(\tilde{\delta} - \hat{\delta}) && (2.8a)\\ &= \tilde{\lambda}'A\tilde{\lambda} && (2.8b)\\ &= (\tilde{\beta} - \hat{\beta})'X'X(\tilde{\beta} - \hat{\beta}) && (2.8c)\\ &= S(\tilde{\beta}) - S(\hat{\beta}). && (2.8d) \end{aligned}$$

In fact these were the statistics proposed by Wolak (1987) for this purpose.

Similarly $\check{\lambda} = -B\hat{\delta}$ is normally distributed with mean $\lambda = -B\delta$ and variance $\sigma^2 B = \sigma^2 A^{-1}$ so that the results of the previous paragraph suggest that we should use Q_M/σ^2 or $Q_M/\hat{\sigma}^2$, as appropriate, to test K_I: $\lambda \geqslant 0$ against K_U: $\lambda \in R^m$, where

$$Q_M = (\tilde{\lambda} - \check{\lambda})' A (\tilde{\lambda} - \check{\lambda}) \tag{2.9a}$$
$$= \tilde{\delta}' B \tilde{\delta} \tag{2.9b}$$
$$= (\tilde{\beta} - \check{\beta})' X' X (\tilde{\beta} - \check{\beta}) \tag{2.9c}$$
$$= S(\check{\beta}) - S(\tilde{\beta}). \tag{2.9d}$$

Further Wolak's (1988) duality principle suggests that we should also use Q_M/σ^2 or $Q_M/\hat{\sigma}^2$ to test H_E: $\delta = 0$ against H_I: $\delta \geqslant 0$. This is the context in which these tests were first proposed by GHM (1982) and Kodde and Palm (1986).

The duality between δ and λ, A and B, and Q_W and Q_M illustrated in this section offers a very convenient foundation for the development of the dual finite-induced tests and directed t-tests in Sections 3 and 4.

Finally we note that the sum of Q_M and Q_W

$$Q_M + Q_W = \hat{\delta}' B \hat{\delta} = \check{\lambda}' A \check{\lambda} \tag{2.10}$$

yields the expression which appears in the numerator of the conventional F-test of H_E: $\delta = 0$ against H_U: $\delta \in R^m$.

3. FINITE-INDUCED TESTS

Unfortunately the likelihood-based tests of the previous section are impractical if m is at all large as their distribution functions become increasingly difficult to compute as the number of constraints increases. Kodde and Palm (1986) have offered a partial solution to this problem by means of a bounds procedure, but, as in the case of the corresponding procedure for the Durbin and Watson test, we still need to have recourse to the full procedure when the bounds procedure is inconclusive. We shall therefore now propose two "finite-induced" variants of the $Q_W/\hat{\sigma}^2$ and $Q_M/\hat{\sigma}^2$ statistics which do not suffer from these defects.

Let c^2 be the chosen critical value of $Q_W/\hat{\sigma}^2$ then the null hypothesis H_I: $\delta \geqslant 0$ is not rejected if $\tilde{\lambda}' A \tilde{\lambda} \leqslant c^2 \hat{\sigma}^2$. This condition is clearly satisfied if $\tilde{\lambda} = h e_i$ where e_i is the i-th column of the $m \times m$ identity matrix I_m and h is a scalar satisfying $0 \leqslant h \leqslant c\hat{\sigma}/a_{ii}^{1/2}$. Further $\tilde{\delta}_i = 0$ when $\tilde{\lambda}_i > 0$ so that, from equation (2.6), we have $\hat{\delta}_i = -h a_{ii} \geqslant -c\hat{\sigma} a_{ii}^{1/2}$. Finally we note that if $\hat{\delta}_i/(\hat{\sigma} a_{ii}^{1/2})$ is expressed as a function of $\tilde{\delta}$ and $\tilde{\lambda}$, and minimised subject to $\tilde{\lambda}' A \tilde{\lambda} \leqslant c^2 \hat{\sigma}^2$, then its minimum value is $-c$ and this minimum value will be attained when $\tilde{\lambda} = (c\hat{\sigma}/a_{ii}^{1/2}) e_i$.

Thus our finite-induced test of H_I: $\delta \geqslant 0$ against H_U: $\delta \in R^m$ is the test induced by the m single element failures of the null hypothesis $\delta \geqslant 0$ and we reject this hypothesis if

$$s_W/\hat{\sigma} = \min_i \hat{\delta}_i/\left(\hat{\sigma}^2 a_{ii}\right)^{1/2} \tag{3.1}$$

takes a value less than the predetermined critical value $-c_W$.

Applying a similar argument to $Q_M/\hat{\sigma}2$ we have a finite-induced test of H_E: $\delta = 0$ against H_I: $\delta \geqslant 0$ which rejects the null hypothesis $\delta = 0$ if

$$-s_M/\hat{\sigma} = \min_i \check{\lambda}_i/\left(\hat{\sigma}^2 b_{ii}\right)^{1/2} \tag{3.2}$$

takes a value less than $-c_M$. In other words, it rejects H_E: $\delta = 0$ if

$$s_M/\hat{\sigma} = \max_i b_i'\hat{\delta}/\left(\hat{\sigma}^2 b_{ii}\right)^{1/2} \tag{3.3}$$

takes a value greater than c_M. This second finite-induced test is formally due to (Rogers, 1986a, 1986b), but our notation reveals that it is also an analogue of the much older one-sided Bonferroni-box test based on expression (3.1), see (Goldberger, 1986, n2) for details.

Under the (least favourable) null hypothesis H_E: $\delta = 0$, the elements on the right of equations (3.1) or (3.3) have a joint multivariate t distribution so that exact or approximate critical values for the associated tests may readily be obtained by numerical integration.

For the case $m = 2$, $R = I$ and $\sigma^2 = 1$, Goldberger (1986) compared the power of the test based on Q_W with that of the test based on s_W. He considered the values $\rho = -0.6$, 0.0, and +0.6 where $\rho = a_{12}/(a_{11}a_{22})^{1/2}$, and found that the power contours of the two tests are very similar when $\rho = 0.6$ but that the Q_W test becomes more powerful for values of δ in the third quadrant ($\delta \leqslant 0$) as ρ falls below zero whilst the s_W test becomes more powerful in the second and fourth quadrants. Our notation reveals that these findings also apply for the Q_M and s_M tests with λ substituted for δ.

In fact (Goldberger, 1986), did not compare the power of the Q_M and s_M tests but instead compared the power of the Q_M test of H_E: $\delta = 0$ against H_I: $\delta \geqslant 0$ with that of the Q_W and s_W tests ($Q_{\bar{W}}$ and $s_{\bar{W}}$) of H_J: $\delta \leqslant 0$ against H_U: $\delta \in R^m$. For $\rho = 0.6$ he found that the $Q_{\bar{W}}$ and $s_{\bar{W}}$ tests are more powerful than the Q_M test for a wide range of values of δ in the first quadrant (which is the only quadrant of interest for the Q_M test), that the $Q_{\bar{W}}$ statistic is identical to the Q_M statistic when $\rho = 0$ (as $Q_{\bar{W}}$ actually tests K_I: $\lambda \geqslant 0$ against K_U: $\lambda \in R^m$ in this case), and that the $Q_{\bar{W}}$ test is less powerful than the Q_M test when $\rho = -0.6$.

4. FURTHER SIMPLIFICATION

Let e_i represent the ith column of the $m \times m$ identity matrix, then equation (3.3) may be rewritten as

$$t_0 = \max_i e_i' B\hat{\delta} / \left(\hat{\sigma}^2 e_i' B e_i\right)^{1/2}. \tag{4.1}$$

Further, suppose that B is close to the $m \times m$ rank one matrix $B_0 = gg'$, where g is an $m \times 1$ nonzero matrix, then equation (4.1) may be approximated by

$$t_1 = g' B\hat{\delta} / \left(\hat{\sigma}^2 g' B g\right)^{1/2} \tag{4.2}$$

if g is nonnegative or nonpositive, and by

$$t_2 = \left|g' B\hat{\delta}\right| / \left(\hat{\sigma}^2 g' B g\right)^{1/2} \tag{4.3}$$

if g possesses positive and negative elements.

Statistics (4.2) and (4.3) form the basis of simple directed tests of H_E: $\delta = 0$ against H_I: $\delta \geqslant 0$ as t_1 is distributed as t with $n - k$ degrees of freedom under the null hypothesis $\delta = 0$ whether or not g is related to B in the manner described, and t_2 is equal to the absolute value of t_1.

Similar tests of H_I: $\delta \geqslant 0$ against H_U: $\delta \in R^m$ may be developed from equation (3.l) by writing its negation in the form

$$t_0^* = \max_i e_i' A\check{\lambda} / \left(\hat{\sigma}^2 e_i' A e_i\right)^{1/2} \tag{4.4}$$

and approximating A by $A_0 = ff'$. Under the particular (least favourable) null hypothesis $\delta = 0$

$$t_1^* = f' A\check{\lambda} / \left(\hat{\sigma}^2 f' A f\right)^{1/2} \tag{4.5}$$

is again distributed as t with $n - k$ degrees of freedom, and

$$t_2^* = \left|f' A\check{\lambda}\right| / \left(\hat{\sigma}^2 f' A f\right)^{1/2} \tag{4.6}$$

as the absolute value of t_1^*. Note that alternative forms of these statistics may be obtained by writing $f' A\check{\lambda} = -f'\hat{\delta}$ and $g' A\hat{\delta} = -g'\check{\lambda}$.

5. DISCUSSION

Let $X = [X_1 \;\; X_2]$ and $R = [0 \;\; I_m]$ be partitioned by their first $k - m$ columns and the remaining m columns and let $r = 0$, then equation (4.2) may be rewritten as

$$t_1 = z' \bar{P}_1 y / \left(\hat{\sigma}^2 z' \bar{P}_1 z\right)^{1/2} \tag{5.1}$$

where

$$z = X_2 g \quad \text{and} \quad \bar{P}_1 = I_n - X_1(X_1'X_1)^{-1}X_1'.$$

Thus our test of $\delta = 0$ against $\delta \geqslant 0$ based on equation (4.2) includes the test proposed by King and Smith (1986) as a special case.

In their paper (King and Smith ,1986) used Hillier's (1986) results to compare the power of their own test with that of GHM's likelihood ratio (LR) test (based on a transformation of our $Q_M/\hat{\sigma}^2$ statistic) when $m = 2$ and $\delta > 0$. However the argument of the previous section suggests that King and Smith's use of $g = [1 \ 1]'$ with pairs of positively correlated economic variables ($b_{12} > 0$) would tend to produce t_0, t_1, and LR tests with similar rejection regions and thus with similar powers. But this argument will not hold true for pairs of negatively correlated variables unless $g = [1 \ 1]'$ is replaced by $g = [1 \ -1]'$ and t_1 by t_2, contrary to King and Smith's assumption that g is strictly positive. Further (Hillier, 1986, p. 668) has shown that the directed t-test t_1 will have a rejection region which is close to that of the LR test when both tests are close to the uniformly most powerful similar test. But he also noted that the t_1 test can have very much lower power than the LR test when this condition is not satisfied.

Moreover, a referee has pointed out that the t_2 test can be inconsistent and that the (Kodde and Palm, 1986) bounds procedure will provide a better approximation to the LR test than t_2 precisely when the t_2 test provides its best approximation, that is, when it is close to t_0.

In view of these observations it seems reasonable to discard the directed tests developed in Section 4 and to recommend the finite-induced tests of Section 3 or the Kodde and Palm bounds procedure when the full LR procedure is impractical. Both of these procedures are easy to use but their relative merits will depend on the nature of the action taken by the practitioner when the Kodde and Palm bounds procedure is inconclusive.

6. CONCLUDING REMARKS

In this paper we have used the concept of duality discussed by Wolak (1988) to illuminate the relationship between two classes of one-sided and inequality tests. For convenience of exposition we have assumed that the constraint matrix R has full row rank, but the statistics of Sections 3 and 4 may readily be generalised to the rank deficient case provided that r is linearly dependent on the columns of R, and that one is prepared to use simulation techniques to determine the distribution of one's chosen statistic. In this context it is worth noting that Wolak (1989, p. 215) has suggested that it may be necessary to use (direct or indirect) simulation techniques to determine the distribution of $Q_W/\hat{\sigma}^2$ in the full rank case when $m \geqslant 8$.

REFERENCES

Farebrother, R. W. (1986). Testing linear inequality constraints in the standard linear model. *Comm. Statist. A* **15**, 7–31.

Farebrother, R. W. (1987). A further remark on testing linear inequality constraints in the standard linear model. *Comm. Statist. A* **16**, 2003–2005.

Farebrother, R. W. (1988). A survey of some recent econometric work on tests for inequality constraints. *J. Quantit. Econ.* **4**, 227–238.

Goldberger, A. S. (1986). One-sided and inequality tests for a pair of means. University of Wisconsin – Madison, Discussion Paper.

Gouriéroux, C., Holly, A. and Monfort, A. (1982). Likelihood ratio test, Wald test, and Kuhn-Tucker test in linear models with inequality constraints on the regression parameters. *Econometrica* **50**, 63–80.

Hillier, G. H. (1986). Joint tests for zero Rrstrictions on nonnegative regression coefficients. *Biometrika* **73**, 657–669.

King, M. L. and Smith, M. D. (1986). Joint one-sided test of linear regression coefficients. *J. Econometrics* **32**, 367–383.

Kodde, D. and Palm, F. C. (1986). Wald criteria for jointly testing equality and inequality constraints. *Econometrica* **56**, 1243–1248.

Rogers, A. J. (1986a). Wald, likelihood ratio and infinite induced test statistics for joint one-sided hypotheses. *Economics Letters* **21**, 261-264.

Rogers, A. J. (1986b). Modified Lagrange multiplier tests for models with one-sided alternatives. *J. Econometrics* **31**, 341-361.

Wolak, F. A. (1987). An exact test for multiple inequality and equality constraints in the linear regression model. *J. Amer. Statist. Assoc.* **82**, 782–793.

Wolak, F. A. (1988). Duality in testing multivariate hypotheses. *Biometrika* **75**, 611–615.

Wolak, F. A. (1989). Testing inequality constraints in linear econometric models. *J. Econometrics* **41**, 205-235.

Multivar. Statist., pp. 75–83
E.–M. Tiit *et al.* (Eds)

INCREASING THE CORRELATIONS WITH THE RESPONSE VARIABLE MAY NOT INCREASE THE COEFFICIENT OF DETERMINATION: A PCA INTERPRETATION

C. M. CUADRAS

Departament d'Estadistica, Universitat de Barcelona,
Diagonal 645, 08028 Barcelona, Spain

ABSTRACT

Let us suppose that a response variable is correlated with a set of s explanatory variables and uncorrelated with another set of t variables. Let $s + t = k$ be fixed. Tiit (1984) proved that it is possible that $R^2(s_1) > R^2(s_2)$ for $s_1 < s_2$, where $R^2(s_1)$ and $R^2(s_2)$ are the coefficients of determination for all k variables. Following (Cuadras, 1993), we use principal components analysis to explain this quirk in multiple regression. It is proved that the fact that $R^2(s)$ is large when s is small is due to the influence on the response variable of the principal components with the smaller variances. A generalization and three examples are also given.

1. INTRODUCTION

Tiit (1984) studied several regression models for which the multiple correlation coefficient and the regression parameters can be formally computed. One of these models deals with the regression of a response variable y on k equicorrelated variables $x_1, \dots, x_k$, i.e, the correlation matrix between the explanatory variables is

$$\mathbf{R} = (1 - c)\mathbf{I} + c\mathbf{J}, \tag{1}$$

where $\mathbf{I}$ is the $k \times k$ identity matrix, $\mathbf{J}$ is the $k \times k$ matrix of ones and $-(k-1)^{-1} < c < 1$. It is supposed in this model that the vector of correlations between y and $x_1, \dots, x_k$ is

$$\rho_s = (r, \dots, r, 0, \dots, 0)', \tag{2}$$

i.e., the first s entries of ρ_s are constants $r \neq 0$ and the last $k - s$ are zeros. The coefficient of determination, i.e., the squared multiple correlation is given by

$$R^2(s) = \rho_s' \mathbf{R}^{-1} \rho_s,$$

(Mardia *et al.*, 1979, p. 168) and can be expressed as

$$R^2(s) = \frac{sr^2\,(1 + (k - s - 1)c)}{(1 + (k - 1)c)\,(1 - c)}. \tag{3}$$

The number of non-zero regressors giving maximum $R^2(s)$ is the closest integer to

$$s' = \big(1 + (k-1)c\big)/2c. \tag{4}$$

Assuming $c > 1/k$, then $s' < k$ so that increasing the number s of correlated explanatory variables, does not necessarily increase $R^2(s)$. For example, for $k = 8$, $c = 0.5$, $r = 0.4$ we find $s' = 4.5$ and $R^2(4) = 0.7111 > R^2(8) = 0.2844$ (see Table 1 below). Thus, it is rather surprising that having $s = 4$ correlated variables and $k - s = 4$ uncorrelated variables with y, we obtain a larger coefficient of determination than having $s = 8$ correlated variables. (Note that for $c \leqslant 1/k$, $R^2(s)$ cannot decrease with increasing s).

The aim of this contribution is to give a Principal Component Analysis interpretation to this quirk in multiple regression.

2. AN INEQUALITY IN MULTIPLE REGRESSION

Let $\mathbf{R}$ be a $k \times k$ non-singular correlation matrix and $\rho_{yx} = (r_{yx_1}, \ldots, r_{yx_k})'$ the vector of correlations between y and the x variables. Hamilton (1987) and Routledge (1990) studied and interpreted the inequality

$$R^2 > r_{yx_1}^2 + \cdots + r_{yx_k}^2 \tag{5}$$

for $k = 2$, where R^2 is the coefficient of determination. Cuadras (1993) studied the case $k > 2$ expressing (5) as

$$\rho_{yx}'\mathbf{R}^{-1}\rho_{yx} > \rho_{yx}'\rho_{yx}$$

and showing that this inequality is equivalent to

$$\rho_{yz}'(\mathbf{I} - \Lambda)\rho_{yz} > 0, \tag{6}$$

where $\mathbf{R} = \mathbf{U}\Lambda\mathbf{U}'$ is the spectral decomposition of $\mathbf{R}$ and $\rho_{yz} = (r_{yz_1}, \ldots, r_{yz_k})'$ contains the correlations between y and the principal components (PC's) $z_1, \ldots, z_k$ extracted from $\mathbf{R}$. (See (Jolliffe, 1986) for the derivation and properties of PC's.)

Inequality (6) can be written in scalar form as

$$\sum_{i=1}^{k} r_{yz_i}^2(1 - \lambda_i) > 0, \tag{7}$$

showing that (5) is satisfied when y is highly correlated with the PC's with variance smaller than 1.

Inequality (5) contravenes the general belief that correlated variables contain redundant information and that x variables poorly (or null) correlated with y should be rejected. The equivalent inequality (7) reveals that (5) holds due to the influence on y of some PC's with small variance. Recently, Atkinson *et al.* (1994) have evaluated the difference between both sides of (5), also using PC's.

If **R** has the structure (1) and $\rho_{yx} = \rho_s$, as $\rho_s' \rho_s = sr^2$, the statement "a high number of x variables correlated with y" is equivalent to "$\rho_s' \rho_s = sr^2$ is large". The surprising cases are: (a) $R^2(s) > sr^2$, (b) $R^2(s_1) > R^2(s_2)$ for $s_1 < s_2$. An explanation for (a) has been given above. We shall prove that (b) can also be justified by studying the influence of the PC's on y.

3. EVALUATING THE INFLUENCE OF THE PRINCIPAL COMPONENTS

For the equicorrelation matrix (1) it is well known that the eigenvalues are $\lambda_1 = 1 + (k-1)c$ and $\lambda_2 = \cdots = \lambda_k = 1 - c$. Although strictly only the first PC exists, let us consider the set of PC's

$$z_i = \mathbf{u}_i' X, \qquad i = 1, \ldots, k,$$

where $X = (x_1, \ldots, x_k)'$ and

$$\begin{aligned}
\mathbf{u}_1 &= (1, 1, \ldots, 1, 1)'/\sqrt{k},\\
\mathbf{u}_2 &= (1, -1, 0, \ldots, 0)'/\sqrt{2},\\
&\vdots\\
\mathbf{u}_k &= (1, 1, \ldots, 1, -(k-1))'/\sqrt{k(k-1)}.
\end{aligned}$$

It is showed below that the choice of $z_2, \ldots, z_k$ does not affect the total influence of the PC's on y.

The vector of correlations between y and $z_1, \ldots, z_k$ is given by

$$\rho_{yz} = \Lambda^{-1/2} \mathbf{U}' \rho_{yx}.$$

Hence, if $\rho_{yx} = \rho_s = (r, \ldots, r, 0, \ldots, 0)'$, we find

$$\rho_{yz}(s) = \Lambda^{-1/2} \mathbf{U}' \rho_s,$$

where $\Lambda = \operatorname{diag}(1 + (k-1)c, 1 - c, \ldots, 1 - c)$ and $\mathbf{U} = (\mathbf{u}_1, \ldots, \mathbf{u}_k)$. Explicitly, we have

$$\begin{aligned}
\rho_{yz}(1) &= (d_1 r, d_2 r, \ldots, d_k r)',\\
\rho_{yz}(2) &= (2d_1 r, 0, 2d_3 r, \ldots, 2d_k r)',\\
&\vdots\\
\rho_{yz}(k-1) &= ((k-1)d_1 r, 0, 0, \ldots, (k-1)d_k r)',\\
\rho_{yz}(k) &= (kd_1 r, 0, 0, \ldots, 0)',
\end{aligned}$$

where

$$d_1 = (k(1+(k-1)c))^{-1/2},$$
$$d_i = (i(i-1)(1-c))^{-1/2}, \qquad i = 2, \ldots, k.$$

The coefficient of determination is

$$R^2(k) = \rho'_{yz}(k)\rho_{yz}(k) = d_1^2 k^2 r^2 = \frac{kr^2}{1+(k-1)c} \qquad (s = k),$$

$$R^2(s) = \frac{s^2r^2}{(1+(k-1)c)k} + \frac{1}{1-c}\sum_{i=s+1}^{k} \frac{s^2r^2}{i(i-1)} \tag{8}$$
$$= r^2\left[\frac{s^2}{k(1+(k-1)c)} + \frac{s^2}{1-c}\left(\frac{1}{s} - \frac{1}{k}\right)\right]$$
$$= r^2\left[A(s) + B(s)\right] \qquad (s \leqslant k),$$

where $A(s) > 0$ and $B(s) \geqslant 0$ are the first and the second expression between brackets.

Comparing $R^2(s_1)$ with $R^2(s_2)$ for $s_1 < s_2$, we can see that $A(s_1) < A(s_2)$, and that $B(s_1) > B(s_2)$ is possible. Hence, $R^2(s_1) > R^2(s_2)$ is also possible, and it happens when y is strongly influenced by the PC's with variance smaller than 1.

In particular, when $s_2 = k$, y is correlated with z_1 only and considering $s_1 = s = [s']$ (see (4)), we find a rational explanation of $R^2(k) < R^2(s)$ and $k > s$ or, equivalently, $R^2(k) < R^2(s)$ and $\rho'_k\rho_k > \rho'_s\rho_s$.

These results do not depend on the PC's chosen. $A(s)$ is related to the first PC, which is unique, while $B(s)$ is only related to the PC's with variances $1-c$. Given c and k, $B(s)$ is function of s, and, as $R^2(s)$ is the same, the second, third, etc., components of $\rho_{yz}(s)$ can change, but not $A(s)$ and $B(s)$, the only relevant quantities in our relation between y and the PC's.

Furthermore, note that $R^2(s)$ (see (3)) is an increasing function of k whichever is s, $1 \leqslant s \leqslant k$. Hence, discarding $(k-s)$ explanatory variables because they are uncorrelated with y, makes the regression model worse, i.e., the forward selection procedure is inappropriate. See Example 3 in Section 4.

Example 1. To illustrate the results obtained, we take $k = 8$, $c = 0.5$, and $r = 0.4$. Table 1 reports the values of $R^2(s)$ and $\rho'_s\rho_s = sr^2$. The difference between the two quantities is a measure of the influence of the PC's with small variance. The columns in Table 2 give some correlation vectors $\rho_{yz}(s)$ between y and the PC's. For $s = 4$, z_1 and four PC's with variance less than 1 are correlated with y, giving $R^2(4)$=0.7111. For $s = 8$ y is correlated with z_1 only and $R^2(8)$=0.2844. (Note that both the squared correlation of y with

Table 1.
Comparisons between $R^2(s)$ and sr^2 for equicorrelated variables

s	1	2	3	4	5	6	7	8
$R^2(s)$	0.2844	0.4978	0.6400	0.7111	0.7111	0.6400	0.4978	0.2844
sr^2	0.16	0.32	0.48	0.64	0.80	0.96	1.12	1.28

Table 2.
Correlations between the response variable y and the principal components for s=1,4,5 and 8

	$s = 1$	$s = 4$	$s = 5$	$s = 8$
z_1	0.0667	0.2667	0.3333	0.5333
z_2	0.4000	0	0	0
z_3	0.2309	0	0	0
z_4	0.1633	0	0	0
z_5	0.1265	0.5060	0	0
z_6	0.1033	0.4131	0.5164	0
z_7	0.0873	0.3491	0.4364	0
z_8	0.0756	0.3024	0.3780	0

z_1 and the sum of squared correlations between the other PC's are invariant, see (8)).

An extreme case is attained at r=0.4743, k and c being the same. Then y is a linear combination of the explanatory variables for $s = 4$, while the multiple correlation is poor for $s = 8$. It is found that

$$\rho_{yz}(4) = (0.3162, 0, 0, 0, 0.6, 0.49, 0.414, 0.3586)',$$
$$\rho_{yz}(8) = (0.6324, 0, 0, 0, 0, 0, 0, 0)',$$

and

$$R^2(8) = 0.4 < R^2(4) = 1.$$

4. A GENERALIZATION

Our explanation why $R^2(s_1) > R^2(s_2)$ for $s_1 < s_2$ has been made on the basis of the equicorrelation matrix (1), in order to obtain closed formulas when relating PC's with the response variable y. However, this approach essentially holds in a more general context.

Let us suppose that both the entries in $\mathbf{R} = (r_{ij})$ and $\rho_{yx} = (r_{yx_1}, \ldots, r_{yx_k})'$ are non-negative. Let y_1 and y_2 be two response variables with related correlation vectors ρ_{y_1x} and ρ_{y_2x}, respectively. Variable y_1 is apparently more correlated with the x variables than y_2 if

$$\rho'_{y_1x}\rho_{y_1x} > \rho'_{y_2x}\rho_{y_2x}. \tag{9}$$

If $\mathbf{u}$, $\mathbf{v}$ are unitary vectors and $\rho_{y_1x} = \alpha\mathbf{u}$, $\rho_{y_2x} = \beta\mathbf{v}$, then (9) is satisfied whenever $\alpha > \beta > 0$.

Suppose now that $\rho_{yx} = \alpha\mathbf{u}$, where

$$\mathbf{u} = \sum_{i=1}^{k} a_i\mathbf{u}_i$$

is a linear combination of the ordered eigenvectors of $\mathbf{R}$. Then

$$R^2 = \rho'_{yx}\mathbf{R}^{-1}\rho_{yx} = \alpha^2\left(\sum_{i=1}^{k} a_i\mathbf{u}_i\right)' \cdot \left(\sum_{i=1}^{k} a_i\mathbf{u}_i/\lambda_i\right) = \alpha^2\left(\sum_{i=1}^{k} a_i^2/\lambda_i\right).$$

The vector of correlations between y and the PC's is given by

$$\rho_{yz} = \alpha\big(a_1/\sqrt{\lambda_1}, \ldots, a_k/\sqrt{\lambda_k}\,\big)'.$$

In particular, for $\mathbf{u} = \mathbf{u}_1$, we have

$$R^2 = \rho'_{yx}\mathbf{R}^{-1}\rho_{yx} = \alpha^2/\lambda_1 \leqslant \alpha^2\left(\sum_{i=1}^{k} a_i^2/\lambda_i\right),$$

showing that, given α, the minimum value of R^2 is attained at $\rho_{yx} = \alpha\mathbf{u}_1$. Then y is only correlated with z_1.

Finally, if $\rho_{y_1x} = \alpha\mathbf{u} = \alpha\Sigma a_i\mathbf{u}_i$ (where, e.g., $\mathbf{u}$ is "close" to $\mathbf{u}_1$) and $\rho_{y_2x} = \beta\mathbf{v} = \beta\Sigma b_i\mathbf{u}_i$, we can indeed find

$$R_1^2 = \alpha^2\left(\sum_{i=1}^{k} a_i^2/\lambda_i\right) < R_2^2 = \beta^2\left(\sum_{i=1}^{k} b_i^2/\lambda_i\right)$$

in spite of y_1 being apparently more correlated with the x variables than y_2

$$\rho'_{y_1x}\rho_{y_1x} = \alpha^2 > \rho'_{y_2x}\rho_{y_2x} = \beta^2.$$

Example 2. Let $k = 4$, the correlation matrix

$$\mathbf{R} = \begin{pmatrix} 1 & 0.31 & 0.40 & 0.52 \\ 0.31 & 1 & 0.52 & 0.40 \\ 0.40 & 0.52 & 1 & 0.31 \\ 0.52 & 0.40 & 0.31 & 1 \end{pmatrix}$$

and y_1, y_2 two response variables with $\rho_{y_1x} = (0.6,\ 0.5,\ 0.4,\ 0.3)'$ and $\rho_{y_2x} = (0.6,\ 0.5,\ 0.1,\ 0.1)'$, respectively. Since

$$\rho'_{y_1x}\rho_{y_1x} = 0.86 > \rho'_{y_2x}\rho_{y_2x} = 0.63,$$

Table 3.
Data used for illustrating that forward selection and principal component regression can fail

	y	x_1	x_2	x_3	x_4
1	5.50772	2.79998	1.33707	2.66431	1.24548
2	4.57777	3.99666	2.57340	3.95719	2.87856
3	5.41479	2.79532	1.59693	2.59463	2.05869
4	8.04209	2.41676	1.47571	2.18515	1.12845
5	5.20152	2.37721	1.63292	2.65330	1.62447
6	8.52368	2.89874	1.36172	2.12835	1.25327
7	7.24278	2.83364	1.90155	2.67437	1.78585
8	8.35068	4.03046	2.50581	3.47559	2.15076
9	4.95802	2.00729	0.99020	2.31781	0.59171
10	6.39466	3.62448	1.99389	2.79650	2.93772
11	7.49485	3.73982	3.38669	3.85941	3.37282
12	6.68145	2.94157	2.46221	3.36319	1.81006
13	7.83918	4.01075	2.69206	3.68645	2.31579
14	0.77292	2.73620	1.44737	3.13476	2.48003
15	5.81083	5.07923	3.79308	5.03852	3.85448
16	6.37814	3.81549	2.88508	3.69216	3.16257
17	7.34863	1.25729	0.38500	0.97631	0.40236
18	3.83970	2.64711	2.05160	3.23590	2.19994
19	3.14950	0.86863	0.35331	1.71441	0.10867
20	9.43909	4.26512	3.24486	3.97550	2.47006

y_1 is apparently more correlated with the x variables than y_2. However, the coefficients of determination are

$$R_1^2 = 0.4833 < R_2^2 = 0.7162.$$

Inequality (5) is satisfied for y_2 but not for y_1, indicating that y_2 is influenced by the PC's with the smallest variances. In fact, it is found that $\rho(y_1, z_1)=$ 0.6027 and $\rho(y_2, z_4) = 0.7206$.

Finally, let us illustrate with an example that two well-known methods in regression, the forward selection and the principal component regression, may be inappropriate when the data follows a correlation structure as described above.

Example 3. Using MATLAB, a $n \times k$ matrix $\mathbf{X}$ has been generated with $k = 4$, $n = 20$, and transformed to have the x variables coming from a normal population, with equicorrelation matrix (1) for $c = 0.9$. The variable y, a linear combination of the columns of $\mathbf{X}$ plus an independent random error, is forced to have a small correlation with x_3 and x_4. The $n \times (k+1)$ data matrix (Table 3) so obtained provides the simple correlations:

	x_1	x_2	x_3	x_4	y
x_1	1	0.9284	0.9078	0.8844	0.3509
x_2	0.9284	1	0.9442	0.8899	0.3425
x_3	0.9078	0.9442	1	0.8832	0.0692
x_4	0.8844	0.8899	0.8832	1	0.0551

The coefficient of determination taking all variables is $R^2 = 0.9995$, but if we reject variables x_3, x_4 and regress y on x_1, x_2, then this coefficient decreases to 0.1252. Furthermore, the forward selection (using STATGRAPHICS) *rejects* all variables. Note that

$$R^2 = 0.9995 > r^2_{yx_1} + \cdots + r^2_{yx_k} = 0.2483,$$

indicating a clear influence of the PC's with small variances. The correlations between y and the ordered PC's are $\rho_{yz} = (0.2129, 0.4129, 0.6162, 0.6356)'$. Hence, performing a regression on the first PC, which accounts for 93% of the total variation, is clearly inappropriate. These computations have been repeated using SAS/Proc Reg, reaching similar conclusions. Finally, note that it is also possible to generate non-normal data following the same correlation structure (Tiit, 1986; Cuadras, 1992).

5. CONCLUSIONS

We have proved that poor or null simple correlations between the response variable y and the explanatory variables, do not necessarily yield a small multiple correlation. Moreover, if we decrease the simple correlations, the whole correlation can increase.

We have shown that the forward selection can discard variables with a high joint influence on y. This anomaly in regression is often attributed to the multicollinearity (see (Hamilton, 1987) for a discussion). Actually, as it happens with inequality (5), both the above quirk and this misbehaviour of the forward selection, are due to the influence on y of the principal components with small variances. Then, in spite of having x variables apparently poor correlated with y, we can find a high ceefficient of determination. To study this influence, we simply must compare the vector of simple correlations between y and the x variables, to several eigenvectors of the matrix $\mathbf{R}$. Thus the multicollinearity is not a necessary, but only a sufficient condition.

The principal component regression can be inefficient for similar reasons. While several authors hope that the first principal components will be predictive (Mardia *et al.*, 1979, p. 244; De Veaux and Ungar, 1994), other authors are less optimistic (Kshirsagar, 1972, p. 462; Jolliffe, 1986, p. 238; Soofi, 1988; Cuadras and Arenas, 1990). Example 3 and real data examples in (Routledge, 1990; Cuadras, 1993) are illustrations of the predictability of the last principal components.

Finally, the users of regression in the presence of the quirks here discussed should rely on the relation between y and principal components with small variances, only if they are interpretable (e.g., as shape factors in biometry). But if they are regarded as "noise" or as axes explaining a part of the variability which is not worth considering, the user should be careful with this relation.

Acknowledgements

Work supported in part by DGICYT grant PB93-0784. The author thanks R. Atkinson, A. Bartkowiak, M. Greenacre, I. Olkin, C. R. Rao, G. Styan, E. Tiit and two anonymous reviewers for useful comments. Thanks are also due to F. Oliva for checking some of the computations independently.

REFERENCES

Atkinson, R. A., Bertrand, P. V. and Critchley, F. (1994). Regression: The whole can be greater than the sum of the parts. How much? How likely? Working paper, University of Birmingham.

Cuadras, C. M. (1992). Probability distributions with given multivariate marginals and given dependence structure. *J. Multivariate Anal.* **42**, 51–66.

Cuadras, C. M. (1993). Interpreting an inequality in multiple regression. *Amer. Statist.* **47**, 256–258.

Cuadras, C. M. and Arenas, C. (1990). A distance based regression model for prediction with mixed variables. *Comm. Statist. Theory Methods* **19**, 2261–2279.

De Veaux, R. D. and Ungar, L. H. (1994). Multicollinearity: a tale of two nonparametric regressions. In: *Selecting Models from Data: AI and Statistics IV*, pp. 393–402, P. Cheeseman and R.W. Oldford (Eds), Springer, New York.

Hamilton, D. (1987). Sometimes $R^2 > r^2_{yx_1} + r^2_{yx_2}$. Correlated variables are not always redundant. *Amer. Statist.* **41**, 129–132.

Jolliffe, I. T. (1986). *Principal Components Analysis.* Springer, New York.

Kshirsagar, A. M. (1972). *Multivariate Analysis.* Marcel Dekker, New York.

Mardia, K. V., Kent, J. T. and Bibby, J. M. (1979). *Multivariate Analysis.* Academic Press, London.

Routledge, R. D. (1990). When stepwise regression fails: correlated variables some of which are redundant. *Int. J. Math. Educ. Sci. Technol.* **21**, 403–410.

Soofi, E. S. (1988). Principal component regression under exchangeability. *Comm. Statist. Theory Methods* **17**, 1717–1733.

Tiit, E. M. (1984). Formal computation of regression parameters. In: *Proceedings Sixth Symposium COMPSTAT 1984*, pp. 497–502, T. Havraneek *et al.* (Eds), Physica-Verlag, Vienna.

Tiit, E. M. (1986). Random vectors with given arbitrary marginals and given correlation matrix. *Acta Comm. Univ. Tartuensis* **733**, 14–39.

Multivar. Statist., pp. 85–92
E.-M. Tiit *et al.* (Eds)

DISCUSSION ON THE PAPER OF C. M. CUADRAS "INCREASING THE CORRELATIONS WITH THE RESPONSE VARIABLE MAY NOT INCREASE THE COEFFICIENT OF DETERMINATION: A PCA INTERPRETATION"

E.-M. TIIT
Department of Mathematical Statistics, University of Tartu,
Liivi 2, EE2400 Tartu, Estonia

ABSTRACT

The assumptions used in the paper of C. M. Cuadras are discussed, and their admissibility is proved. The results gained by Cuadras are explained and somewhat elaborated. The Hamilton–Rotledge–Cuadras inequality is strenghtened and illustrated with examples.

1. INTRODUCTION

After the paper of C. M. Cuadras on the 5th Tartu Conference on Multivariate Statistics an active discussion started in the conference room. Unfortunately, the time-schedule of the conference was quite tight, so there was no time for a thorough discussion, and several questions remained unanswered. The author of the present discussion has studied similar problems during several years (see (Tiit, 1980, 1981, 1982, 1983a, 1983b)). Using the right of an editor to highlight some questions more deeply, an explanation will be given to seemingly contradictory or questionable features connected with the report of Professor C. M. Cuadras.

The questions asked the speaker can be divided into three groups, concerning admissibility of the assumptions, correctness of the conclusions and practical value of the results.

2. THE QUESTIONS

2.1. *Is the set of assumptions in the report admissible or contradictory?*

The assumptions are the following: A random vector $X = (X_1, \ldots, X_k)'$ has equicorrelated components, $r(X_i, X_j) = c$, $i, j = 1, \ldots, k$, $i \neq j$, i.e., the correlation matrix $\mathbf{R}_X$ has the following form

$$\mathbf{R}_X = \begin{pmatrix} 1 & c & \cdots & c \\ c & 1 & \cdots & c \\ \cdots & \cdots & \cdots & \cdots \\ c & c & \cdots & 1 \end{pmatrix}, \tag{1}$$

where c satisfies the restriction $-1/k < c < 1$. A random variable Y has nonzero correlations (equal to r) with several components $X_1, \ldots, X_s$ of X, and zero correlations with the remaining components $X_{s+1}, \ldots, X_k$.

a) The main objection of the opponents of the report was, that as the components of the vector X were *exchangeable*, so all components X_i should have the same correlations with Y.

b) The second version of the objection is that it is impossible, that a part of the equicorrelated components are uncorrelated with a random variable when at the same time the remaining part of the components are (substantially) correlated with the same variable.

2.2. *Is the principal inequality, explained by C. Cuadras, correct at all?*

Let $R^2(s)$ denote the coefficient of determination (squared multiple correlation) characterizing the linear regression model where Y is the dependent variable, explanatory vector X has the correlation matrix (1) and Y is correlated with s components of X and uncorrelated with $k - s$ components of X. Cuadras explained that for some values of k, s_1, s_2, c, and r the inequality

$$R^2(s_1) > R^2(s_2), \qquad \text{where } s_1 < s_2, \tag{2}$$

is possible. The inequality seems to be questionable for several reasons:

a) The subspace $E(X_j)$, defined by a component X_j, uncorrelated with Y, is orthogonal to Y. Therefore, the projection of Y into $E(X_j)$ has zero length, and including X_j into the regression will not improve the model.

b) It is a well-known fact, that as a result of adding an argument to the regression model, the (theoretical) value of the coefficient of determination R^2 cannot decrease. The inequality (2) contradicts this fact.

c) Maybe, Cuadras has used an unusual definition of the coefficient of determination, and therefore the conclusions made by him had some unusual meaning.

2.3. *The result is a very special, artificial case of regression model that has no theoretical and practical value*

Probably in empirical studies similar effects will almost never appear, and hence the results obtained by Cuadras *et al.* are useless.

3. THE ANALYSIS OF OBJECTIONS

3.1. *The correctness of assumptions*

The existence of the $(k+1)$-variate random vector $(Y, X_1, \ldots, X_k)'$, having the correlation matrix

$$\mathbf{R} = \begin{pmatrix} 1 & \rho' \\ \rho & \mathbf{R}_X \end{pmatrix}, \tag{3}$$

where $\mathbf{R}_X$ has the shape (1), depends on the regularity of the matrix $\mathbf{R}$. The matrix $\mathbf{R}_X$ is regular for all values of k and c, satisfying the following inequalities (Tiit, 1980)

$$-\frac{1}{k-1} \leqslant c \leqslant 1. \tag{4}$$

Using this, it is easy to find the conditions for the vector of correlations $\rho = (\rho_1, \rho_2, \ldots, \rho_k)'$ between Y (the dependent variable) and X (vector of arguments) to guarantee the regularity of the matrix $\mathbf{R}$.

a) In the case of constant correlation vector, i.e., when $s = k$, and $\rho_j = r$ for $j = 1, \ldots, k$, the sufficient condition for r is the following (see (Tiit, 1981, 1982))

$$r^2 < c + \frac{1-c}{k}. \tag{5}$$

In this case the coefficient of determination R^2 equals to the following ratio (see also equation (3) from the paper of C. Cuadras, (C3) in future)

$$R^2(r) = \frac{k\,r^2}{1+(k-1)c}. \tag{6}$$

b) If the vector ρ is not constant, then the coefficient of determination R^2 has the following form

$$R^2 = R^2(\bar{r}) + \frac{1}{1-c}\sum_{j=1}^{k}(\rho_i - \bar{r})^2 = R^2(\bar{r}) + D, \tag{7}$$

where

$$\bar{r} = \frac{1}{k}\sum_{j=1}^{k}\rho_j$$

(see (Tiit, 1981, 1983b)). The formula (7) is derived immediately, using the usual definition of R^2 and the expression of $\mathbf{R}^{-1}$. From here follows, that *the greater the variability in components of vector ρ is, the larger is the value of R^2*. The similarity of this formula to the formulae derived by Cuadras is obvious. The formula (7) lets us deduce the *restrictions for the vector* ρ. It is obvious, that the inequality

$$D \leqslant 1 - R^2(r),$$

where D is defined by (7), should be fulfilled.

c) If the vector ρ has the block-structure used by Cuadras,

$$(\underbrace{r, \ldots, r}_{s \text{ times}}, 0, \ldots, 0)', \tag{8}$$

then the formula (C3) follows immediately from the formula (7). If the value r is assumed to be positive, then it must fulfill the following condition

$$r^2 \leqslant \frac{(1+(k-1)c)(1-c)}{s(1+(k-s-1)c)}. \tag{9}$$

In the case of considerably small values of c the condition (9) is not very restrictive; e.g, in the case $c = 0.5$, $k = 8$, and $s = 4$ the maximal admissible value of r is 0.4743 (the same result has been gained by Cuadras, using the PCs, see the paper). It is obvious, that all matrices $\mathbf{R}$, having the shape (3) and satisfying the conditions $0 \leqslant R^2 < 1$, where R^2 is expressed by (7), are regular. The existence of a normally distributed vectors having given correlation matrix, follows from the regularity of the correlation matrix. Consequently, the first question is answered. Concerning the exchangeability of the components of the vector X, it should be mentioned, that the equicorrelatedness of components is *a necessary but not sufficient condition for the components to be exchangeable*. It is sufficient only in the limiting case when $c = 1$, provided that all marginal distributions are identical. When an arbitrary $(p+1)$-st component is appended to a p-component exchangeable vector, then the resulting $(p+1)$-st component vector is, in general, not exchangeable.

3.2. *The seeming contradiction of the inequality* $R^2(s_1) > R^2(s_2)$

The seeming contradiction of the above inequality when $s_1 < s_2$ can be explained, analysing the correlation model.

a) The step-wise construction of a model using the projections. Let us regard the process of building a linear model

$$Y = \sum_{i=1}^{k} b_i \cdot X_i + \epsilon.$$

At the first step the argument X_i cannot be orthogonal to Y, that means the condition $s_1 \geqslant 1$ must be fulfilled. Let p steps be made. Then *the residual* ϵ_p is

$$\epsilon_p = Y - \sum_{i=1}^{p} b_i \cdot X_i. \tag{10}$$

To improve the model for the next step there are following (simplified) possibilities:

- X_{p+1} is correlated with Y, but not with X_i, $i = 1, \ldots, p$. Then the corresponding regression coefficient b_{p+1} has the same sign as $r(Y, X_{p+1})$.
- X_{p+1} is not correlated with Y, but is correlated with (some or all) explanatory variables X_i, included into the model. Then the sign of the following regression coefficient b_{p+1} is opposite to the correlation $r(X_i, X_{p+1})$. *As X_{p+1} is not correlated with* Y, the next term can reduce the ϵ_p without any influence to Y.

In practice the situation is usually more complicated. It follows from the equation (10) that the best next argument is the one that has *different signs of correlations with dependent variable Y and independent variables X_i*. At the same time, when including into the model the argument X_{p+1} having almost equal correlations with Y and arguments $X_1, \ldots, X_p$, cannot reduce the residual ϵ_p essentially. It follows, that after several steps the best choice would be iclusion into the model of an argument X_{p+1} uncorrelated with Y.

b) Adding the components should increase the coefficient of determination. The answer to this objection is simple – in the report of C. Cuadras the *general number of arguments was constant*, equal to (k). The problem was connected with the number of *correlated versus uncorrelated* arguments.

c) Which version of the coefficient of determination was used?

The coefficient of determination used in all calculations is the usual *theoretical* squared multiple coefficient of correlation R^2, where R is defined as the correlation between Y and the best possible linear forecast $\tilde{Y}$ by $X_1, \ldots, X_k$ without any modification.

3.3. *The background and practical use of the problem*

The background of the considered problem is connected with the complicated dependence structure in the given example. Two main types of similar models have been used in practical tasks.

a) The linear regression equations. In this case the correlations between explanatory variables have been taken as something indicating the *ill-conditioned* problem, something that should be avoided. The influence of different correlation structures of variables has been investigated considerably superficially.

b) The time-dependent structures, where *a priori* all variables, including dependent one, are positively correlated, and other dependence structures do not have any realistic interpretation.

From here it follows that most current models assume (either openly or in hidden form) that the variables are positively correlated.

As a matter of fact, several practical tasks exist, where the dependence structure is much more complicated. In these tasks it is convenient to use the uncorrelated or poorly correlated arguments for building regression models, provided that the argumants are correlated among themselsves. It is true, that the variance of estimates increases due to using correlated arguments, but this is not a problem, when the number of observations n is very large. The most appropriate strategy should be used, accounting the losses due poor model versus strongly correlated arguments. It is obvious that in the case of high-dimensional multivariate distributions the scope of different dependence structures is huge. If we regard only the signs of correlations, then we get 2^{k-1} different structures for the k-dimensional matrices. E.g., in the case of $k = 8$ the number of different dependence structures equals to 256. Including all possible zero correlations (i.e., independent subvectors) of a 8-dimensional vector, the number of different dependence structures will be about 80 000 (see (Tiit, 1986)).

4. THE STUDY OF THE HAMILTON–ROTLEDGE–CUADRAS INEQUALITY

Cuadras has mentioned in his report that it is possible to build an example, when the common inequality

$$R^2 \leqslant \sum_{i=1}^{k} r_i^2 \tag{11}$$

does not hold, and instead of it the inequality (in future HRC inequality)

$$R^2 > \sum_{i=1}^{k} r_i^2 \tag{C5}$$

is valid, which seems to be quite surprising for most practical statisticians.

Example 1. Let us demonstrate that it is possible to find the random vector, having such correlation matrix, that a stronger version of the HRC inequality,

$$\frac{R^2}{\sum_{i=1}^{k} r_i^2} > C \tag{12}$$

will be satisfied, where C is a constant fulfilling the condition

$$1 < C < k.$$

The construction of the vector $(Y, X_1, \ldots, X_k)$ can be reduced to the construction of the correlation matrix $\mathbf{R}$, which is defined by (1) and (3), where

$$c = r = -\frac{1}{k}, \qquad \text{and} \qquad s = k. \tag{13}$$

In this case we get from the equation (6)

$$R^2 = \frac{k(1/k)^2}{1 + (k-1) \cdot (-1/k)} = 1, \tag{14}$$

and, as it follows,

$$R^2 = k \sum_{i=1}^{k} r_i^2.$$

In fact, the correlation matrix $\mathbf{R}$ is singular in this case. But for each C satisfying (12) it is possible to find such positive η, that if we take $r = c = -1/k + \eta$, then the matrix $\mathbf{R}_\eta$ is *regular* and the coefficient of determination R^2, calculated by $\mathbf{R}_\eta$, satisfies the inequality (12).

The result is well-known, connected with the representation of a vector having multinomial distribution via dependent dichotomeous variables. If we take k events A_i, $P(A_i) = 1/k$, where $A_i \cap A_j = \varnothing$, if $i \neq j$, and take X_i equal to the indicator of event A_i $(i = 1, \ldots, k)$, then the correlation matrix of the vector $X = (X_1, \ldots, X_k)'$ has the shape given in the last example, while the last variable X_k can be exactly expressed (prognosed) by all others – the fact that is expressed in the equation (14).

Next, the more exiting example, using variability of components of the vector ρ, will be constructed.

Example 2. For given $k > 1$ and arbitrary C there exists a random vector $(Y, X_1, \ldots, X_k)'$ so that the inequality (12) is fulfilled. Let us regard the equation (7) and design the vector ρ in such a way that $\bar{r} = 0$. Then we get from (7) the following expression for R^2:

$$R^2 = \frac{1}{1-c} \sum_{i=1}^{k} {\rho_i}^2.$$

If we choose c and r in such a way, that $(1-c)^{-1} > C$ and $R^2 < 1$, then the task is solved. The admissibility of this choice follows from the conditions (4) and (5), i.e., c should be as close to 1 as necessary to fulfill the first condition, and r should be as close to 0 as necessary to fulfill the second one.

REFERENCES

Tiit, E. (1980). The comparison of theoretical models in linear regression analysis. In: *The Proceedings of Conference 'Theoretical and Applied Problems in Mathematics'*, Tartu, pp. 233–235 (in Estonian).

Tiit, E.-M. (1981). The choice of models in linear regression analysis. *Proceedings of Computer Centre of Tartu State University* **46**, 60–84 (in Russian).

Tiit, E. (1982). Class of admissible empirical models in regression analysis. *Computing Systems* **88**, 99–106 (in Russian).

Tiit, E. (1983). About stability of the linear regression model. *Proceedings of Computer Centre of Tartu State University* **49**, 76–95 (in Russian).

Tiit, E. (1983). About *a priori* choice of regressors into linear model. *Zavodskaya Laboratoriya* **7**, 44–49 (in Russian).

Tiit, E.-M. (1986). Random vectors with given arbitrary marginals and given correlation matrix. *Acta et Comment Universitatis Tartuensis* **733**, 14–39.

Multivar. Statist., pp. 93–106
E.–M. Tiit *et al.* (Eds)

TESTING LACK OF FIT IN MIXED EFFECT MODELS FOR LONGITUDINAL DATA*

M. PALTA and R. P. QU
Department of Preventive Medicine, University of Wisconsin-Madison, 504 N. Walnut St., Madison, Wisconsin 53705, U.S.A.

ABSTRACT

Recent years have seen much development in analyzing longitudinal and repeated measures data with multivariate methods. However, relatively little attention has been directed toward checking the fit of such models. We consider the mixed effect model of Laird and Ware (1982). Motivated by the common occurrence of cohort and period effects in longitudinal studies, we model an alternative arising from such omitted covariates. We present two approaches to testing for this alternative, and apply these to a data set generated by enviromental monitoring of mobile homes.

1. INTRODUCTION

Clinical and epidemiologic studies often involve follow-up of individuals with observation of risk factors and outcomes over time. For example, a recent study (Newschaffer *et al.*, 1992) examined the decline in total serum cholesterol with age in persons age 65 and older. As a cross-sectional analysis led to an apparently stronger decline in cholesterol level than did a longitudinal approach, the authors concluded that lower cholesterol levels in earlier birth cohorts may have exaggerated the decline. Such cohort effects (Dwyer *et al.*, 1992), can be viewed as omitted covariates.

Period effects are also commonly found. In one study (Hanrahan *et al.*, 1985), mean formaldehyde emission (in kppm) in mobile homes was found to decline linearly with the natural logarithm of home age (in months) across 127 homes, with a between home regression slope of $\beta = \mathbf{SL}_b = -0.22$. However, fitting regression lines longitudinally within each home yielded a substantially stronger decline, with a within home slope $\beta = \mathbf{SL}_w = -0.45$. As many homes entered the study in spring and were followed through fall and formaldehyde emission decreases with lower temperature and humidity, these variables may have counfounded the estimator of within home decline.

We have previously considered omitted cohort and period effects and examined their impact on different regression estimators (Palta and Yao, 1991).

*Supported by grant CA-53786 from the National Cancer Institute.

Other investigators have demonstrated that differences in regression coefficients may arise from non-linearity (Louis *et al.*, 1986). Comparison of regression coefficients can therefore detect several forms of model misspecification. Although many methods exist to analyze longitudinal data, few methods for examining model fit have been developed.

In this paper, we use a mixed effects regression model (Laird and Ware, 1982) to model dependence of a response on known covariates under the null hypothesis. To formulate an alternative hypothesis, we generalize our previous model for confounding by an omitted covariate to a multivariate setting. We also generalize our previous tests (Palta *et al.*, 1994) for misspecification of the regression model, which are sensitive to the resulting alternative, and apply these tests to the formaldehyde data introduced above.

2. MODEL AND NOTATION

2.1. *Null model*

For $i = 1, 2, \ldots, n$, let $\mathbf{y}_i = (y_{i1}, y_{i2}, \ldots, y_{ik_i})'$ be the k_i responses and $\widetilde{\mathbf{X}}_i = (\mathbf{X}_i, \mathbf{Z}_i) = (\tilde{\mathbf{x}}_{i1}, \tilde{\mathbf{x}}_{i2}, \ldots, \tilde{\mathbf{x}}_{i(s+p)})$ the covariate matrix for individual i, where $\mathbf{X}_i = (\mathbf{x}_{i1}, \mathbf{x}_{i2}, \ldots, \mathbf{x}_{is})$ and $\mathbf{Z}_i = (\mathbf{z}_{i1}, \mathbf{z}_{i2} \ldots, \mathbf{z}_{ip})$ are $k_i \times s$ and $k_i \times l$ matrices, respectively, and $\mathbf{W}_i = (\mathbf{1}, \widetilde{\mathbf{X}}_i)'$. Let $\boldsymbol{\gamma} = (\gamma_1, \gamma_2, \ldots, \gamma_{s+p})'$, $\boldsymbol{\beta} = (\beta_1, \beta_2, \ldots, \beta_{s+p})'$. Let $\mathbf{I}$ be the identity matrix, $\mathbf{1} = (1, 1, \ldots, 1)'$, and $\mathbf{0} = (0, 0, \ldots, 0)'$ of proper orders according to their respective context. The general model for longitudinal data is

$$\mathbf{y}_i \mid \widetilde{\mathbf{X}}_i, a_i, \mathbf{b}_i \sim N\big(\beta_0 \mathbf{1} + \widetilde{\mathbf{X}}_i \boldsymbol{\beta} + a_i \mathbf{1} + \mathbf{X}_i \mathbf{b}_i, \sigma_0^2 \mathbf{I}\big), \tag{1}$$

for $i = 1, 2, \ldots, n$. Here σ_0^2 is the unknown error variance and $(a_i, \mathbf{b}_i')'$ are random effects. We assume that

$$\begin{pmatrix} a_i \\ \mathbf{b}_i \end{pmatrix} \overset{\text{i.i.d.}}{\sim} N\left[\begin{pmatrix} 0 \\ \mathbf{0} \end{pmatrix}, \mathbf{D} = \begin{pmatrix} \sigma_a^2 & \boldsymbol{\sigma}'_{ab} \\ \boldsymbol{\sigma}_{ab} & \mathbf{D}_b \end{pmatrix}\right].$$

Integrating out a_i and $\mathbf{b}_i$, we obtain the marginal multivariate normal model

$$\mathbf{y}_i \mid \widetilde{\mathbf{X}}_i \sim N\big(\beta_0 \mathbf{1} + \widetilde{\mathbf{X}}_i \boldsymbol{\beta}, \Sigma_i\big)$$

with variance matrix

$$\Sigma_i = \sigma_0^2 \mathbf{I} + (\mathbf{1}, \mathbf{X}_i) \mathbf{D} (\mathbf{1}, \mathbf{X}_i)'.$$

2.2. *Alternative model*

Let $\mathbf{U}_i = (\mathbf{u}_{i1}, \mathbf{u}_{i2}, \ldots, \mathbf{u}_{il})$ denote a matrix of unknown covariates which are omitted from the model fitting, and $\boldsymbol{\eta} = (\eta_0, \eta_1, \ldots, \eta_l)'$ the corresponding coefficients. Assume that the correct model with both known and unknown covariates is

$$\mathbf{y}_i \mid \widetilde{\mathbf{X}}_i, \mathbf{U}_i, a_i, \mathbf{b}_i \sim N\big(\beta_0 \mathbf{1} + \widetilde{\mathbf{X}}_i \boldsymbol{\beta} + \mathbf{U}_i \boldsymbol{\eta} + a_i \mathbf{1} + \mathbf{X}_i \mathbf{b}_i, \sigma_0^2 \mathbf{I}\big)$$

for $i = 1, 2, \ldots, n$.

We show in Appendix 1 that under certain conditions, the alternative model conditional on only observed covariates is given by

$$\mathbf{y}_i \mid \widetilde{\mathbf{X}}_i, a_i, \mathbf{b}_i \sim N\big[\beta_0 \mathbf{1} + \widetilde{\mathbf{X}}_i \boldsymbol{\beta}^* + \big(\bar{\bar{\mathbf{x}}}_i' \otimes \mathbf{1}\big)\boldsymbol{\gamma} + a_i \mathbf{1} + \mathbf{X}_i \mathbf{b}_i, \widetilde{\Sigma}\big], \qquad (2)$$

where $\bar{\mathbf{x}}_i = k_i^{-1} \mathbf{X}_i' \mathbf{1} = (\bar{x}_{i1}, \bar{x}_{i2}, \ldots, \bar{x}_{is})'$ and $\bar{\bar{\mathbf{x}}}_i = k_i^{-1} \widetilde{\mathbf{X}}_i' \mathbf{1} = (\bar{\bar{x}}_{i1}, \bar{\bar{x}}_{i2}, \ldots, \bar{\bar{x}}_{i(s+p)})'$ the individual specific means of the covariates and $\widetilde{\Sigma}$ is a covariance matrix with exchangeable correlation structure (i.e., the correlation between any two residuals on the same person is constant), and $\boldsymbol{\beta}^*$ contains the coefficients of $\widetilde{\mathbf{X}}_i$ in the new model.

For the situation of only one covariate $\mathbf{x}$ in the model, Palta and Yao (1991) derived the alternative model (2) from a model of form (1). The derivation for the more general multivariate case is given in Appendix 1. Integrating out the random effects in (2) leads to the marginal model

$$\mathbf{y}_i \mid \widetilde{\mathbf{X}}_i \sim N\big[\beta_0 \mathbf{1} + \widetilde{\mathbf{X}}_i \boldsymbol{\beta}^* + \big(\bar{\bar{\mathbf{x}}}_i' \otimes \mathbf{1}\big)\boldsymbol{\gamma}, \Sigma_i^*\big],$$

where Σ_i^* has the same form as Σ_i.

3. GOODNESS OF FIT TESTS

We present two testing procedures: (a) a comparison of unweighted within and between individual regression coefficients; and (b) testing $\mathbf{H}_0$: $\boldsymbol{\gamma} = \mathbf{0}$ in model (2) by fitting a mixed effects model (Laird and Ware, 1982). The two tests are closely related, but involve different weighting of the data.

3.1. *Comparison of unweighted within and between regression coefficients*

This procedure compares the regression coefficients obtained by averaging within individual slopes and by regressing $\bar{y}_i$ on $\bar{\mathbf{x}}_i$ across individuals. For each i, by least squares

$$(\hat{\beta}_{0i}, \hat{\boldsymbol{\beta}}_i')' = \big(\mathbf{W}_i' \mathbf{W}_i\big)^{-1} \mathbf{W}_i' \mathbf{y}_i.$$

We denote the unweighted average of these individual estimates of coefficients of $\widetilde{\mathbf{X}}_i$ by $\mathbf{SL}_w$. It is easy to see that $\mathbf{SL}_w$ is unbiased under both (1) and (2).

The between individual estimate of the regression coefficients is

$$(\tilde{\beta}_0, \tilde{\boldsymbol{\beta}}')' = \left(\overline{\mathbf{W}}'\overline{\mathbf{W}}\right)^{-1}\overline{\mathbf{W}}'\bar{\mathbf{y}},$$

where $\bar{\mathbf{y}} = (\bar{y}_1, \bar{y}_2, \ldots, \bar{y}_n)'$ and $\overline{\mathbf{W}} = \mathbf{1}'(k_1^{-1}\mathbf{W}_1', k_2^{-1}\mathbf{W}_2', \ldots, k_n^{-1}\mathbf{W}_n')'$ is the covariate matrix for the between individual model. Let $\mathbf{SL}_b$ be the sub-vector $\tilde{\boldsymbol{\beta}}$. Under the null hypothesis, $\mathrm{E}(\mathbf{SL}_b \mid \{\widetilde{\mathbf{X}}_i\}) = \mathrm{E}(\mathbf{SL}_w \mid \{\widetilde{\mathbf{X}}_i\})$ and under the alternative hypothesis formulated as in (2),

$$E\left(\mathbf{SL}_b \mid \{\widetilde{\mathbf{X}}_i\}\right) = \mathrm{E}\left(\mathbf{SL}_w \mid \{\widetilde{\mathbf{X}}_i\}\right) + \gamma.$$

Other options for obtaining a between-individual regression coefficient exist (e.g., Wei and Stram, 1988).

Let $\hat{\mathbf{q}} = \mathbf{SL}_w - \mathbf{SL}_b$. Being a linear combination of ordinary least squares regression estimators, $\hat{\mathbf{q}}$ is normally distributed and has covariance matrix

$$\begin{aligned}&\mathrm{Cov}\left(\hat{\mathbf{q}} \mid \{\widetilde{\mathbf{X}}_i\}\right)\\ &\quad = \mathrm{Cov}\left(\mathbf{SL}_w \mid \{\widetilde{\mathbf{X}}_i\}\right) + \mathrm{Cov}\left(\mathbf{SL}_b \mid \{\widetilde{\mathbf{X}}_i\}\right) - 2\mathrm{Cov}\left(\mathbf{SL}_w, \mathbf{SL}_b \mid \{\widetilde{\mathbf{X}}_i\}\right).\end{aligned}$$

Hence $\mathrm{F} = \hat{\mathbf{q}}'[\mathrm{Cov}(\hat{\mathbf{q}} \mid \{\widetilde{\mathbf{X}}_i\})]^{-1}\hat{\mathbf{q}}$ follows a chi-squared distribution with degrees of freedom $s+p$ and can be used for testing lack-of-fit of the underlying model (1).

Denote by $\mathbf{M}_{(1)}$ the second diagonal block of any $1+s+p$ square matrix $\mathbf{M}$ partitioned into orders 1 and $s+p$. Let $\mathrm{diag}\{C_i\}$ denote any matrix with only elements or submatrices $C_1, C_2, \ldots, C_m$ on the main diagonal. Let

$$\mathbf{D}_b^* = \begin{pmatrix} \mathbf{D}_b & \mathbf{0} \\ \mathbf{0} & \mathbf{0} \end{pmatrix}.$$

In Appendix 2, we show that

$$\mathrm{Cov}\left(\mathbf{SL}_w \mid \{\widetilde{\mathbf{X}}_i\}\right) = n^{-1}\mathbf{D}_b^* + n^{-2}\sigma_0^2\Sigma(\mathbf{W}_i'\mathbf{W}_i)_{(1)}^{-1}$$

and

$$\begin{aligned}\mathrm{Cov}\left(\mathbf{SL}_w, \mathbf{SL}_b \mid \{\widetilde{\mathbf{X}}_i\}\right) &= n^{-1}\mathbf{D}_b^*,\\ \mathrm{Cov}\left(\mathbf{SL}_b \mid \{\widetilde{\mathbf{X}}_i\}\right) &= \left\{\mathrm{Cov}\left[(\tilde{\beta}_0, \tilde{\boldsymbol{\beta}}')' \mid \{\widetilde{\mathbf{X}}_i\}\right]\right\}_{(1)},\end{aligned}$$

where

$$\begin{aligned}&\mathrm{Cov}\left[(\tilde{\beta}_0, \tilde{\boldsymbol{\beta}}')' \mid \{\widetilde{\mathbf{X}}_i\}\right]\\ &\quad = \left(\overline{\mathbf{W}}'\overline{\mathbf{W}}\right)^{-1}\overline{\mathbf{W}}'\mathrm{diag}\left\{k_i^{-1}\sigma_0^2 + \sigma_a^2 + 2\boldsymbol{\sigma}_{ab}'\bar{\mathbf{x}}_i + \bar{\mathbf{x}}_i'\mathbf{D}_b\bar{\mathbf{x}}_i\right\}\overline{\mathbf{W}}\left(\overline{\mathbf{W}}'\overline{\mathbf{W}}\right)^{-1}.\end{aligned}$$

There are $s(s+1)/2+1$ unknown parameters in $\mathrm{Cov}(\mathbf{SL}_w, \mathbf{SL}_b \mid \{\widetilde{\mathbf{X}}_i\})$. Following the suggestion of Swamy (1970) which is based on the least squares estimates and residuals $\mathbf{r}_i = \mathbf{y}_i - \mathbf{W}_i(\hat{\beta}_{0i}, \hat{\boldsymbol{\beta}}_i')'$:

$$\hat{\sigma}_0^2 = n^{-1}\sum_{i=1}^{n}\left[k_i - (1+s+p)\right]^{-1}\mathbf{r}_i'\mathbf{r}_i$$
$$= n^{-1}\sum_{i=1}^{n}\left[k_i - (1+s+p)\right]^{-1}\mathbf{y}_i\left[\mathbf{I} - \mathbf{W}_i(\mathbf{W}_i'\mathbf{W}_i)^{-1}\mathbf{W}_i'\right]\mathbf{y}_i',$$
$$\widehat{\mathbf{D}} = (n-1)^{-1}\sum_{i=1}^{n}\left[\begin{pmatrix}\hat{\beta}_{0i}\\ \hat{\boldsymbol{\beta}}_i\end{pmatrix} - n^{-1}\sum_{j=1}^{n}\begin{pmatrix}\hat{\beta}_{0j}\\ \hat{\boldsymbol{\beta}}_j\end{pmatrix}\right]\left[\begin{pmatrix}\hat{\beta}_{0i}\\ \hat{\boldsymbol{\beta}}_i\end{pmatrix} - n^{-1}\sum_{j=1}^{n}\begin{pmatrix}\hat{\beta}_{0j}\\ \hat{\boldsymbol{\beta}}_j\end{pmatrix}\right]'$$
$$- n^{-1}\hat{\sigma}_0^2\sum_{i=1}^{n}(\mathbf{W}_i'\mathbf{W}_i)^{-1}.$$

Swamy (1970) proved that both $\hat{\sigma}_0^2$ and $\widehat{\mathbf{D}}$ are consistent estimators of σ_0^2 and $\mathbf{D}$ respectively when k_i and n tend to infinity. Thus substituting $\hat{\sigma}_0^2$ and $\widehat{\mathbf{D}}$ into F gives a legitimate test statistic for practical application.

This testing approach has the advantage of involving inspection of the simple unweighted within and between estimators. In situations with non-compliance bias or survival bias, individuals with longer follow-up may display different trends than those with shorter follow-up. Weighted estimators may therefore be biased. The test statistic F has the potential for detecting misspecification other than model (2). Louis *et al.* (1986) showed that $\mathbf{SL}_w$ and $\mathbf{SL}_b$ differ when non-linearity is present. However, the lack of efficient weighting may decrease power. Outliers among the individual slopes $\hat{\boldsymbol{\beta}}_i$ present another problem. In the example discussed below we found it necessary to remove such outliers to avoid overestimating the variance of $\mathbf{SL}_w$ and thereby under-emphasizing the lack of fit.

The above approach to testing lack of fit is similar to that of Hausman (1978), but we do not present this latter approach here, due to space limitations. In our applications, the test of Hausman was generally less powerful than the above test.

3.2. *Directly testing regression coefficients in model* (2)

In the second approach we test the hypothesis $\mathbf{H}_0$: $\gamma = 0$ (i.e., whether model (1) holds) by applying maximum likelihood procedures to model (2). A simple version of this procedure is to use the test statistic $\hat{\gamma}'[\widehat{\mathrm{Cov}}(\gamma \mid \widetilde{\mathbf{X}}_i)]^{-1}\hat{\gamma}$ where the covariance is obtained from the information matrix. The approach uses efficient weighting and the covariance matrix of $\hat{\gamma}$ under the alternative hypothesis, which is expected to be smaller than the variances used in the previous test.

4. APPLICATION

The study of Hanrahan *et al.* (1985), sought to determine the health effects and trends in formaldehyde levels in mobile homes. To this end, the investigators collected data in 129 mobile homes in Wisconsin during 1980/81. The number of observations in each home, k_i ranged from 2 to 10, with 9 as the modal value. The follow-up period of each home was short relative to the range of

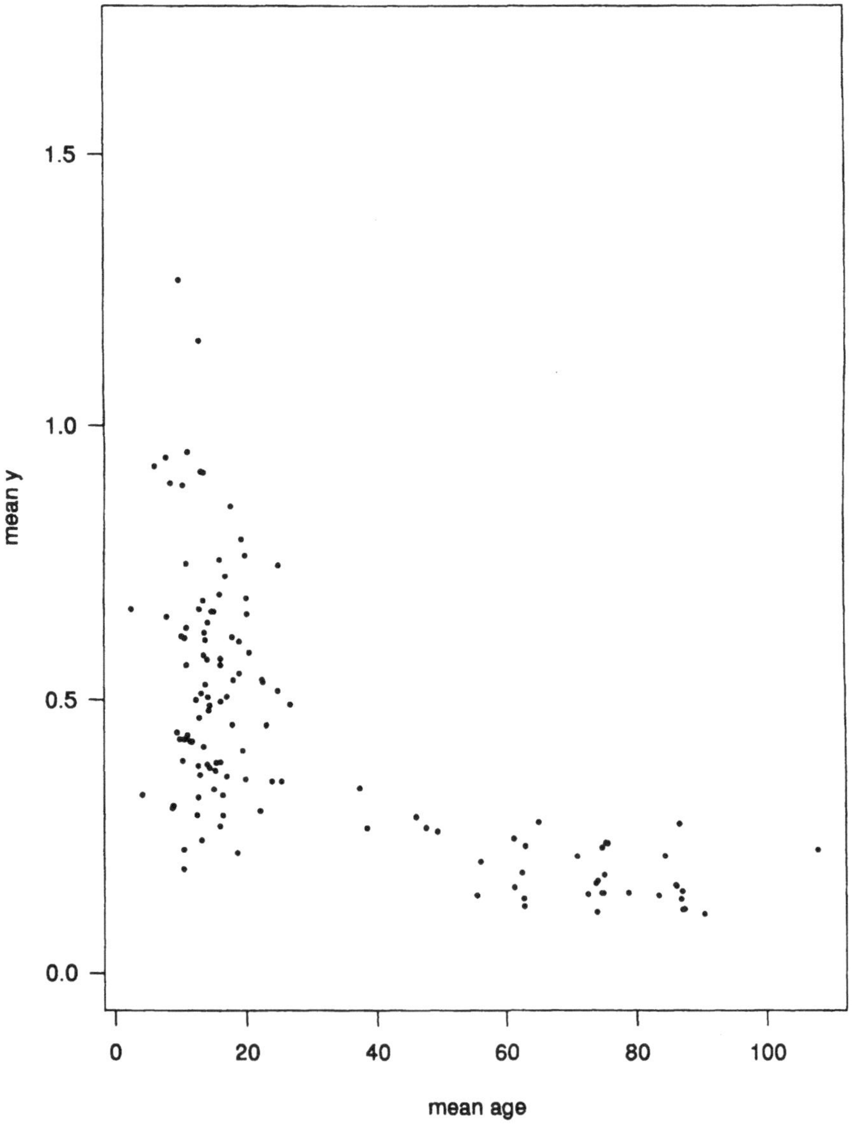

Figure 1. Mean home formaldehyde level versus home age in months.

home ages across homes. We study changes in formaldehyde level with home aging.

Figure 1 shows a plot of mean formaldehyde level $\bar{y}_i$ versus mean age in months $\bar{x}_i$. Non-linearity is obvious. We found that the transformation $\log(x)$ removed the non-linearity. Figure 2 is the counterpart of Figure 1 using $\log(x)$ instead of x and Figure 3 shows, as a further linearity check, $\hat{\beta}_i$ versus mean of $\log(x)$. We removed two outliers ($|\hat{\beta}_i| > 5$) before performing further goodness of fit testing. Table 1 below therefore includes only 127 homes.

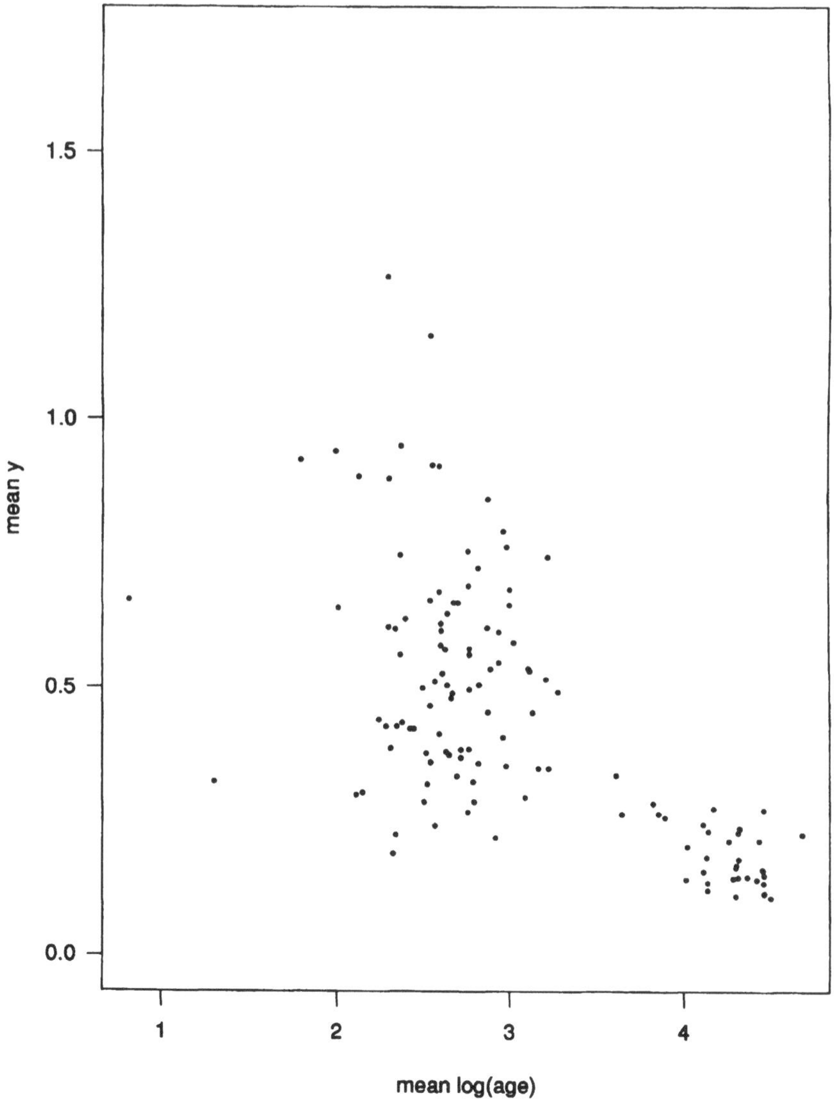

Figure 2. Mean home formaldehyde level versus mean log of home age.

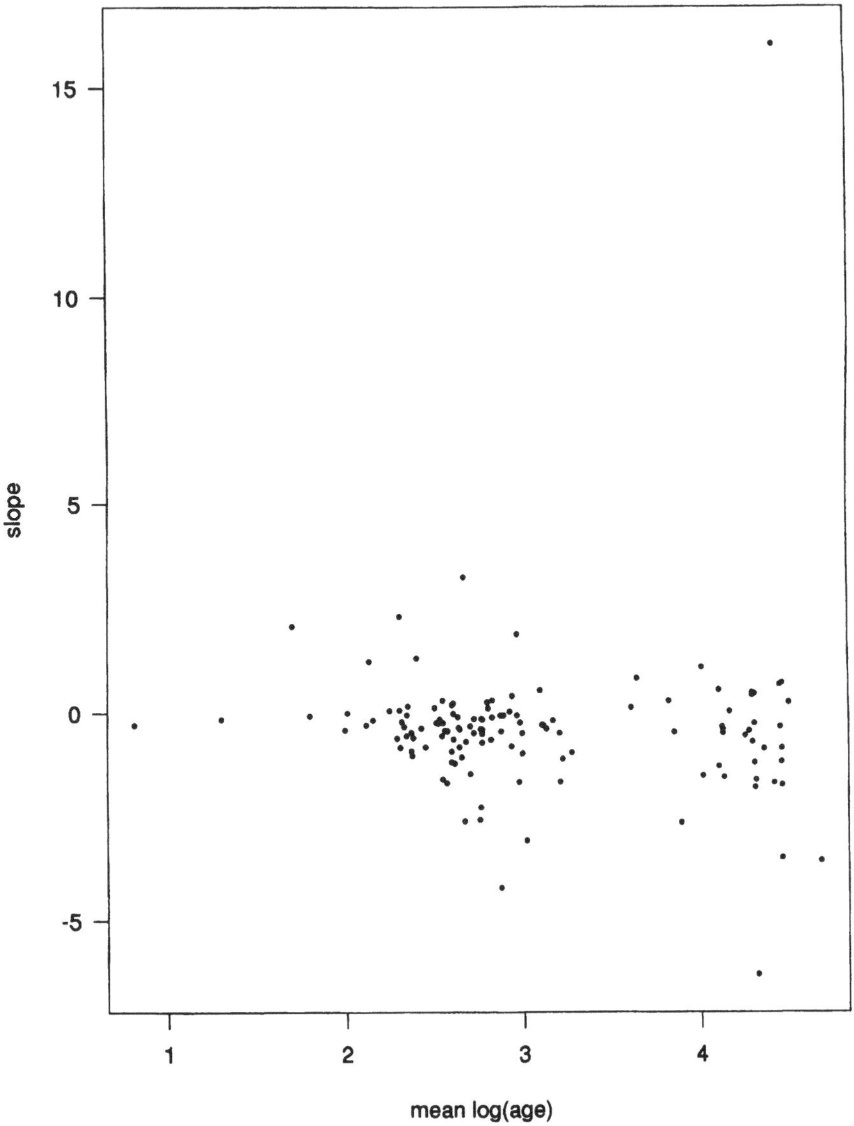

Figure 3. Individual slope in formaldehyde level versus log of home age.

Table 1 shows considerable differences between the cross-sectional and longitudinal regression coefficients. Both statistical tests of the difference in Table 1 indicate the presence of omitted covariates, with the test for $\gamma = 0$ yielding a more significant result. As there was more spread in home ages between than within homes, $\mathbf{SL}_w$ differs more from the maximum likelihood estimator than does $\mathbf{SL}_b$.

Table 1.
Goodness of fit testing for model describing formaldehyde level with home aging

Estimates	Intercept	Slope (standard deviation)		
Within regression	2.03	−0.454(0.093)		
Between regression	1.13	−0.218(0.090)		
ML for model (1)	1.22	−0.247(0.019)		
Tests	**Numerator**	**Denominator**	**Statistic**	**p-value**
Unweighted test	0.236	0.082	2.86	< 0.005
ML test for $\gamma = 0$	0.192	0.038	5.11	< 0.001

The decline in formaldehyde and the calendar month of the final follow-up were related ($r = 0.23$, $p = 0.01$). Limiting analysis to the summer months gave a mean within home slope of −0.19.

We conclude that the within home estimators in this study are confounded by period effects and that the cross-sectional estimates of formaldehyde decline are likely to be the least biased. Because period effects were primary confounders in the above study, and as the design was more cross-sectional than longitudinal (i.e., more spread in ages across homes than within), the multivariate random effects approach also gives reasonable results.

5. DISCUSSION

Cohort and period effects are often of concern in analyzing longitudinal data, but tend to be ignored when longitudinal models are fit. Fortunately, longitudinal data give rise to opportunities to detect such effects as well as other confounders that are associated with risk factors of interest differently within individuals from between individuals. We have found comparison of regression coefficients arising from different weighting of cross-sectional and longitudinal approaches to analysis to be especially useful.

Of the two approaches to goodness of fit testing we considered, the second, i.e., testing coefficients of functions of $\bar{x}_i$ in the random effects model, appears more flexible and possibly the most powerful. We have pointed out the advantage of also inspecting the unweighted coefficients, especially when x_{ij} are highly unbalanced between individuals.

Considering the biases expected in different regression coefficients with confounding also reaffirms the importance of a well planned study design. A well-informed choice of study design can greatly help in avoiding biases arising from cohort and period effects. If cohort effects are of major concern emphasis on achieving a wide range of follow-up within each individual leads to a multivariate model that is essentially unbiased. Conversely, if period effects are of major concern, choosing individuals with a wide range of the risk factors leads to minimal bias in multivariate modelling.

APPENDIX

1. *Conditional Distribution of* $\mathbf{y}_i$ *Given* $\widetilde{\mathbf{X}}_i$, a_i, *and* $\mathbf{b}_i$

We will derive model (2) for equal sample sizes $k_i = k$. For unequal k_i, model (2) holds approximately with k as the harmonic mean of k_i.

Let $g = s + p$, $\tilde{\mathbf{x}}_{i\bullet j} = (\tilde{x}_{i1j}, \tilde{x}_{i2j}, \ldots, \tilde{x}_{igj})'$ and $\mathbf{u}_{i\bullet j} = (u_{i1j}, u_{i2j}, \ldots, u_{ilj})'$, $\boldsymbol{\mu}_{x_i} = (\mu_{x_{i1}}, \mu_{x_{i2}}, \ldots, \mu_{x_{ig}})'$, $\boldsymbol{\mu}_{u_i} = (\mu_{u_{i1}}, \mu_{u_{i2}}, \ldots, \mu_{u_{il}})'$, $\mathbf{e}_{x_{ij}} = (e_{x_{i1j}}, e_{x_{i2j}}, \ldots, e_{x_{igj}})'$ and $\mathbf{e}_{u_{ij}} = (e_{u_{i1j}}, e_{u_{i2j}}, \ldots, e_{u_{ilj}})'$ for $i = 1, 2, \ldots, n$ and $j = 1, 2, \ldots, k$. Assume that $\{\mathbf{U}_i\}$ and $\{(a_i, \mathbf{b}_i')'\}$ are independent and

$$\begin{pmatrix} \tilde{\mathbf{x}}_{i\bullet j} \\ \mathbf{u}_{i\bullet j} \end{pmatrix} = \begin{pmatrix} \boldsymbol{\mu}_{x_i} \\ \boldsymbol{\mu}_{u_i} \end{pmatrix} + \begin{pmatrix} \mathbf{e}_{x_{ij}} \\ \mathbf{e}_{u_{ij}} \end{pmatrix},$$

$$\begin{pmatrix} \boldsymbol{\mu}_{x_i} \\ \boldsymbol{\mu}_{u_i} \end{pmatrix} \overset{\text{i.i.d.}}{\sim} N\left[\begin{pmatrix} \mathbf{0} \\ \mathbf{0} \end{pmatrix}, \Lambda_\mu = \begin{pmatrix} \Lambda_{\mu_{xx}} & \Lambda_{\mu_{xu}} \\ \Lambda_{\mu_{ux}} & \Lambda_{\mu_{uu}} \end{pmatrix}\right],$$

and

$$\begin{pmatrix} \mathbf{e}_{x_{ij}} \\ \mathbf{e}_{u_{ij}} \end{pmatrix} \overset{\text{i.i.d.}}{\sim} N\left[\begin{pmatrix} \mathbf{0} \\ \mathbf{0} \end{pmatrix}, \Lambda_e = \begin{pmatrix} \Lambda_{e_{xx}} & \Lambda_{e_{xu}} \\ \Lambda_{e_{ux}} & \Lambda_{e_{uu}} \end{pmatrix}\right].$$

Let $\tilde{\mathbf{x}}_i = (\tilde{\mathbf{x}}_{i1}', \tilde{\mathbf{x}}_{i2}', \ldots, \tilde{\mathbf{x}}_{ig}')'$ and $\mathbf{u}_i = (\mathbf{u}_{i1}', \mathbf{u}_{i2}', \ldots, \mathbf{u}_{il}')'$. Then

$$\begin{pmatrix} \tilde{\mathbf{x}}_i \\ \mathbf{u}_i \end{pmatrix} \overset{\text{i.i.d.}}{\sim} N\left[\begin{pmatrix} \mathbf{0} \\ \mathbf{0} \end{pmatrix}, \mathbf{A} = \begin{pmatrix} \mathbf{A}_{xx} & \mathbf{A}_{xu} \\ \mathbf{A}_{ux} & \mathbf{A}_{uu} \end{pmatrix}\right].$$

It follows that

$$\mathbf{u}_i \mid \tilde{\mathbf{x}}_i \sim N\big(\mathbf{A}_{ux}\mathbf{A}_{xx}^{-1}\tilde{\mathbf{x}}_i, \mathbf{A}_{uu} - \mathbf{A}_{ux}\mathbf{A}_{xx}^{-1}\mathbf{A}_{xu}\big).$$

Note that $\tilde{\mathbf{x}}_{it} = \mu_{x_{it}}\mathbf{1} + (e_{x_{it1}}, e_{x_{it2}}, \ldots, e_{x_{itk}})'$ and $\mathbf{u}_{it} = \mu_{u_{it}}\mathbf{1} + (e_{u_{it1}}, e_{u_{it2}}, \ldots, e_{u_{itk}})'$. Thus

$$\operatorname{Cov}\left(\tilde{\mathbf{x}}_{it}, \tilde{\mathbf{x}}_{it'}\right) = \mathbf{1}\operatorname{Cov}\left(\mu_{x_{it}}, \mu_{x_{it'}}\right)\mathbf{1}' + \operatorname{Cov}\left(e_{x_{it1}}, e_{x_{it'1}}\right)\mathbf{I},$$

for $i = 1, 2, \ldots, n$ and $t, t' = 1, 2, \ldots, g$, and $\mathbf{0}$ otherwise. Hence

$$\mathbf{A}_{xx} = \Lambda_{\mu_{xx}} \otimes \mathbf{1}\mathbf{1}' + \Lambda_{e_{xx}} \otimes \mathbf{I},$$

where $\otimes$ is the Kronecker product. Let $\mathbf{B} = \mathbf{I} \otimes \mathbf{1}$. Then $\Lambda_{\mu_{xx}} \otimes \mathbf{1}\mathbf{1}' = \mathbf{B}\Lambda_{\mu_{xx}}\mathbf{B}'$ and (see (Rao, 1985, Exer. 2.9, p. 33))

$$\mathbf{A}_{xx}^{-1} = (\Lambda_{e_{xx}} \otimes \mathbf{I})^{-1} - (\Lambda_{e_{xx}} \otimes \mathbf{I})^{-1}\mathbf{B}\big[\mathbf{B}'(\Lambda_{e_{xx}} \otimes \mathbf{I})^{-1}\mathbf{B} + \Lambda_{\mu_{xx}}^{-1}\big]^{-1} \times \mathbf{B}'\left(\Lambda_{e_{xx}} \otimes \mathbf{I}\right)^{-1},$$

$$\left(\Lambda_{e_{xx}} \otimes \mathbf{I}\right)^{-1}\mathbf{B} = \left(\Lambda_{e_{xx}}^{-1} \otimes \mathbf{I}\right)\mathbf{B} = \left(\mathbf{I} \otimes \mathbf{1}\right)\Lambda_{e_{xx}}^{-1},$$
$$\mathbf{B}'\left(\Lambda_{e_{xx}} \otimes \mathbf{I}\right)^{-1}\mathbf{B} = \left(\mathbf{I} \otimes \mathbf{1}'\right)\left(\mathbf{I} \otimes \mathbf{1}\right)\Lambda_{e_{xx}}^{-1} = k\Lambda_{e_{xx}}^{-1}.$$

Thus

$$\mathbf{A}_{xx}^{-1} = \Lambda_{e_{xx}}^{-1} \otimes \mathbf{I} - \left(\mathbf{I} \otimes \mathbf{1}\right)\Lambda_{e_{xx}}^{-1}\left(k\Lambda_{e_{xx}}^{-1} + \Lambda_{\mu_{xx}}^{-1}\right)^{-1}\Lambda_{e_{xx}}^{-1}\left(\mathbf{I} \otimes \mathbf{1}'\right),$$

and

$$\mathbf{A}_{ux}\mathbf{A}_{xx}^{-1}\tilde{\mathbf{x}}_i = \mathbf{B}_1\tilde{\mathbf{x}}_i + \mathbf{B}_2\bar{\tilde{\mathbf{x}}}_i,$$

where $\mathbf{B}_1 = \mathbf{A}_{ux}(\Lambda_{e_{xx}}^{-1} \otimes \mathbf{I}) = \mathbf{A}_{ux}\Lambda_{e_{xx}}^{-1} \otimes \mathbf{I}$ and

$$\mathbf{B}_2 = \mathbf{A}_{ux}\left(\mathbf{I} \otimes \mathbf{1}\right)\Lambda_{e_{xx}}^{-1}\left(\Lambda_{e_{xx}}^{-1} + k^{-1}\Lambda_{\mu_{xx}}^{-1}\right)^{-1}\Lambda_{e_{xx}}^{-1}.$$

The fact that k components in each $\mathbf{u}_{it}$ depend in the same way on $\bar{\tilde{\mathbf{x}}}_i$ leads to that $\mathbf{B}_2 = \widetilde{\mathbf{B}}_2 \otimes \mathbf{1}$, for some $\widetilde{\mathbf{B}}_2 = (\tilde{b}_{2lt})_{l\times g}$ which is an $l \times g$ matrix. Let $\widetilde{\mathbf{B}}_1 = \mathbf{A}_{ux}\Lambda_{e_{xx}}^{-1} = (\tilde{b}_{1it})_{l\times g}$. Then

$$\mathrm{E}\left(\mathbf{u}_{it} \mid \tilde{\mathbf{x}}_i\right) = \sum_{\alpha=1}^{g} \tilde{b}_{1t\alpha}\tilde{\mathbf{x}}_{i\alpha} + \Sigma_{\alpha=1}^{g}\tilde{b}_{2t\alpha}\bar{\tilde{x}}_{i\alpha}\mathbf{1}$$

for $i = 1, 2, \ldots, n$ and $t = 1, 2, \ldots, l$. Hence

$$\mathrm{E}\left(\mathbf{y}_i \mid \widetilde{\mathbf{X}}_i, a_i, \mathbf{b}_i\right) = \beta_0\mathbf{1} + \widetilde{\mathbf{X}}_i\boldsymbol{\beta}^* + \left(\bar{\tilde{\mathbf{x}}}_i' \otimes \mathbf{1}\right)\boldsymbol{\gamma} + a_i\mathbf{1} + \mathbf{X}_i\mathbf{b}_i,$$

where $\boldsymbol{\beta}^*$ is the new coefficient of $\widetilde{\mathbf{X}}_i$. The conditional covariance matrix of $\mathbf{y}_i$ given $\widetilde{\mathbf{X}}_i$, a_i, and $\mathbf{b}_i$ is

$$\begin{aligned}\mathrm{Cov}\left(\mathbf{y}_i \mid \widetilde{\mathbf{X}}_i, a_i, \mathbf{b}_i\right) &= \mathrm{Cov}\left[\mathrm{E}(\mathbf{y}_i \mid \widetilde{\mathbf{X}}_i, \mathbf{U}_i, a_i, \mathbf{b}_i) \mid \widetilde{\mathbf{X}}_i, a_i, \mathbf{b}_i\right] \\ &\quad + \mathrm{E}\left[\mathrm{Cov}\left(\mathbf{y}_i \mid \widetilde{\mathbf{X}}_i, \mathbf{U}_i, a_i, \mathbf{b}_i\right) \mid \widetilde{\mathbf{X}}_i, a_i, \mathbf{b}_i\right] \\ &= \mathrm{Cov}\left(\mathbf{U}_i\boldsymbol{\eta} \mid \widetilde{\mathbf{X}}_i, a_i, \mathbf{b}_i\right) + \sigma_0^2\mathbf{I}\end{aligned}$$

which is independent of $\widetilde{\mathbf{X}}_i$, a_i, and $\mathbf{b}_i$ by the conditional distribution of $\mathbf{u}_i$ given $\widetilde{\mathbf{X}}_i$. Considering the fact that the components of $\mathbf{y}_i$ have the same relationship with one another, we see that the resulting covariance matrix of $\mathbf{y}_i$ given $\widetilde{\mathbf{X}}_i$, a_i, and $\mathbf{b}_i$ is of compound symmetry. Heavy algebra can also be used to prove this result. Hence the resulting model from omitting $\mathbf{U}_i$ is of the form of model (2).

2. *Variance and covariance matrices of* $\mathbf{SL}_b$ *and* $\mathbf{SL}_w$

2.1. *Variance matrix of* $\mathbf{SL}_w$

$$\mathbf{SL}_w = n^{-1}\Sigma\hat{\boldsymbol{\beta}}_i, \mathrm{Cov}\left(\mathbf{SL}_w \mid \{\widetilde{\mathbf{X}}_i\}\right) = n^{-2}\Sigma\mathrm{Cov}\left(\hat{\boldsymbol{\beta}}_i \mid \{\widetilde{\mathbf{X}}_i\}\right),$$

$$\mathrm{Cov}\left[(\hat{\beta}_{0i}, \hat{\boldsymbol{\beta}}_i')' \mid \widetilde{\mathbf{X}}_i, a_i, \mathbf{b}_i\right] = \sigma_0^2(\mathbf{W}_i'\mathbf{W}_i)^{-1}.$$

Denote

$$\boldsymbol{\sigma}_{ab}^* = \begin{pmatrix} \boldsymbol{\sigma}_{ab} \\ \mathbf{0} \end{pmatrix} \quad \text{and} \quad \mathbf{D}^* = \begin{pmatrix} \mathbf{D} & \mathbf{0} \\ \mathbf{0} & \mathbf{0} \end{pmatrix}.$$

Then

$$\begin{aligned}\mathrm{Cov}\left[(\hat{\beta}_{0i}, \hat{\boldsymbol{\beta}}_i')' \mid \widetilde{\mathbf{X}}_i\right] &= \mathrm{E}\left\{\mathrm{Cov}\left[(\hat{\beta}_{0i}, \hat{\boldsymbol{\beta}}_i')' \mid \widetilde{\mathbf{X}}_i, a_i, \mathbf{b}_i\right]\right\} \\ &\quad + \mathrm{Cov}\left\{\mathrm{E}\left[(\hat{\beta}_{0i}, \hat{\boldsymbol{\beta}}_i')' \mid \widetilde{\mathbf{X}}_i, a_i, \mathbf{b}_i\right]\right\} \\ &= \sigma_0^2(\mathbf{W}_i'\mathbf{W}_i)^{-1} + \mathbf{D}^*.\end{aligned}$$

So

$$\mathrm{Cov}\left(\mathbf{SL}_w \mid \{\widetilde{\mathbf{X}}_i\}\right) = n^{-1}\mathbf{D}_b^* + n^{-2}\sigma_0^2\Sigma(\mathbf{W}_i'\mathbf{W}_i)_{(1)}^{-1}.$$

2.2. *Variance matrix of* $\mathbf{SL}_b$

Note that $(\tilde{\beta}_0, \tilde{\boldsymbol{\beta}}')' = (\overline{\mathbf{W}}'\overline{\mathbf{W}})^{-1}\overline{\mathbf{W}}'\bar{\mathbf{y}}$. So

$$\mathrm{Cov}\left[(\tilde{\beta}_0, \tilde{\boldsymbol{\beta}}')' \mid \{\widetilde{\mathbf{X}}_i\}\right] = (\overline{\mathbf{W}}'\overline{\mathbf{W}})^{-1}\overline{\mathbf{W}}'\mathrm{Cov}\left(\bar{\mathbf{y}} \mid \{\widetilde{\mathbf{X}}_i\}\right)\overline{\mathbf{W}}(\overline{\mathbf{W}}'\overline{\mathbf{W}})^{-1},$$

$$\begin{aligned}\mathrm{Cov}\left(\bar{y}_i \mid \widetilde{\mathbf{X}}_i\right) &= \mathrm{Cov}\left(k_i^{-1}\mathbf{1}'\mathbf{y}_i \mid \widetilde{\mathbf{X}}_i\right) = k_i^{-2}\mathbf{1}'\left[\sigma_0^2\mathbf{I} + (\mathbf{1}, \mathbf{X}_i)\mathbf{D}(\mathbf{1}, \mathbf{X}_i)'\right]\mathbf{1} \\ &= k_i^{-1}\sigma_0^2 + \sigma_a^2 + 2\boldsymbol{\sigma}_{ab}'\bar{\mathbf{x}}_i + \bar{\mathbf{x}}_i'\mathbf{D}_b\bar{\mathbf{x}}_i.\end{aligned}$$

Thus

$$\begin{aligned}&\mathrm{Cov}\left[(\tilde{\beta}_0, \tilde{\boldsymbol{\beta}}')' \mid \{\widetilde{\mathbf{X}}_i\}\right] \\ &= (\overline{\mathbf{W}}'\overline{\mathbf{W}})^{-1}\overline{\mathbf{W}}'\mathrm{diag}\left\{k_i^{-1}\sigma_0^2 + \sigma_a^2 + 2\boldsymbol{\sigma}_{ab}'\bar{\mathbf{x}}_i + \bar{\mathbf{x}}_i'\mathbf{D}_b\bar{\mathbf{x}}_i\right\}\overline{\mathbf{W}}(\overline{\mathbf{W}}'\overline{\mathbf{W}})^{-1}\end{aligned}$$

and $\mathrm{Cov}(\mathbf{SL}_b \mid \{\widetilde{\mathbf{X}}_i\}) = [\mathrm{Cov}(\tilde{\boldsymbol{\beta}} \mid \{\widetilde{\mathbf{X}}_i\})]_{(1)}$.

2.3. *Covariance matrix of* $\mathbf{SL}_w$ *and* $\mathbf{SL}_b$

For each i

$$\begin{aligned}\mathrm{Cov}\left[(\hat{\beta}_{0i}, \hat{\boldsymbol{\beta}}_i')', \bar{y}_i \mid \widetilde{\mathbf{X}}_i\right] &= \mathrm{Cov}\left[(\mathbf{W}_i'\mathbf{W}_i)^{-1}\mathbf{W}_i'\mathbf{y}_i, k_i^{-1}\mathbf{1}'\mathbf{y}_i \mid \widetilde{\mathbf{X}}_i\right] \\ &= k_i^{-1}(\mathbf{W}_i'\mathbf{W}_i)^{-1}\mathbf{W}_i'\mathrm{Cov}\left(\mathbf{y}_i \mid \widetilde{\mathbf{X}}_i\right)\mathbf{1} \\ &= k_i^{-1}(\mathbf{W}_i'\mathbf{W}_i)^{-1}\mathbf{W}_i'\left[\sigma_0^2\mathbf{I} + (\mathbf{1}, \mathbf{X}_i)\mathbf{D}(\mathbf{1}, \mathbf{X}_i)'\right]\mathbf{1} \\ &= k_i^{-1}(\mathbf{W}_i'\mathbf{W}_i)^{-1}\mathbf{W}_i'\left[\sigma_0^2\mathbf{I} + \mathbf{W}_i\mathbf{D}^*\mathbf{W}_i'\right]\mathbf{1} \\ &= k_i^{-1}\sigma_0^2(\mathbf{W}_i'\mathbf{W}_i)^{-1}\mathbf{W}_i'\mathbf{1} + k_i^{-1}\mathbf{D}^*\mathbf{W}_i'\mathbf{1}.\end{aligned}$$

So

$$\text{Cov}\left(\hat{\beta}_i, \bar{y}_i \mid \widetilde{\mathbf{X}}_i\right) = \boldsymbol{\sigma}^*_{ab} + \mathbf{D}^*_b \bar{\bar{\mathbf{x}}}_i,$$

since $\mathbf{W}'_i \mathbf{W}_i (\mathbf{W}'_i \mathbf{W}_i)^{-1} = \mathbf{I}$. By independence,

$$\text{Cov}\left[\left(\hat{\beta}'_i, \hat{\beta}'_2, \ldots, \hat{\beta}'_n\right)', \bar{\mathbf{y}} \mid \{\widetilde{\mathbf{X}}_i\}\right] = \text{diag}\left\{\boldsymbol{\sigma}^*_{ab} + \mathbf{D}^*_b \bar{\bar{\mathbf{x}}}_i\right\}.$$

Thus

$$\begin{aligned}
&\text{Cov}\left(\mathbf{SL}_w, \mathbf{SL}_b \mid \{\widetilde{\mathbf{X}}_i\}\right) \\
&= \left\{n^{-1}(\mathbf{I}, \ldots, \mathbf{I}) \text{Cov}\left[\left(\hat{\beta}'_1, \hat{\beta}'_2, \ldots, \hat{\beta}'_n\right)', \bar{\mathbf{y}} \mid \{\widetilde{\mathbf{X}}_i\}\right] \overline{\mathbf{W}}\left(\overline{\mathbf{W}}'\overline{\mathbf{W}}\right)^{-1}\right\}_{(1)} \\
&= \left\{n^{-1}(\mathbf{I}, \ldots, \mathbf{I}) \,\text{diag}\left\{\boldsymbol{\sigma}^*_{ab} + \mathbf{D}^*_b \bar{\bar{\mathbf{x}}}_i\right\} \overline{\mathbf{W}}\left(\overline{\mathbf{W}}'\overline{\mathbf{W}}\right)^{-1}\right\}_{(1)} = \text{I} + \text{II},
\end{aligned}$$

where

$$\begin{aligned}
\text{I} &= n^{-1}\left\{(\mathbf{I}, \ldots, \mathbf{I}) \,\text{diag}\left\{\boldsymbol{\sigma}^*_{ab}, \boldsymbol{\sigma}^*_{ab}, \ldots, \boldsymbol{\sigma}^*_{ab}\right\} \overline{\mathbf{W}}\left(\overline{\mathbf{W}}'\overline{\mathbf{W}}\right)^{-1}\right\}_{(1)} \\
&= n^{-1}\left\{\left(\boldsymbol{\sigma}^*_{ab} \otimes \mathbf{1}'\right) \overline{\mathbf{W}}\left(\overline{\mathbf{W}}'\overline{\mathbf{W}}\right)^{-1}\right\}_{(1)} = n^{-1}\left\{\boldsymbol{\sigma}^*_{ab} \otimes \mathbf{1}'\overline{\mathbf{W}}\left(\overline{\mathbf{W}}'\overline{\mathbf{W}}\right)^{-1}\right\}_{(1)} \\
&= n^{-1}\left\{\left(\boldsymbol{\sigma}^*_{ab}, \mathbf{0}, \ldots, \mathbf{0}\right)_{s \times (1+s+p)}\right\}_{(1)} = \mathbf{0}, \\
\text{II} &= n^{-1}\left\{(\mathbf{I}, \ldots, \mathbf{I}) \,\text{diag}\left\{\mathbf{D}^*_b \bar{\bar{\mathbf{x}}}_i\right\} \overline{\mathbf{W}}\left(\overline{\mathbf{W}}'\overline{\mathbf{W}}\right)^{-1}\right\}_{(1)} \\
&= \left\{\mathbf{D}^*_b \left(\bar{\bar{\mathbf{x}}}_1, \bar{\bar{\mathbf{x}}}_2, \ldots, \bar{\bar{\mathbf{x}}}_n\right) \overline{\mathbf{W}}\left(\overline{\mathbf{W}}'\overline{\mathbf{W}}\right)^{-1}\right\}_{(1)} = n^{-1}\mathbf{D}^*_b.
\end{aligned}$$

Again, the last equality is due to $\mathbf{I} = \overline{\mathbf{W}}'\overline{\mathbf{W}}(\overline{\mathbf{W}\mathbf{W}})^{-1}$. Hence $\text{Cov}(\mathbf{SL}_w, \mathbf{SL}_b \mid \{\widetilde{\mathbf{X}}_i\}) = n^{-1}\mathbf{D}^*_b$.

REFERENCES

Dwyer, J. H., Feinleib, M., Lippert, P. and Hoffmeister, H. (1992). *Statistical Models for Longitudinal Studies of Health* (*Chapter 1*). Oxford University Press, New York.

Hanrahan, L. P., Anderson, H. A., Dally, K. A., Eckman, A. D. and Kanarek, M. S. (1985). Formaldehyde concentrations in Wisconsin mobile homes. *J. Control Association* **35**, 1164–1167.

Hausman, J. (1978). Specification tests in econometrics. *Econometrica* **46**, 1251–1271.

Laird, N. M. and Ware, J. H. (1982). Random effects models for longitudinal data. *Biometrics* **38**, 963–974.

Louis, T. A., Robins, J., Dockery, D. W., Spiro, A. III. and Ware, J. H. (1986). Explaining discrepancies between longitudinal and cross-sectional models. *J. Chronie Dis.* **39**, 831–839.

Newschaffer, C. J., Bush, T. L. and Hale, W. E. (1992). Aging and total cholesterol level: Cohort, period and survivorship effect. *Amer. J. Epidemiol.* **136**, 23–34.

Palta, M. and Yao, T. J. (1991). Analysis of longitudinal data with unmeasured confounders. *Biometrics* **47**, 1355–1369.

Palta, M., Yao, T. J. and Velu, R. (1994). Testing for omitted variables andnon-linearity in regression models for longitudinal data. *Stat. Med.* **13**, in press.

Rao, C. R. (1985). *Linear Statistical Inference and Its Application.* Wiley, New York.

Swamy, P. A. V. B. (1970). Efficient inference in a random coefficient regression model. *Econometrica* **38**, 311–323.

Wei, L. J. and Stram, D. O. (1988). Analyzing repeated measurements withpossibly missing observations by modelling marginal distributions. *Stat. Med.* **7**, 139–148.

Part Three

TESTS IN MULTIVARIATE STATISTICS

Multivar. Statist., pp. 109–122
E.-M. Tiit *et al.* (Eds)

A PROJECTION NT-TYPE TEST FOR SPHERICAL SYMMETRY OF A MULTIVARIATE DISTRIBUTION*

LI-XING ZHU[1], KAI-TAI FANG[2] and JIN-TING ZHANG[3]
[1,3]Probability Laboratory, Institute of Applied Mathematics,
Academia Sinica, Beijing,

[2]Hong Kong Baptist University and Institute of Applied Mathematics,
Academia Sinica, Beijing

ABSTRACT

The test of spherical symmetry of a multivariate distribution plays an important role in high-dimensional data analysis and has attracted more and more attention. Several tests have been suggested for past years. In this paper, we propose a new test, the so-called projection NT-type test which is consistent against any fixed alternative and is easy to be computed. The limit distribution of the test statistic is obtained. Simulations and examples in the paper show that the proposed test is powerful. It can be seen that our approach could be similarly applied to many projection pursuit tests.

1. INTRODUCTION

In past years the multivariate analogies of univariate symmetric distribution, so-called spherical distribution or spherically symmetric distribution including the multivariate normal distribution, multivariate t-distribution and the uniform distribution on a unit sphere in high dimensional space, have attracted much attention (Fang *et al.*, 1990). It is well known that these distributions possess many properties which are similar to those of the multivariate standard normal distribution. Hence, if one has known that the underlying distribution of the collected data is spherical, some tools of traditional multivariate analysis, as Friedman (1987) pointed out, can be applied to analyze the data, especially in exploratory stage of data analysis. Otherwise, as well known, it is often necessary to choose delicately a method or several methods to analyze the data so that the trustful conclusions can be obtained. Thus, testing the spherical symmetry of a multivariate distribution is of some interest in the area of high-dimensional data analysis. Kariya and Eaton (1977) considered this issue.

*This work was supported by the Hong Kong UPGC grant and the NFS of China.

Recently, Baringhaus(1991) proposed a Cramer–von Mises type test for this task. His test is consistent against any fixed alternative. Fang *et al.* (1993) suggested a necessary test that can be applied to test whether or not d marginal distributions at d orthogonal directions are identical where d is the dimension of the multivariate distribution. Later, Zhu *et al.* (1993) developed a test such that it is consistent against any fixed alternative. However, the above tests mentioned are not flawless. The test of Fang *et al.* (1993) has a simple limit null distribution for its statistic but it is only a necessary test. On the other hand, although the tests of Baringhaus (1991) and Zhu *et al.* (1993) are some consistent tests, the limit distributions of their statistics are not tractable. Thus, to determine the critical values for both tests, they proposed, respectively, approaches. One is based on the approximation in terms of the orthogonal decomposition of the limit distribution, the other depends on the bootstrap method. Both approaches need heavy computing workload and are not easily used in practice.

Let $F(\boldsymbol{x})$ be a underlying d-dimensional distribution and we want to test H_0: $F(\boldsymbol{x})$ is spherical against H_1: $F(\boldsymbol{x})$ is not spherical. In this paper we shall develop a new test, the so-called projection NT-type test for this purpose. Firstly we employ the idea of the projection pursuit (PP) test and find a univariate version of the test based on all the directions on the unit sphere S^d in R^d. Secondly we use the number-theoretic method (NTM) to choose a set of representative directions $A_s = \{\boldsymbol{a}_1, \boldsymbol{a}_2, \ldots, \boldsymbol{a}_s\}$ on S^d to construct a projection NT-type statistic. We shall prove that the test is consistent against any fixed alternative and its critical values can be determined by looking simply at the χ^2-distribution (chi-square distribution) table. In Section 2, we shall motivate our approach for the test of spherical symmetry and present the main result. Section 3 contains several simulation experiments and an example in which the presented test is applied to a real-life data set. Some further remarks are given in Section 4. The theoretic proof of main result is given in Section 5. It will be seen that the approach proposed in the paper could be similarly applied to many PP tests, for example, the test for multivariate normality (see (Zhu *et al.*, 1994)). Throughout this paper, the boldface symbols will represent vectors or matrices.

2. THE PROJECTION NT-TYPE TEST

2.1. *Motivation for constructing a new test*

Suppose that $\boldsymbol{x}_1, \boldsymbol{x}_2, \ldots, \boldsymbol{x}_n$ is an i.i.d. sample from the population distribution $F(\boldsymbol{x})$ in R^d. We are now required to test the hypothesis of the spherical symmetry of $F(\boldsymbol{x})$, i.e.,

$$H_0\colon\ F(\boldsymbol{x}) \text{ is spherical, } \quad \text{versus} \quad H_1\colon\ F(\boldsymbol{x}) \text{ is not spherical.} \tag{2.1}$$

Using the results concerning the theory of spherically symmetric distributions, we can transform the hypothesis H_0 into its univariate version:

$$H_0\colon \text{ all } \boldsymbol{a}'\boldsymbol{x} \text{ have the same distribution,} \tag{2.2}$$

where $\boldsymbol{a}$ is a unit vector on the unit sphere $S^d = \{\boldsymbol{a}\colon \|\boldsymbol{a}\| = \mathbf{1},\ \boldsymbol{a} \in R^d\}$ and $\boldsymbol{x} \sim F(\boldsymbol{x})$. Denote by $F_{\boldsymbol{a}}(\cdot)$ the marginal distribution of $\boldsymbol{a}'\boldsymbol{x}$ at the direction $\boldsymbol{a} \in S^d$. For convenience, we denote the distribution of $\boldsymbol{a}_0'\boldsymbol{x}$ by F_0 throughout this paper where $\boldsymbol{a}_0$ is a unit vector fixed arbitrarily on S^d. Obviously, the above transformation (2.2) relies on the fact that $\boldsymbol{x}$ has a spherical distribution if and only if for each $\boldsymbol{a} \in S^d$, $\boldsymbol{a}'\boldsymbol{x}$ has the same distribution as F_0, i.e., $F_{\boldsymbol{a}} \equiv F_0$. The reader can refer to (Fang *et al.*, 1990, Theorem 2.5.4) for details.

Furthermore, suppose that $F_{\boldsymbol{a}}(t)$ is continuous with respect to t for any fixed $\boldsymbol{a} \in S^d$. Then as (Zhu *et al.*, 1994, Theorem 1) showed that, $F_{\boldsymbol{a}}(\cdot)$ is also continuous with respect to $\boldsymbol{a}$. That is, for any fixed $\varepsilon > 0$, there exists $\delta(\varepsilon) > 0$ such that $\sup_t |F_{\boldsymbol{a}}(t) - F_{\boldsymbol{b}}(t)| < \varepsilon$ provided $\|\boldsymbol{a} - \boldsymbol{b}\| \leqslant \delta(\varepsilon)$ where $t \in R^1$. Indeed, the assumption that $F_{\boldsymbol{a}}(t)$ is continuous with respect to t for any fixed $\boldsymbol{a} \in S^d$ is very mild. As (Zhu *et al.*, 1994, corollary) pointed out that if $F(\cdot)$ has a density function, then the above assumption can be satisfied. Then a necessary and sufficient condition that $\boldsymbol{x}$ has a spherical distribution is as follows: for any countable dense set $A = \{\boldsymbol{a}_1, \ldots, \boldsymbol{a}_s, \ldots\}$ of S^d (i.e., for given $\boldsymbol{a} \in R^d$ and δ there exits a $\boldsymbol{a}_i \in A$ such that $\|\boldsymbol{a} - \boldsymbol{a}_i\| < \delta$), $F_{\boldsymbol{a}_i} \equiv F_0$ for each $\boldsymbol{a}_i \in A$. Indeed, necessity is obvious. Now we prove the sufficiency. By the continuity of $F_{\boldsymbol{a}}(t)$ with respect to $\boldsymbol{a}$, for any fixed $\boldsymbol{a} \in S^d$ and any given $\varepsilon > 0$, we can find an $\boldsymbol{a}_i \in A$ such that

$$\sup_t |F_{\boldsymbol{a}}(t) - F_{\boldsymbol{a}_i}(t)| < \varepsilon. \tag{2.3}$$

As $F_{\boldsymbol{a}_i}(t) = F_0(t)$ for all $t \in R^1$, we have

$$\sup_t |F_{\boldsymbol{a}}(t) - F_0(t)| < \varepsilon. \tag{2.4}$$

Letting $\varepsilon \to 0$, we obtain $F_{\boldsymbol{a}}(t) = F_0(t)$ for each t. The proof for the sufficiency is completed. Hence, the null hypothesis can be turned into a simpler version:

$$H_0\colon F_{\boldsymbol{a}_i} \equiv F_0 \qquad \text{for each} \quad \boldsymbol{a}_i \in A. \tag{2.5}$$

Let $A_s = \{\boldsymbol{a}_1, \boldsymbol{a}_2, \ldots, \boldsymbol{a}_s\} \subset A$ be a set of representative directions of A in a certain sense below. When s is large, the following hypothesis is an approximation to (2.5):

$$H_{0s}\colon F_{\boldsymbol{a}_i} \equiv F_0 \qquad \text{for each} \quad \boldsymbol{a}_i \in A_s. \tag{2.6}$$

This motivates us to construct a Wilcoxon type statistic for testing the hypothesis (2.6) and then the hypothesis (2.2). For any $\boldsymbol{a} \in A_s$, define

$$V_n(\boldsymbol{a}) = \frac{1}{n^2}\sum_{i=1}^{n}\sum_{j=1}^{n} I(\boldsymbol{a}_0'\boldsymbol{x}_i < \boldsymbol{a}'\boldsymbol{x}_j), \tag{2.7}$$

where $I(x) = 0$ if $x \leqslant 0$ and $I(x) = 1$ if $x > 0$. It is well known that this statistic can be applied to test whether or not the two distributions of $\boldsymbol{a}_0'\boldsymbol{x}$ and $\boldsymbol{a}'\boldsymbol{x}$ are identical. Denote an empirical process by $\{\sqrt{n}(V_n(\boldsymbol{a}) - 1/2)\colon \boldsymbol{a} \in S^d\}$. Furthermore, we derive the following assertion whose proof will be given in Section 5.

THEOREM 1. *Under the null hypothesis H_0 which is defined in* (2.2), $\{\sqrt{n}(V_n(\boldsymbol{a}) - 1/2)\colon \boldsymbol{a} \in S^d\}$ *converges weakly to a centered continuous Gaussian process* $\{J(\cdot, \boldsymbol{a}),\ \boldsymbol{a} \in S^d\}$ *with the mean function zero and the covariance function*:

$$R(\boldsymbol{a}, \boldsymbol{b}) = EJ(\cdot, \boldsymbol{a})J(\cdot, \boldsymbol{b}) = E\big(F_0(\boldsymbol{a}_0'\boldsymbol{x}) - F_0(\boldsymbol{a}'\boldsymbol{x})\big)\big(F_0(\boldsymbol{a}_0'\boldsymbol{x}) - F_0(\boldsymbol{b}'\boldsymbol{x})\big) \tag{2.8}$$

for any $\boldsymbol{a}, \boldsymbol{b} \in S^d$.

This conclusion implies that any finite subset of components of $\{J(\cdot, \boldsymbol{a})\colon \boldsymbol{a} \in S^d\}$, say $\{J(\cdot, \boldsymbol{a}_i)\colon \boldsymbol{a}_i \in A_s\}$, has the s-variate normal distribution with mean vector $\mathbf{0}$ and covariance matrix $\boldsymbol{R}_s = (R(\boldsymbol{a}_i, \boldsymbol{a}_j))_{1 \leqslant i,j \leqslant s}$. Then, if $\boldsymbol{R}_s$ is a positive definite matrix, $\boldsymbol{R}_s^{-1/2}(J(\cdot, \boldsymbol{a}_1), \ldots, J(\cdot, \boldsymbol{a}_s))' = \boldsymbol{R}_s^{-1/2}\boldsymbol{J}_s' \sim N_s(\mathbf{0}, \boldsymbol{I}_s)$ and

$$\boldsymbol{J}_s'\boldsymbol{R}_s^{-1}\boldsymbol{J}_s \sim \chi^2(s), \tag{2.9}$$

where $\chi^2(s)$ is the chi-square distribution with s degrees of freedom. Letting $s \to \infty$, we obtain

$$\frac{1}{\sqrt{2s}}\big(\boldsymbol{J}_s'\boldsymbol{R}_s^{-1}\boldsymbol{J}_s - s\big) \longrightarrow N(0, 1) \tag{2.10}$$

in distribution. Using Theorem 1, we can construct an estimate of $\boldsymbol{J}_s'\boldsymbol{R}_s^{-1}\boldsymbol{J}_s$. Denote

$$\begin{aligned}
\hat{J}(\cdot, \boldsymbol{a}) &= \sqrt{n}\big(V_n(\boldsymbol{a}) - 1/2\big),\\
\hat{\boldsymbol{J}}_s &= \big(\hat{J}(\cdot, \boldsymbol{a}_1), \ldots, \hat{J}(\cdot, \boldsymbol{a}_s)\big)',\\
\widehat{R}(\boldsymbol{a}, \boldsymbol{b}) &= \frac{1}{n}\sum_{j=1}^{n}\big(\widehat{F}_0(\boldsymbol{a}_0'\boldsymbol{x}_j) - \widehat{F}_0(\boldsymbol{a}'\boldsymbol{x}_j)\big)\big(\widehat{F}_0(\boldsymbol{a}_0'\boldsymbol{x}_j) - \widehat{F}_0(\boldsymbol{b}'\boldsymbol{x}_j)\big),
\end{aligned}$$

and

$$\widehat{\boldsymbol{R}}_s = \big(\widehat{R}(\boldsymbol{a}_i, \boldsymbol{a}_j)\big)_{1 \leqslant i,j \leqslant s},$$

where $\widehat{F}_0(\cdot)$ is the empirical distribution based on the projection sample $\boldsymbol{a}_0'\boldsymbol{x}_1, \ldots, \boldsymbol{a}_0'\boldsymbol{x}_n$. Due to the consistency of $\hat{\boldsymbol{J}}_s$ and $\widehat{\boldsymbol{R}}_s$ to $\boldsymbol{J}_s$ and $\boldsymbol{R}_s$, $\hat{\boldsymbol{J}}_s'\widehat{\boldsymbol{R}}_s^{-1}\hat{\boldsymbol{J}}_s$ is also a consistent estimate of $\boldsymbol{J}_s'\boldsymbol{R}_s^{-1}\boldsymbol{J}_s$ as $n \to \infty$. However, $\widehat{\boldsymbol{R}}_s^{-1}$ is not a robust estimate of $\boldsymbol{R}_s^{-1}$ if $\boldsymbol{R}_s$ is very close to be degenerate. In this case we might suggest to use

$$\widehat{\boldsymbol{R}}_{s1} = \widehat{\boldsymbol{R}}_s + k\boldsymbol{I}_s \tag{2.11}$$

as an estimate of $\boldsymbol{R}_s$ where $\boldsymbol{I}_s$ is an identity matrix of order s and k can be regarded as a ridge parameter similarly as in the ridge regression and can be taken as $k = n^{-2}$, $n^{-3/2}$ or n^{-1}. We now recommend the following test statistic for testing the hypothesis (2.2):

$$\mathrm{PNT} = \hat{\boldsymbol{J}}_s'\widehat{\boldsymbol{R}}_{s1}^{-1}\hat{\boldsymbol{J}}_s. \tag{2.12}$$

For large s, we use the following statistic:

$$\mathrm{PNT}_1 = \frac{1}{\sqrt{2s}}\left(\hat{\boldsymbol{J}}_s'\widehat{\boldsymbol{R}}_{s1}^{-1}\hat{\boldsymbol{J}}_s - s\right) \tag{2.13}$$

which approximately has a standard normal distribution. This fact will be presented in the next subsection.

2.2. *The main result*

THEOREM. *Suppose that $\boldsymbol{R}_s$ is a positive definite matrix for each $s \geqslant 2$ under the hypothesis H_0. Then we have*

$$\lim_{s\to\infty}\lim_{n\to\infty}\frac{1}{\sqrt{2s}}\left(\hat{\boldsymbol{J}}_s'\widehat{\boldsymbol{R}}_{s1}^{-1}\hat{\boldsymbol{J}}_s - s\right) \sim N(0,1), \tag{2.14}$$

in distribution and for any fixed s,

$$\lim_{n\to\infty}\hat{\boldsymbol{J}}_s'\widehat{\boldsymbol{R}}_{s1}^{-1}\hat{\boldsymbol{J}}_s \sim \chi^2(s) \tag{2.15}$$

in distribution. If the hypothesis H_0 is false and $\sup_{\boldsymbol{a}}|EF_0(\boldsymbol{a}'\boldsymbol{x}) - 1/2| > 0$, then with probability one we have

$$\lim_{s\to\infty}\lim_{n\to\infty}\hat{\boldsymbol{J}}_s'\widehat{\boldsymbol{R}}_{s1}^{-1}\hat{\boldsymbol{J}}_s = \infty \tag{2.16}$$

and thus

$$\lim_{s\to\infty}\lim_{n\to\infty}\frac{1}{\sqrt{2s}}\left(\hat{\boldsymbol{J}}_s'\widehat{\boldsymbol{R}}_{s1}^{-1}\hat{\boldsymbol{J}}_s - s\right) = \infty. \tag{2.17}$$

2.3. *How to choose the representative points on S^d*

In the above two subsections, we have pointed out that to test approximately the hypothesis H_0 we can test the hypothesis H_{0s}. For testing the hypothesis H_{0s}, it is enough to test whether or not $\boldsymbol{a}_0'\boldsymbol{x}, \boldsymbol{a}_1'\boldsymbol{x}, \ldots, \boldsymbol{a}_s'\boldsymbol{x}$ have the same distribution. To this end, we construct the test statistics (2.12) and (2.13). From the process of constructing the statistics, we see that it is very important to choose properly $s+1$ projection directions $\boldsymbol{a}_0, \boldsymbol{a}_1, \ldots, \boldsymbol{a}_s$ such that they can represent S^d effectively, especially in the case of small sample size. In the literature, the projection directions $\boldsymbol{a}_0, \boldsymbol{a}_1, \ldots, \boldsymbol{a}_s$ are often called the representative points of S^d. How to choose these directions? We meet two aspects: on the one hand, we wish that the $s+1$ chosen points do have a good representation for S^d. On the other hand, we hope that $\boldsymbol{R}_s$ defined in terms of these projection directions is not degenerate under H_0. Obviously, we need a method for choosing the representative points so that these points can easily fit the above two aspects. Fortunately, the number-theoretic method (NTM) can meet these two requirements perfectly. The essence of NTM is to find a set of points that is uniformly scattered over an s-dimensional unit cube $[0,1]^s$ (it is called a NT-net on $[0,1]^s$) and sometimes this set can be used instead of random numbers in Monte Carlo method with a better result. Recently, Fang and Wang (1993) gave a systematic study on applications of NTM in statistics. In particular, they proposed a way to generate a set of representative points on S^d and gave the corresponding algorithm that is the so-called TFWW algorithm. They pick up the good lattice point (glp) method, one of NTM, to generate a NT-net on $[0,1]^s$ as follows: For given integer s $(s > d^{-2})$, choose an integer vector $\boldsymbol{h} = (h_1, \ldots, h_{d-1})$ satisfying $1 \leqslant h_i < s+1$, $h_i \neq h_j$, for $i \neq j$. Denote

$$c_{ki} = \{(2kh_i - 1)/2(s+1)\}, \qquad k = 1, \ldots, s+1, \quad i = 1, \ldots, d-1,$$

and

$$\boldsymbol{c}_k = (c_{k1}, \ldots, c_{k,d-1})', \qquad k = 1, \ldots, s+1,$$

where $\{x\}$ denotes the fraction part of x. The choice of $\boldsymbol{h} = (h_1, \ldots, h_{d-1})$ can be found in (Hua and Wang, 1981, Appendix) or (Fang and Wang, 1993). Then we can use the TFWW algorithm to generate the desired points $\{\boldsymbol{a}_0, \ldots, \boldsymbol{a}_s\}$ on S^d corresponding to $\{\boldsymbol{c}_0, \ldots, \boldsymbol{c}_s\}$.

3. SIMULATIONS AND EXAMPLE

To demonstrate powers of the proposed test PNT for the spherical symmetry of a multivariate distribution, we will apply it to several simulated data sets and a real-life example in this section.

The simulated results are listed in Table 1 below. The data sets are generated from 11 different distributions, where

1) $N_d(\boldsymbol{0}, \boldsymbol{I}_d)$: the d-variate standard normal distribution;

2) $Mt_d(7, \mathbf{0}, \boldsymbol{I}_d)$: the d-variate t-distribution with the density function

$$c_d\left(1+\frac{\boldsymbol{x}'\boldsymbol{x}}{10}\right)^{-(5+d/2)},$$

where c_d is a normalizing constant;

3) Ex (1) – 1: the standard exponential distribution with a shift (–1) and mean zero;

4) $\chi^2(2) - 2$: the chi-square distribution with 2 degrees of freedom with a shift (–2) and mean zero;

5) U (–0.5, 0.5): the uniform distribution over [–0.5, 0.5];

6) Ga (2) – 2: the gamma distribution with a density function $xe^{-x}I(x > 0)$ with shift (–2) and mean zero;

7) Beta (5, 1) – 5/6: the beta distribution with parameters 5 and 1 with a shift (–5/6) and mean zero.

Finally, the notation t-i.i.d. $F(\cdot)$ stands for a distribution with t independently and identically distributed components from $F(\cdot)$. When d variables are split into two parts, these parts are independent.

In the simulation we consider the low-dimensional cases of $d = 3, 4, 5$ only. It can be expected that our test is also powerful in high dimensional cases. For these eleven distributions of Table 1, the various sample sizes, say $n = 150, 200$, are investigated. In each case, $k = n^{-2}$, 31 projection directions are chosen by the glp method and TFWW algorithm and 500 duplicates are performed for each population. As the number of directions is not large ($\leqslant 200$), we use only the statistic PNT to treat the testing problem. The critical values, corresponding to the number of directions 31 are 43.77 under the significance level 5%.

Table 1.
Proportion of rejecting H_0 ($s = 31$, $\alpha = 5\%$)

Alternatives	$d=3$ $n=150$	$d=3$ $n=200$	$d=4$ $n=150$	$d=4$ $n=200$	$d=5$ $n=150$	$d=5$ $n=200$
$N(\mathbf{0}, \boldsymbol{I}_d)$	0.068	0.034	0.040	0.020	0.070	0.050
$Mt_d(7, \mathbf{0}, \boldsymbol{I}_d)$	0.050	0.050	0.020	0.090	0.050	0.040
d-i.i.d. U (–0.5, 0.5)	0.154	0.252	0.110	0.090	0.050	0.100
d-i.i.d. (Ex (1) – 1)	0.998	1.000	1.000	1.000	1.000	1.000
d-i.i.d. $(\chi^2(2) - 2)$	1.000	1.000	1.000	1.000	1.000	1.000
d-i.i.d. (Ga (2) – 2)	0.942	1.000	0.990	1.000	0.970	1.000
d-i.i.d. (Beta (5, 1) – 5/6)	0.930	0.998	1.000	1.000	0.970	1.000
$\{N(\mathbf{0}, \boldsymbol{I}_{d-1}), (\text{Ex}(1) - 1)\}$	0.680	0.958	0.520	0.840	0.330	0.580
$\{N(\mathbf{0}, \boldsymbol{I}_{d-2}), 2 \text{ i.i.d. } (\text{Ex}(1) - 1)\}$	0.946	0.996	0.960	1.000	0.560	0.780
$\{N(\mathbf{0}, \boldsymbol{I}_{d-1}), (\chi^2(2) - 2)\}$	0.968	0.994	1.000	1.000	1.000	1.000
$\{N(\mathbf{0}, \boldsymbol{I}_{d-2}), 2 \text{ i.i.d. } (\chi^2(2) - 2)\}$	1.000	1.000	1.000	1.000	1.000	1.000

Table 1 lists the simulation results. It can be seen that there are small probabilities of rejecting H_0 for cases of $N_d(\mathbf{0}, \mathbf{I}_d)$ and $Mt_d(7, \mathbf{0}, \mathbf{I}_d)$ because these two distributions are spherical. Powers of other nine alternatives are very good. Except the alternative d-i.i.d. U$(-0.5, 0.5)$. Furthermore, it seems that powers of test will be better with increasing the ratio n/s.

The following is a real-life example in which the new test is employed. The data set pertains to the derivation of the standardization of dress of Chinese men in 1976 and is taken from (Fang *et al.*, 1992). There are 12 measurements of the body considered, including

X_1: the height from the waist up,
X_2: the arm length,
X_3: the bust,
X_4: the neck,
X_5: the shoulder length,
X_6: the width of the front part of chest,
X_7: the width of the back part of chest,
X_8: the height,
X_9: the height without head and neck,
X_{10}: the height from the waist down,
X_{11}: the waist circumference,
X_{12}: the buttocks

with 100 observations. We now test whether or not some of variates are distributed spherically after the transformation

$$\boldsymbol{y} = \widehat{\boldsymbol{\Sigma}}^{-1/2}(\boldsymbol{x} - \bar{\boldsymbol{x}}) \tag{3.1}$$

where $\boldsymbol{y} = (y_1, y_2, \ldots, y_{12})'$, $\boldsymbol{x} = (X_1, \ldots, X_{12})'$, $\bar{\boldsymbol{x}}$ and $\widehat{\boldsymbol{\Sigma}}$ are the sample mean and the sample covariance matrix of $\boldsymbol{x}$, respectively. Suppose that we are interested in testing spherical symmetry of the subset $\boldsymbol{y}^{(8)} = (y_1, y_3, y_8, y_{10}, y_{12})'$, then we denote the observations by $\{\boldsymbol{y}_i^{(8)},\ i = 1, \ldots, 100\}$. When the number of projection directions is $s = 31$ the test statistic value is calculated to be PNT $= 25.76$. Comparing this value with the corresponding critical value 43.77, we obtain the conclusion 'Accepted'. That is, we recommend that the subset $\boldsymbol{y}^{(8)} = (y_1, y_3, y_8, y_{10}, y_{12})'$ is spherical. Similarly, several other subsets are considered. The results are presented in Table 2.

The conclusions listed in Table 2 tell us that all considered subsets of $\boldsymbol{y}$ are recommended to be spherical. It seems that some conclusions obtained here would be not consistent with the conclusions obtained by Zhu *et al.* (1993). However, they investigated the multinormality of some subsets for the same data set and drew a conclusion that all corresponding subsets of $\boldsymbol{x}$ possess good multinormality except $\boldsymbol{x}^{(2)} = (X_2, X_5, X_6)'$, $\boldsymbol{x}^{(4)} = (X_2, X_4, X_6, X_{11})'$ and $\boldsymbol{x}^{(10)} = (X_3, X_4, X_5, X_6, X_{11})'$. On this issue, our explanation is that some subsets of $\boldsymbol{y}$ can be spherical even if the corresponding subsets of $\boldsymbol{x}$ may be nonnormal originally.

Table 2.
Testing the Sphericity of Several Subsets of $\boldsymbol{y}$

$\boldsymbol{y}^{(i)}$	PNT	Conclusions	Normality
$\boldsymbol{y}^{(1)} = \{y_1, y_3, y_{11}\}$	24.18	Accepted	Accepted
$\boldsymbol{y}^{(2)} = \{y_2, y_5, y_6\}$	29.88	Accepted	Rejected
$\boldsymbol{y}^{(3)} = \{y_8, y_{10}, y_{12}\}$	24.50	Accepted	Accepted
$\boldsymbol{y}^{(4)} = \{y_2, y_4, y_6, y_{11}\}$	25.55	Accepted	Rejected
$\boldsymbol{y}^{(5)} = \{y_1, y_3, y_8, y_{10}\}$	40.40	Accepted	Accepted
$\boldsymbol{y}^{(6)} = \{y_1, y_8, y_{10}, y_{12}\}$	41.37	Accepted	Accepted
$\boldsymbol{y}^{(7)} = \{y_3, y_8, y_{10}, y_{12}\}$	23.05	Accepted	Accepted
$\boldsymbol{y}^{(8)} = \{y_1, y_3, y_8, y_{10}, y_{12}\}$	25.76	Accepted	Accepted
$\boldsymbol{y}^{(9)} = \{y_1, y_2, y_7, y_8, y_9\}$	34.57	Accepted	Accepted
$\boldsymbol{y}^{(10)} = \{y_3, y_4, y_5, y_6, y_{11}\}$	27.07	Accepted	Rejected
$\boldsymbol{y}^{(11)} = \{y_2, y_4, y_5, y_9, y_{11}\}$	37.52	Accepted	Rejected

4. FURTHER DISCUSSION

Although the results of our empirical illustration are quite encouraging, many questions can be posed when regarding practical implementation of the proposed test. For instance, are there automatic methods for setting the ridge parameter in (2.11) that assures good performance of the test for small-sample cases? We have tried 3 parameters n^{-1}, $n^{-3/2}$, and n^{-2} in our simulations, effects of the test are similar. Which ratio is moderate for assuring good performance of the test for fixed sample size n and the dimension of distributions? These open problems will be in the future research.

5. THE PROOF OF THEOREMS

Before proving Theorem 1 we first introduce two lemmas. For any fixed $\boldsymbol{a} \in S^d$, let

$$F_{n\boldsymbol{a}}(x) = \frac{1}{n}\sum_{j=1}^{n} I(\boldsymbol{a}^\tau \boldsymbol{x}_j \leqslant x), \qquad F_{n0}(x) = F_{n\boldsymbol{a}_0}(x),$$

$F_{\boldsymbol{a}}(x) = P\{\boldsymbol{a}^\tau \boldsymbol{x} \leqslant x\}$, and $F_0 \equiv F_{\boldsymbol{a}_0}$. As a special case of Dudley and Philipp (1983, Theorem 7.2.1), we have the invariant principle which is needed in the following proof.

LEMMA 5.1. *For any fixed $\theta > 0$ we have*

$$\sup_{\boldsymbol{a},x} \sqrt{n}\left|\frac{1}{n}\sum_{j=1}^{n}\Big(I\big(\boldsymbol{a}^\tau \boldsymbol{x}_j \leqslant x\big) - F_{\boldsymbol{a}}(x)\Big) - W(\cdot, \boldsymbol{a}, x)\right| = \mathrm{O}_p\big((\log n)^{-\theta}\big)$$

where $\boldsymbol{W} = \{W(\cdot, \boldsymbol{a}, x)\colon\ \boldsymbol{a} \in S^d,\ x \in R^1\}$ *is a centered continuous Gaussian process with the covariance function*

$$\begin{aligned} &E\{W(\cdot, \boldsymbol{a}, x)W(\cdot, \boldsymbol{a}_1, x_1)\} \\ &\quad = E\{I(\boldsymbol{a}^\tau \boldsymbol{x} \leqslant x)I(\boldsymbol{a}_1^\tau \boldsymbol{x} \leqslant x_1)\} - F_{\boldsymbol{a}}(x)F_{\boldsymbol{a}_1}(x_1). \end{aligned} \tag{5.1}$$

The following result describes the relationship between $\mathcal{L}^2$-seminorm continuity and almost sure continuity for the Gaussian process.

LEMMA 5.2. (Pollard, 1984, Theorem IV. 13). *For the Gaussian process defined in Lemma* 5.1, *there exists a constant* $c > 0$ *(depending only on the dimension of* S^d*) such that if*

$$\Big(P\big(I(\boldsymbol{a}^\tau \boldsymbol{x} \leqslant x) - I(\boldsymbol{a}_1^\tau \boldsymbol{x} \leqslant x_1)\big)^2\Big)^{1/2} \leqslant \delta(w),$$

then

$$|W(w, \boldsymbol{a}, x) - W(w, \boldsymbol{a}_1, x_1)| \leqslant c\delta(w), \tag{5.2}$$

where $\delta(w)$ *is finite for every* w.

The proof of Theorem 1. Note that we can express (2.7), as

$$\begin{aligned} V_n(\boldsymbol{a}) &= \int F_{n0}(x)\,\mathrm{d}F_{n\boldsymbol{a}}(x), \\ \sqrt{N}\Big(V_n(\boldsymbol{a}) - \frac{1}{2}\Big) &= \sqrt{n}\Big\{\int F_{n0}(x)\,\mathrm{d}F_{n\boldsymbol{a}}(x) - \int F_0(x)\,\mathrm{d}F_{\boldsymbol{a}}(x)\Big\} \\ &= \sqrt{n}\int \big(F_{n0}(x) - F_0(x)\big)\,\mathrm{d}F_{\boldsymbol{a}}(x) \\ &\quad + \sqrt{n}\int \big(F_{n0}(x) - F_0(x)\big)\,\mathrm{d}\big(F_{n\boldsymbol{a}}(x) - F_{\boldsymbol{a}}(x)\big) \\ &\quad - \sqrt{n}\int \big(F_{n\boldsymbol{a}}(x) - F_{\boldsymbol{a}}(x)\big)\,\mathrm{d}F_0(x) \\ &\triangleq I_{n1}(\boldsymbol{a}) + I_{n2}(\boldsymbol{a}) - I_{n3}(\boldsymbol{a}). \end{aligned}$$

By Lemmas 5.1 and 5.2, for any fixed $\theta > 0$, we have

$$\begin{aligned} \sup_{\boldsymbol{a}} |I_{n2}(\boldsymbol{a})| &\leqslant \sup_{\boldsymbol{a}} \Big|\int W(\cdot, \boldsymbol{a}_0, x)\,\mathrm{d}\big(F_{n\boldsymbol{a}}(x) - F_{\boldsymbol{a}}(x)\big)\Big| + \mathrm{O}\big((\log n)^{-\theta}\big) \quad \text{a.s.} \\ &\triangleq I'_{n2} + \mathrm{O}\big((\log n)^{-\theta}\big) \quad \text{a.s.} \end{aligned}$$

We now show that $I'_{n2} \to 0$ a.s. as $n \to \infty$. It suffices to show that for any $\varepsilon > 0$

$$\limsup_{n\to\infty} I'_{2n} \leqslant \varepsilon \quad \text{a.s.} \tag{5.3}$$

For any $\varepsilon > 0$, there exists an $M > 0$ such that $P\{\|\boldsymbol{x}\| \geqslant M\} \leqslant \varepsilon$. Hence it is easy to see that for all $\boldsymbol{a} \in S^d$

$$\begin{cases} 1 - F_{\boldsymbol{a}}(x) \leqslant \varepsilon & \text{for} \quad x \geqslant M, \\ F_{\boldsymbol{a}}(x) \leqslant \varepsilon & \text{for} \quad x \leqslant -M. \end{cases}$$

Meanwhile, since $\sup\limits_{\boldsymbol{a},x} |F_{n\boldsymbol{a}}(x) - F_{\boldsymbol{a}}(x)| \to 0$ a.s. as $n \to \infty$, then for large n

$$\sup_{\boldsymbol{a}} \big(1 - F_{n\boldsymbol{a}}(x)\big) \leqslant 2\varepsilon \quad \text{a.s.} \quad \text{for} \quad x \geqslant M,$$

$$\sup_{\boldsymbol{a}} F_{n\boldsymbol{a}}(x) \leqslant 2\varepsilon \quad \text{a.s.} \quad \text{for} \quad x \leqslant -M.$$

We rewrite I'_{n2} by $I'_{n2}(w)$ for emphasizing I'_{n2} as a function on sample space. Furthermore, as is known, W is uniformly bounded. Hence for almost all w there exists a constant $c(w)$, by Lemma 5.2, so that

$$\begin{aligned} |I'_{n2}(w)| &\leqslant \sup_{\boldsymbol{a}} \left| \int_{-M}^{M} W(w, \boldsymbol{a}_0, x)\, \mathrm{d}\big(F_{n\boldsymbol{a}}(x) - F_{\boldsymbol{a}}(x)\big) \right| + c(w)\varepsilon \\ &\triangleq I''_{n2}(w) + c(w)\varepsilon. \end{aligned} \tag{5.4}$$

Divide the interval $[-M, M]$ into l subintervals $[c_0, c_1), \ldots, [c_{l-1}, c_l]$, $c_0 = -M$ and $c_l = M$ such that

$$\begin{aligned} &\sup_{\boldsymbol{a}} \Big(P\big(I(\boldsymbol{a}^\tau \boldsymbol{x} \leqslant c_{i+1}) - I(\boldsymbol{a}^\tau \boldsymbol{x} \leqslant c_i)\big)^2\Big)^{1/2} \\ &\quad = \big(F\big(c_{i+1}\big) - F(c_i)\big)^{1/2} \leqslant \varepsilon \end{aligned} \tag{5.5}$$

for $i = 1, \ldots, l-1$. Consequently $I''_{n2}(w)$ can be bounded by

$$\begin{aligned} I''_{n2}(w) &\leqslant \sup_{\boldsymbol{a}} \left| \sum_{i=0}^{l-1} \int_{c_i}^{c_{i+1}} \big(W(w, \boldsymbol{a}_0, x) - W(w, \boldsymbol{a}_0, c_i)\big) \mathrm{d}F_{n\boldsymbol{a}}(x) \right| \\ &\quad + \sup_{\boldsymbol{a}} \left| \sum_{i=0}^{l-1} \int_{c_i}^{c_{i+1}} \big(W(w, \boldsymbol{a}_0, x) - W(w, \boldsymbol{a}_0, c_i)\big) \mathrm{d}F_{\boldsymbol{a}}(x) \right| \\ &\quad + \sup_{\boldsymbol{a}} \sum_{i=0}^{l-1} \left| \int_{c_i}^{c_{i+1}} W(w, \boldsymbol{a}_0, c_i)\, \mathrm{d}\big(F_{n\boldsymbol{a}}(x) - F_{\boldsymbol{a}}(x)\big) \right| \\ &\triangleq I^{(1)}_{n2}(w) + I^{(2)}_{n2}(w) + I^{(3)}_{n2}(w). \end{aligned} \tag{5.6}$$

By Lemma 5.2 and the equation (5.5), for some $c > 0$ we have

$$I^{(1)}_{n2}(w) \leqslant c\varepsilon \int_{-M}^{M} \mathrm{d}F_{n\boldsymbol{a}}(x) \leqslant c\varepsilon$$

and

$$I_{n2}^{(2)}(w) \leqslant c\varepsilon.$$

Combining (5.4) and (5.6), we finish the proof of (5.3).

On the other hand, by Lemma 5.1, we have

$$\begin{aligned}&\{I_{n1}(\boldsymbol{a}) - I_{n3}(\boldsymbol{a}):\ \boldsymbol{a} \in S^d\}\\ &\Longrightarrow \Big\{ \int W(\cdot, \boldsymbol{a}_0, x)\,\mathrm{d}F_{\boldsymbol{a}}(x) - \int W(\cdot, \boldsymbol{a}, x)\,\mathrm{d}F_0(x):\ \boldsymbol{a} \in S^d \Big\}.\end{aligned}$$

This process is nothing but the same one defined in (2.8). The proof of Theorem 1 is completed.

Proof of Theorem 2. Theorem 1 yields that for any fixed s,

$$\boldsymbol{J}_s = \{J(\cdot, \boldsymbol{a}_i):\ i = 1, \ldots, s\} \sim N_s(\boldsymbol{0}, \boldsymbol{R}_s). \tag{5.7}$$

Thus, as $n \to \infty$ we have $\hat{\boldsymbol{J}}_s \Rightarrow N_s(\boldsymbol{0}, \boldsymbol{R}_s)$ and $\widehat{\boldsymbol{R}}_{s1} \Rightarrow \boldsymbol{R}_s$ in probability. Consequently we have

$$\hat{\boldsymbol{J}}_s' \widehat{\boldsymbol{R}}_{s1}^{-1} \hat{\boldsymbol{J}}_s \Longrightarrow \chi^2(s) \qquad \text{in probability.}$$

Let s tend to infinity, we can easily derive , via the central limit theorem, that

$$\lim_{s\to\infty} \lim_{n\to\infty} \frac{1}{\sqrt{2s}} (\hat{\boldsymbol{J}}_s' \widehat{\boldsymbol{R}}_{s1}^{-1} \hat{\boldsymbol{J}}_s - s) \sim N(0, 1).$$

So (2.14) and (2.15) are proved. Now we turn to prove (2.16) and (2.17). We only prove (2.16), and (2.17) can be proved similarly.

Under H_1, we can find at least one direction $\boldsymbol{a}$ such that

$$E(F_0(\boldsymbol{a}'\boldsymbol{x})) \neq \frac{1}{2}. \tag{5.8}$$

Furthermore, by the continuity of $F_{\boldsymbol{a}}(\cdot)$ with respect to $\boldsymbol{a}$ and the properties of the chosen representation points, when s is large, we can find a direction $\boldsymbol{a}_i$ such that

$$E(F_0(\boldsymbol{a}_i'\boldsymbol{x})) \neq \frac{1}{2}. \tag{5.9}$$

Thus

$$\hat{\boldsymbol{J}}_s' = \Big\{ \sqrt{n}\big(V_n(\boldsymbol{a}_i) - E(F_0(\boldsymbol{a}_i'\boldsymbol{x}))\big),\ i = 1, \ldots, s \Big\} + \sqrt{n}\boldsymbol{b}_s$$

is asymptotically multinormal where $\boldsymbol{b}_s = \{E(F_0(\boldsymbol{a}_i'\boldsymbol{x})) - 1/2,\ i = 1, 2, \ldots, s\}$. This means that

$$\hat{\boldsymbol{J}}_s' - \sqrt{n}\boldsymbol{b}_s = \mathrm{O}(1) \qquad \text{in probability.} \tag{5.10}$$

Let $\widehat{\boldsymbol{R}}_{s2} = \widehat{\boldsymbol{R}}_{s1} + \varepsilon \boldsymbol{I}_s$ for a fixed positive constant ε, then as $n \to \infty$

$$\widehat{\boldsymbol{R}}_{s2}^{-1} \Longrightarrow \left(\boldsymbol{R}_s + \varepsilon \boldsymbol{I}_s\right)^{-1} \quad \text{in probability.}$$

By a well-known result, due to the nonnegative definiteness of $\boldsymbol{R}_s$, we have

$$\hat{\boldsymbol{J}}_s' \widehat{\boldsymbol{R}}_{s1}^{-1} \hat{\boldsymbol{J}}_s \geqslant \hat{\boldsymbol{J}}_s' \widehat{\boldsymbol{R}}_{s2}^{-1} \hat{\boldsymbol{J}}_s.$$

On the other hand, the nonnegative definiteness of $\boldsymbol{R}_s$ implies that

$$\boldsymbol{b}_s' \left(\boldsymbol{R}_s + \varepsilon \boldsymbol{I}_s\right)^{-1} \boldsymbol{b}_s > 0.$$

Consequently

$$\begin{aligned} \lim_{n\to\infty} \frac{1}{n} \hat{\boldsymbol{J}}_s' \widehat{\boldsymbol{R}}_{s1}^{-1} \hat{\boldsymbol{J}}_s &\geqslant \lim_{n\to\infty} \frac{1}{n} \hat{\boldsymbol{J}}_s' \widehat{\boldsymbol{R}}_{s2}^{-1} \hat{\boldsymbol{J}}_s \\ &= \lim_{n\to\infty} \left\{ \frac{1}{n} (\hat{\boldsymbol{J}}_s - \sqrt{n}\boldsymbol{b}_s)' \widehat{\boldsymbol{R}}_{s2}^{-1} (\hat{\boldsymbol{J}}_s - \sqrt{n}\boldsymbol{b}_s) \right. \\ &\quad \left. + \frac{2}{\sqrt{n}} \boldsymbol{b}_s' \widehat{\boldsymbol{R}}_{s2}^{-1} (\hat{\boldsymbol{J}}_s - \sqrt{n}\boldsymbol{b}_s) + \boldsymbol{b}_s' \widehat{\boldsymbol{R}}_{s2}^{-1} \boldsymbol{b}_s \right\} \\ &= \boldsymbol{b}_s' \left(\boldsymbol{R}_s + \varepsilon \boldsymbol{I}_s\right)^{-1} \boldsymbol{b}_s > 0 \qquad \text{in probability.} \end{aligned}$$

This means that $\hat{\boldsymbol{J}}_s' \widehat{\boldsymbol{R}}_s^{-1} \hat{\boldsymbol{J}}_s$ converges to infinity with n rate. (2.16) is proved. Following this, (2.17) is an obvious consequence. The proof is completed.

Acknowlegdments

The authors are grateful to the editor and referees for their comments which improve the representation of this paper. We also thank Mr. Run-Ze Li for his computational work and assistance.

REFERENCES

Baringhaus, L. (1991). Testing for spherical symmetry of a multivariate distribution. *Ann. Statist.* **19**, 899–917.

Dudley, R. M. and Philipp, W. (1983). Invariance principle for sums of Banach space valued random elements and empirical processes. *Z. Wahrsch. verw. Geb.* **62**, 509–552.

Fang, K. T., Kotz, S. and Ng, K. W. (1990). *Symmetric Multivariate and Related Distributions.* Chapman and Hall, London.

Fang, K. T., Yuan, K. H. and Bentler, P. M. (1992). Applications of sets of points uniformly distributed on a sphere to testing multinormality and rabust estimation. In: *Probability and Statistics* Z. P. Jiang, S. J. Yan, P. Cheng and R. Wu (Eds), World Scientific, Singapore.

Fang, K. T. and Wang, Y. (1993). *Applications of Number-Theoretic Methods in Statistics.* Chapman and Hall, London.

Fang, K. T., Zhu, L. X. and Bentler, P. M. (1993). A necessary test for sphericity of a high-dimensional distribution. *J. Multiv. Anal.* **44**, 34–55.

Friedman, J. H. (1987). Exploratory projection pursuit. *J. Amer. Statist. Assoc.* **82**, 249–266.

Hua, L. G. and Wang, Y. (1981). *The Applications of Number Theory to Approximation Analysis.* Science Press, Beijing.

Huber, P. J. (1985). Projection pursuit (with discussion). *Ann. Statist.* **13**, 435–475.

Kariya, T. and Eaton, M. L. (1977). Robust tests for spherical symmetry. *Ann. Statist.* **5**, 206–215.

Pollard, D. (1984). *Convergence of Stochastic Processes.* Springer, New York.

Rothman, E. D. and Woodrofe, M. (1972). A Cramér von Mises type statistic for testing symmetry. *Ann. Math. Statist.* **43**, 2035–2038.

Roy, S. N. (1953). On a heuristic method of test constructionand its use in multivariate analysis. *Ann. Math., Statist.* **24**, 220–238.

Wang, Y. and Fang, K. T. (1990). Number-theoretic methods in applied statistics. *Chinese Ann. Math. Ser. B* **11**, 41–55.

Zhu, L. X., Fang, K. T., Bhatti, I. M. and Bentler, P. M. (1993). Testing sphericity of a high-dimensional distribution based on bootstrap approximation, to be submitted for publication.

Zhu, L. X., Wong, H. L. and Fang, K. T. A test formultivariate normality based on sample entropy and projection pursuit. *J. Statist. Plann. Inference* (in press).

Zhu, L. X., Fang, K. T. and Li, R. Z. (1994). A projection NT-type test for multinormality based on the skewness and kurtosis, to be submitted. Technical Report 53, Department of Mathematics, Hong Kong Baptist College.

Multivar. Statist., pp. 123–133
E.-M. Tiit *et al.* (Eds)

SOME APPLICATIONS OF DIRECTIONAL STATISTICS TO ASTRONOMY

P. E. JUPP
School of Mathematical and Computational Sciences,
University of St Andrews, North Haugh,
St Andrews KY16 9SS, United Kingdom

ABSTRACT

Directional statistics is applied to three astronomical problems. These concern cometary orbits, orbits of visual binary stars and spin axes of asteroids. The first involves rotational data while the others require models allowing for mirror-image ambiguity and selection effects. The common feature is the use of score tests to test uniformity.

1. INTRODUCTION

In many areas of astronomy information about distance is not available or is unreliable, so that observations can be reduced to directions and can therefore be treated as points on the celestial sphere. Examples include perihelion directions of comets and positions of binary stars. For handling such observations the techniques of directional statistics are required. Directional statistics is the area of multivariate analysis in which the observations are not the familiar unrestricted vectors but, typically, directions (unit vectors) or axes (unit vectors up to sign). The main texts on directional statistics are those by Mardia (1972) and Fisher *et al.* (1987). A rather technical treatment of some topics appears in (Watson, 1983). A review of the theory of directional statistics was given by Jupp and Mardia (1989). The present paper aims to illustrate the special flavour of directional statistics by considering three non-standard applications to astronomy.

Section 2 is concerned with testing the uniformity of the orbits of long-period comets and illustrates methods for analysing data in which each observation is a rotation matrix. The application in Section 3 is to the orbital planes of visual binary stars. Testing these for uniformity is complicated by the problems of (i) ambiguity of directions, (ii) geometrical selection effects. A solution using the models of Jupp and Cuypers (1994) is presented. Score tests are used to test for uniformity and for selection. In Section 4 the application is to the spin axes of asteroids. Here the observations are axes rather than directions and they too are subject to ambiguity and selection. Models and tests analogous to those of Section 3 are given.

2. COMETS

2.1. *Astronomical problem*

The orbit of a periodic comet is an ellipse with the sun at one focus. Taking the sun as the origin of $\boldsymbol{R}^3$, we can associate two unit vectors with such an orbit. These are the direction $\boldsymbol{u}$ of the perihelion (the point of closest approach to the sun) and the directed unit normal $\boldsymbol{v}$ to the orbit (with direction specified by the sense of rotation). Putting $\boldsymbol{w} = \boldsymbol{u} \wedge \boldsymbol{v}$ yields a third unit vector such that the matrix $\boldsymbol{X} = (\boldsymbol{u}, \boldsymbol{v}, \boldsymbol{w})$ is a rotation matrix, i.e., $\boldsymbol{X}'\boldsymbol{X} = I_3$ and $\det(\boldsymbol{X}) = 1$. Thus the directional aspects of the orbit of a periodic comet can be represented by a rotation matrix. See Figure 1.

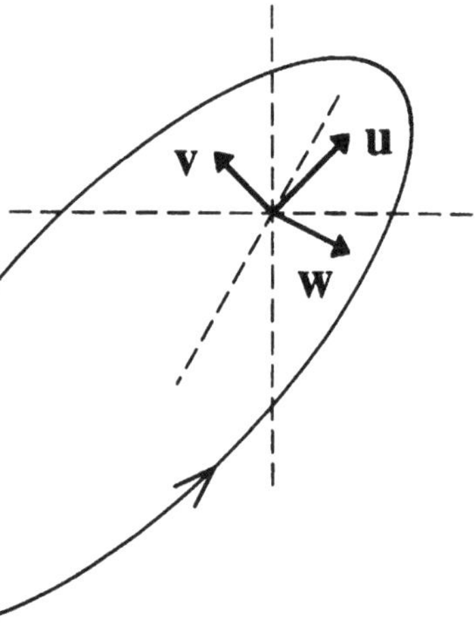

Figure 1. The rotation matrix $\boldsymbol{X} = (\boldsymbol{u}, \boldsymbol{v}, \boldsymbol{w})$ of a cometary orbit.

Three natural questions of astronomical interest are:

(i) are the orbits uniformly distributed?

(ii) are the perihelion directions uniformly distributed?

(iii) are the orbital planes of comets with a given perihelion direction distributed uniformly about that axis?

2.2. *Statistical tools*

Uniformity of the orbits means that the underlying random rotation $\boldsymbol{X}$ has the same distribution as $\boldsymbol{RX}$, for any fixed rotation $\boldsymbol{R}$. Similarly, uniformity of perihelion directions means that $\boldsymbol{u}$ and $\boldsymbol{Ru}$ have the same distribution. Perhaps the simplest test of uniformity on the sphere is the Rayleigh test. (See (Mardia, 1972, Section 9.3.1) or (Fisher *et al.*, 1987, Section 5.3.1)). The intuitive idea is that uniformity is rejected if the sample mean $\bar{\boldsymbol{u}}$ of unit vectors

$\boldsymbol{u}_1, \ldots, \boldsymbol{u}_n$ is far from $\mathbf{0}$. More precisely, uniformity is rejected for large values of $3n\bar{\boldsymbol{u}}'\bar{\boldsymbol{u}}$. Under uniformity, the large-sample asymptotic distribution of $3n\bar{\boldsymbol{u}}'\bar{\boldsymbol{u}}$ is χ^2_3. Similarly, if $\boldsymbol{X}_1, \ldots, \boldsymbol{X}_n$ are rotations with sample mean $\overline{\boldsymbol{X}} = n^{-1}(\boldsymbol{X}_1 + \cdots + \boldsymbol{X}_n)$, the Rayleigh test for rotations rejects uniformity for large values of $3n\,\mathrm{tr}(\overline{\boldsymbol{X}}'\overline{\boldsymbol{X}})$. The large-sample asymptotic distribution of $3n\,\mathrm{tr}(\overline{\boldsymbol{X}}'\overline{\boldsymbol{X}})$ under uniformity is χ^2_9. The Rayleigh tests are score tests of uniformity in the Fisher distribution (see (3.1) below) and the matrix Fisher distribution (see (Jupp and Mardia, 1989, p. 284)), respectively.

The decomposition

$$\boldsymbol{X} = (\boldsymbol{u}, \boldsymbol{Y}),$$

where $\boldsymbol{Y} = (\boldsymbol{v}, \boldsymbol{w})$ yields a decomposition

$$3n\,\mathrm{tr}\left(\overline{\boldsymbol{X}}'\overline{\boldsymbol{X}}\right) = 3n\bar{\boldsymbol{u}}'\bar{\boldsymbol{u}} + 3n\,\mathrm{tr}\left(\overline{\boldsymbol{Y}}'\overline{\boldsymbol{Y}}\right) \tag{2.1}$$

of the Rayleigh test statistic for uniformity of rotations and a corresponding decomposition of the large-sample asymptotic χ^2 distribution under uniformity. The term $3n\,\mathrm{tr}(\overline{\boldsymbol{Y}}'\overline{\boldsymbol{Y}})$ in (2.1) measures the asymmetry of the distribution of orbits about perihelion directions. As its asymptotic distribution depends on $\boldsymbol{u}_1, \ldots, \boldsymbol{u}_n$, it is more useful to test symmetry using the conditional distribution of $\overline{\boldsymbol{Y}}$ given $\boldsymbol{u}_1, \ldots, \boldsymbol{u}_n$. A convenient test is that which rejects symmetry for large values of

$$2\bar{\boldsymbol{v}}'\left\{\sum_{i=1}^{n}(\boldsymbol{I}_3 - \boldsymbol{u}_i\boldsymbol{u}_i')\right\}^{-1}\bar{\boldsymbol{v}}. \tag{2.2}$$

Under symmetry the large-sample asymptotic distribution of (2.2) conditional on $\boldsymbol{u}_1, \ldots, \boldsymbol{u}_n$ is χ^2_3.

There is a large class of tests of uniformity which generalise the Rayleigh test. (See (Jupp and Mardia, 1989, Section 3.1)). Jupp and Spurr (1983) showed that these tests admit a decomposition analogous to (2.1) into the sum of a test of uniformity on the sphere and a test of symmetry.

2.3. *Data analysis*

Applying the above tests to the data set of 204 long period comets in the catalogue of Marsden and Williams (1993) gives $3n\,\mathrm{tr}(\overline{\boldsymbol{X}}'\overline{\boldsymbol{X}}) = 17.9$ and $3n\bar{\boldsymbol{u}}'\bar{\boldsymbol{u}} = 8.0$. Use of the χ^2 approximations suggests that these values are significant at the 5% level, leading to rejection of both uniformity of orbits and uniformity of perihelion directions. The value of the statistic (2.2) is 0.04, which is significantly small as a value of χ^2_3 and so indicates a surprising degree of symmetry. Analyses of related data sets can be found in (Mardia, 1975, pp. 363–365; Jupp and Mardia, 1979; Jupp and Spurr, 1983; Watson, 1983, pp. 28–35; Fisher *et al.*, 1987, pp. 160–161).

Some caution is required in interpreting the above analysis because it has been tacitly assumed that the data form a random sample. In fact, as in many astronomical problems, there are complicated selection effects. A careful discussion of selection in observations on cometary orbits is given by Kresák (1982). Modelling the selection effects which occur in two other contexts is an important aspect of Sections 3 and 4.

3. BINARY STARS

3.1. *Astronomical problem*

A binary star consists of a pair of stars which rotate about their common centre of mass. It is thought that at least 50% of stars are binaries and about 65 000 binary pairs are known. An important class of binary stars are the visual binaries, in which the two stars can be distinguished from one another when seen through an optical telescope. The directional aspects of the observations on a visual binary are its position $\boldsymbol{u}$ on the celestial sphere and the direction $\boldsymbol{z}$ of the axis about which the two stars rotate (with sign specified by the sense of rotation). A hypothesis of astronomical interest is that the directions $\boldsymbol{z}$ are uniformly distributed on the celestial sphere. (See (Berglund, 1938; Gillett, 1988)). Testing the hypothesis of uniformity is complicated by

(i) the *mirror-image ambiguity* of many visual binaries, i.e., for these binaries it is not possible to distinguish between $\boldsymbol{z}$ and its reflection $(2\boldsymbol{u}\boldsymbol{u}' - \boldsymbol{I}_3)\boldsymbol{z}$ in the line of sight along $\boldsymbol{u}$ (see Figure 2);

(ii) possible selection effects, with orbits having $\boldsymbol{z}$ close to $\pm\boldsymbol{u}$ being more likely to be observed than other orbits.

3.2. *Statistical models*

The standard alternative model to uniformity is the Fisher distribution. (See (Mardia, 1972, Section 8.5.4) or (Fisher *et al.*, 1987, Section 4.4.3)). This has

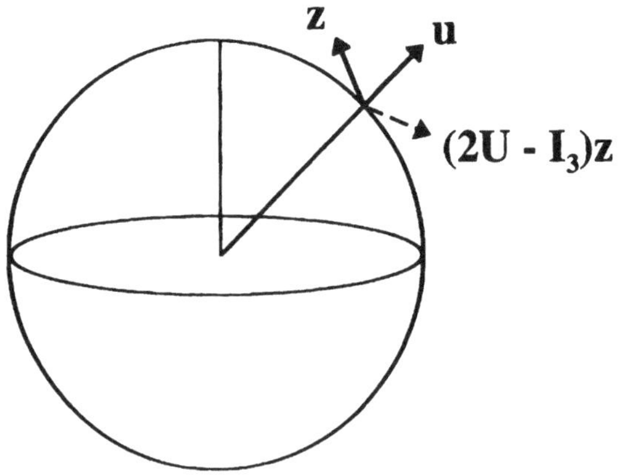

Figure 2. The mirror-image ambiguity for directional data.

probability density function

$$f_0(\boldsymbol{z}; \boldsymbol{\mu}, \kappa) = \frac{\kappa}{4\pi \sinh \kappa} \exp\{\kappa \boldsymbol{\mu}'\boldsymbol{z}\}, \tag{3.1}$$

where $\kappa \geqslant 0$ and $\boldsymbol{\mu}$ is a unit vector. The parameter κ measures the concentration of.the distribution about the mean direction $\boldsymbol{\mu}$. If $\kappa = 0$ then (3.1) reduces to the uniform distribution.

Let $\boldsymbol{U} = \boldsymbol{u}\boldsymbol{u}'$ denote the orthogonal projection onto the line of sight. Because $\boldsymbol{z}$ is subject to mirror-image ambiguity, we cannot use the Fisher distribution (3.1) directly. However, although we cannot distinguish between $\boldsymbol{z}$ and $(2\boldsymbol{U} - \boldsymbol{I}_3)\boldsymbol{z}$, we do know both $\boldsymbol{U}\boldsymbol{z}$, the part of $\boldsymbol{z}$ parallel to the line of sight, and $\pm(\boldsymbol{I}_3 - \boldsymbol{U})\boldsymbol{z}$, the axis determined by the part of $\boldsymbol{z}$ normal to the line of sight. Since

(i) $\boldsymbol{U}\boldsymbol{z}$ and $\pm(\boldsymbol{I}_3 - \boldsymbol{U})\boldsymbol{z}$ depend on $\boldsymbol{u}$,

(ii) the distribution of observable $\boldsymbol{u}$ is influenced by the positions of observatories on the earth's surface,

it is appropriate to consider the conditional distribution given $\boldsymbol{u}$ of the observable parts $\boldsymbol{U}\boldsymbol{z}$ and $\pm(\boldsymbol{I}_3 - \boldsymbol{U})\boldsymbol{z}$ of $\boldsymbol{z}$. If $\boldsymbol{z}$ has the Fisher distribution (3.1) then the conditional density of $(\boldsymbol{U}\boldsymbol{z}, \pm(\boldsymbol{I}_3 - \boldsymbol{U})\boldsymbol{z})$ given $\boldsymbol{u}$ is

$$\begin{aligned} f_M(\boldsymbol{z}; \boldsymbol{\mu}, \kappa \mid \boldsymbol{u}) &= \frac{1}{2}\{f_0(\boldsymbol{z}; \boldsymbol{\mu}, \kappa) + f_0((2\boldsymbol{U} - \boldsymbol{I}_3)\boldsymbol{z}; \boldsymbol{\mu}, \kappa)\} \\ &= \frac{\kappa}{4\pi \sinh \kappa} \cosh\left(\kappa \boldsymbol{\mu}'(\boldsymbol{I}_3 - \boldsymbol{U})\boldsymbol{z}\right) \exp\{\kappa \boldsymbol{\mu}'\boldsymbol{U}\boldsymbol{z}\}. \end{aligned} \tag{3.2}$$

In order to allow for possible selection effects, we need to modify distribution (3.2). A convenient and reasonable model for selection takes the probability of observing direction $\boldsymbol{z}$ at position $\boldsymbol{u}$ to be given by the Watson $W(\boldsymbol{u}, \beta)$ distribution (see (Mardia, 1972, Section 8.5.5a) or (Fisher *et al.*, 1987, Section 4.4.4)). This has probability density function

$$f_S(\boldsymbol{z}; \beta \mid \boldsymbol{u}) = \exp\left\{\beta(\boldsymbol{z}'\boldsymbol{u})^2 - c(\beta)\right\}, \tag{3.3}$$

where β is a parameter measuring the strength of the selection effect and $c(\beta)$ is the logarithm of the norming constant. If $\beta = 0$ then there is no selection; if $\beta > 0$ then we tend to observe those $\boldsymbol{z}$ which are parallel to $\pm\boldsymbol{u}$; if $\beta < 0$ then we tend to observe those $\boldsymbol{z}$ which are normal to $\boldsymbol{u}$. Combining (3.2) and (3.3) yields the conditional density of the observable $(\boldsymbol{U}\boldsymbol{z}, \pm(\boldsymbol{I}_3 - \boldsymbol{U})\boldsymbol{z})$ given $\boldsymbol{u}$ as

$$\begin{aligned} f(\boldsymbol{z}; \boldsymbol{\mu}, \kappa, \beta \mid \boldsymbol{u}) &= f_M(\boldsymbol{z}; \boldsymbol{\mu}, \kappa \mid \boldsymbol{u}) f_S(\boldsymbol{z}; \beta \mid \boldsymbol{u}) \\ &= \cosh\left(\kappa \boldsymbol{\mu}'(\boldsymbol{I}_3 - \boldsymbol{U})\boldsymbol{z}\right) \\ &\quad \times \exp\left\{\kappa \boldsymbol{\mu}'\boldsymbol{U}\boldsymbol{z} + \beta \boldsymbol{z}'\boldsymbol{U}\boldsymbol{z} - c(\boldsymbol{\mu}, \kappa, \beta, \boldsymbol{u})\right\}, \end{aligned} \tag{3.4}$$

where $c(\boldsymbol{\mu}, \kappa, \beta, \boldsymbol{u})$ is the logarithm of the norming constant. Note that $\kappa = 0$ corresponds to uniformity (prior to selection) of $\boldsymbol{z}$.

Model (3.4) can be regarded as a directional regression model in the spirit of that used by Kendall and Young (1984) to analyse a fascinating problem about a possible large-scale asymmetry of the universe. Unfortunately, there is not space to discuss their paper here.

3.3. *Testing uniformity*

The conditional log-likelihood under model (3.4) given by observations $(\boldsymbol{u}_1, \boldsymbol{z}_1), \ldots, (\boldsymbol{u}_n, \boldsymbol{z}_n)$ is

$$l(\boldsymbol{\mu}, \kappa, \beta; \boldsymbol{z}_1, \ldots, \boldsymbol{z}_n \mid \boldsymbol{u}_1, \ldots, \boldsymbol{u}_n)$$
$$= \sum_{i=1}^{n} \log\big(\cosh(\kappa\boldsymbol{\mu}'(\boldsymbol{I}_3 - \boldsymbol{U}_i)\boldsymbol{z}_i)\big) + n\{\kappa\boldsymbol{\mu}'\bar{\boldsymbol{z}} + \beta m\} - \sum_{i=1}^{n} c(\boldsymbol{\mu}, \kappa, \beta, \mathbf{u}_i),$$

where

$$\boldsymbol{U}_i = \boldsymbol{u}_i\boldsymbol{u}_i', \qquad m = \frac{1}{n}\sum_{i=1}^{n}(\boldsymbol{z}_i'\boldsymbol{u}_i)^2$$

and

$$\bar{\boldsymbol{z}} = \frac{1}{n}\sum_{i=1}^{n}\boldsymbol{U}_i\boldsymbol{z}_i.$$

Note that m and $\bar{\boldsymbol{z}}$ can be interpreted as the "line-of-sight moment of inertia" and the mean component of $\boldsymbol{z}$ along the line-of-sight, respectively.

In order to test the hypothesis $\kappa = 0$ in (3.4) one might consider using the likelihood ratio test. However, this is computationally demanding because $c(\boldsymbol{\mu}, \kappa, \beta, \boldsymbol{u})$ is a very complicated function of $\boldsymbol{\mu}$, κ, β, and $\boldsymbol{u}$. It is more practical to use the score test. The score test statistic for testing $\kappa = 0$ in model (3.4) is

$$w_\kappa = n\bar{\boldsymbol{z}}'\left\{m\sum_{i=1}^{n}\boldsymbol{U}_i\right\}^{-1}\bar{\boldsymbol{z}}.$$

If $\kappa = 0$ then the large-sample asymptotic distribution of w_κ is χ_3^2.

The hypothesis of no selection, i.e., that $\beta = 0$, in model (3.3), can also be tested using a score test. The score test statistic is

$$w_\beta = \frac{45n}{4}\left(m - \frac{1}{3}\right)^2$$

and its asymptotic null distribution is χ_1^2.

If it is accepted that $\kappa = 0$ in model (3.4) then it is appropriate to fit model (3.3). The maximum likelihood estimate $\hat{\beta}$ of β in model (3.3) is given by

$$\hat{\beta} = D^{-1}(m), \tag{3.5}$$

where

$$D(\beta) = \frac{\int_{-1}^{1} u^2 e^{\beta u^2} du}{\int_{-1}^{1} e^{\beta u^2} du}.$$

The function $D^{-1}(\cdot)$ is tabulated in (Fisher *et al.*, 1987, Table A10).

The goodness of fit of model (3.3) with $\beta = \dot{\beta}$ can be assessed fairly easily using $c_1, \ldots, c_n$, where $c_i = z_i' u_i$. An informal assessment is given by plotting $c_{(i)}^2$ against $\widehat{F}^{-1}(\frac{i-1/2}{n})$, where $c_{(i)}$ denotes the i-th smallest of $c_1, \ldots, c_n$ and $\widehat{F}$ is the cumulative distribution function of $(z'u)^2$, with $z|u \sim W(u, \dot{\beta})$. This plot is a refinement of the "colatitude plot" of (Fisher *et al.*, 1987, p. 168). The latter uses $-\log(1 - \frac{i-1/2}{n})$ instead of $\widehat{F}^{-1}(\frac{i-1/2}{n})$, an approximation which is reasonable for $\dot{\beta} \geqslant 5$. For $\dot{\beta} < 5$, $\widehat{F}^{-1}(\cdot)$ can be found by numerical integration. Pronounced departure of the plot from a straight line through the origin indicates a poor fit. For a more formal investigation of goodness of fit, standard tests such as the Kolmogorov–Smirnov test can be used to measure the fit of the distribution $\widehat{F}$ to $c_1^2, \ldots, c_n^2$.

3.4. *Data analysis*

A suitable data set is provided by the 170 orbits of grade 1 or 2 in the catalogue of Worley and Heintz (1983). The sample size is presumably large enough to justify comparison of the score statistics with their limiting χ^2 distributions.

First we test for uniformity. The score test of $\kappa = 0$ in (3.4) yields $w_\kappa = 2.12$, which is not significant compared with χ_3^2. Thus there is no evidence of non-uniformity. This conclusion is in agreement with the informal analysis of Berglund (1938) and the more rigorous analysis by Gillett (1988). These analyses did not allow for selection effects and were restricted to orbits not subject to mirror-image ambiguity.

Next we test for selection effects. Since we have accepted that $\kappa = 0$, we can fit the Watson model (3.3). The score test of $\beta = 0$ in (3.3) yields $w_\beta = 9.36$, which is highly significant compared with χ_1^2, so we conclude that there is selection. The maximum likelihood estimate $\dot{\beta}$ of β in (3.3) is $\dot{\beta} = 0.74$. Since $\dot{\beta} > 0$, we conclude that there is higher probability of selection for orbits with z parallel to $\pm u$. This conclusion that there is selection, with orbits having inclination nearer to 0° or 180° being more likely to be selected, agrees with that obtained informally by Pensado and Lahulla (1967).

Finally, we consider goodness of fit. The plot of $c_{(i)}^2$ against $\widehat{F}^{-1}(\frac{i-1/2}{n})$ is almost a straight line, indicating a good fit. The Kolmogorov–Smirnov test statistic is 0.40, which is not significant at the 5% level. Thus the Watson model

$$z_i | u_i \sim W(u_i, 0.74)$$

fits the data well.

4. ASTEROIDS

4.1. *Astronomical problem and statistical model*

The asteroids are a belt of several thousand minor planets orbiting round the sun in orbits between Mars and Jupiter. Many asteroids have an axis around which they are observed to spin. A hypothesis of astronomical interest is that the directed spin axes $\boldsymbol{z}$ are uniformly distributed. For many asteroids the sense of spin is unknown, so that we cannot distinguish between $\boldsymbol{z}$ and $-\boldsymbol{z}$. It follows that models (3.1), (3.2), and (3.4) cannot be used. Because we observe only the axis $\pm\boldsymbol{z}$ instead of the direction $\boldsymbol{z}$, it is appropriate to replace the Fisher distribution (3.1) for $\boldsymbol{z}$ (prior to selection) by a Bingham distribution (see (Mardia, 1972, pp. 234–236) or (Fisher *et al.*, 1987, p. 97)). This is the standard model for axial data and has probability density function

$$g_0(\boldsymbol{z}; \boldsymbol{A}) = \exp\left\{\boldsymbol{z}'\boldsymbol{A}\boldsymbol{z} - c(\boldsymbol{A})\right\}, \tag{4.1}$$

where the parameter $\boldsymbol{A}$ is a symmetric 3×3 matrix. Because $\boldsymbol{z}'\boldsymbol{z} = 1$, we may assume without loss of generality that $\operatorname{tr}(\boldsymbol{A}) = 0$. If $\boldsymbol{A} = \boldsymbol{0}$ then (4.1) reduces to the uniform distribution. Because of the mirror-image ambiguity, the observable parts of $\pm\boldsymbol{z}$ are $\pm\boldsymbol{U}\boldsymbol{z}$ and $\pm(\boldsymbol{I}_3 - \boldsymbol{U})\boldsymbol{z}$. See Figure 3. If $\boldsymbol{z}$ has the Bingham distribution (4.1) then the conditional density of $(\pm\boldsymbol{U}\boldsymbol{z}, \pm(\boldsymbol{I}_3 - \boldsymbol{U})\boldsymbol{z})$ given $\boldsymbol{u}$ is

$$g_M(\boldsymbol{z}; \boldsymbol{A} \mid \boldsymbol{u}) = \exp\left\{\operatorname{tr}\left(\boldsymbol{A}\left[\boldsymbol{U}\boldsymbol{z}\boldsymbol{z}'\boldsymbol{U} + (\boldsymbol{I}_3 - \boldsymbol{U})\boldsymbol{z}\boldsymbol{z}'(\boldsymbol{I}_3 - \boldsymbol{U})\right]\right) - c(\boldsymbol{A}, \boldsymbol{u})\right\}, \tag{4.2}$$

where $c(\boldsymbol{A}, \boldsymbol{u})$ is the logarithm of the norming constant. If model (4.2) is combined with selection of the form (3.3), we obtain as the conditional probability

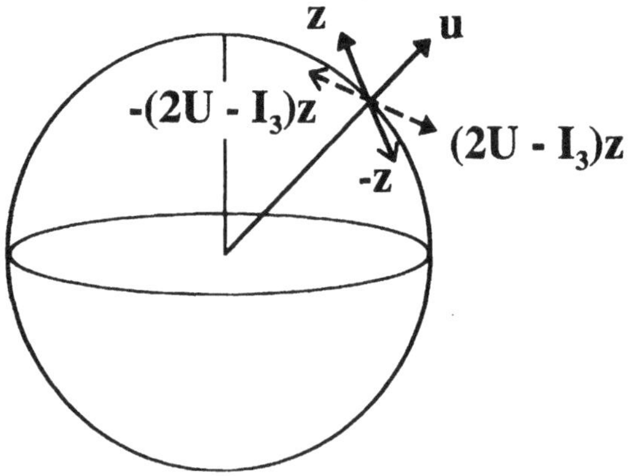

Figure 3. The mirror-image ambiguity for axial data.

density function given u of the observable parts $(\pm Uz, \pm(I_3 - U)z)$ of the selected axes $\pm z$

$$g(z; A, \beta \mid u) = g_M(z; A, \mid u) f_S(z; \beta \mid u)$$
$$= \exp\left\{\operatorname{tr}\left(A\left[Uzz'U + (I_3 - U)zz'(I_3 - U)\right]\right) \quad (4.3)\right.$$
$$\left. + \beta(z'u)^2 - c(A, \beta, u)\right\},$$

where $c(A, \beta, u)$ is the logarithm of the norming constant. Note that $A = 0$ in (4.3) corresponds to uniformity (prior to selection) of $\pm z$.

The conditional log-likelihood under model (4.3) given by observations

$$(u_1, \pm z_1), \ldots, (u_n, \pm z_n)$$

is

$$l(A, \beta; z_1, \ldots, z_n \mid u_1, \ldots, u_n) = \operatorname{tr}(AM) + n\beta m - \sum_{i=1}^{n} c(A, \beta, u_i),$$

where

$$M = \sum_{i=1}^{n} \{U_i z_i z_i' U_i + (I_3 - U_i) z_i z_i' (I_3 - U_i)\}.$$

The matrix M can be interpreted as the "observable moment of inertia". Note that $\operatorname{tr}(M) = n$. Put

$$N = M - \frac{(3m-1)}{2} \sum_{i=1}^{n} U_i - \frac{(1-m)}{2} n I_3$$

and $v = \operatorname{vec}(N)$, the vector obtained by stacking the columns of N on top of one another. Then v is the score vector and the score test statistic of $A = 0$ in model (4.3) is

$$w_A = v' \left\{ \sum_{i=1}^{n} X_i C X_i' \right\}^{-} v.$$

Here C denotes the variance matrix of vec(ww') for w with a Watson $W(n, \dot{\beta})$ distribution, where n is a fixed unit vector, $\dot{\beta}$ is given by (3.5), X_i transforms $\operatorname{vec}(ww')$ to

$$\operatorname{vec}\left(U_i z_i z_i' U_i + (I_3 - U_i) z_i z_i' (I_3 - U_i) - \frac{(3m-1)}{2} U_i - \frac{(1-m)}{2} I_3\right),$$

and $\{\ \}^{-}$ denotes a generalised inverse. The large-sample asymptotic null distribution of w_A is χ^2_5.

4.2. *Data analysis*

A suitable data set is given by the 54 asteroids with known spin axis listed in the catalogue of Magnusson (1988). (For most of these asteroids several determinations of $\pm\boldsymbol{z}$ are available. In the data set used here, for each asteroid an arbitrary choice was made from the determinations of its spin axis $\pm\boldsymbol{z}$.) The score test of $\boldsymbol{A} = \boldsymbol{0}$ in (4.3) yields $w_{\boldsymbol{A}} = 4.59$, so we may accept uniformity of the poles $\pm\boldsymbol{z}$. As the comparison with χ_5^2 may not be valid for such a small sample size, a randomisation test was carried out in which each $\boldsymbol{z}$ was subjected to a rotation through a uniformly-distributed random angle about $\boldsymbol{u}$. Since the observed significance level from the randomisation distribution was 0.16, uniformity of the $\pm\boldsymbol{z}$ can be accepted.

The score test of $\beta = 0$ in (3.3) yields $w_\beta = 0.6$. Comparison with χ_1^2 indicates that there is no selection. Since we have accepted that $\boldsymbol{A} = \boldsymbol{0}$ and $\beta = 0$ in (4.3), the uniform distribution may be fitted to this data set.

Acknowledgements

This research has made use of the Simbad database, operated at CDS, Strasbourg, France. I am grateful to Jan Cuypers for bringing the problem of the mirror-image ambiguity to my attention and to Catherine Thomson for computational assistance.

REFERENCES

Berglund, F. (1938). The orientation of the orbit planes of visual binary stars. *Lunds Universitets Årsskrift* **34**, 2–12.

Fisher, N. I., Lewis, T. and Embleton, B. J. J. (1987). *Statistical Analysis of Spherical Data*. Cambridge University Press, Cambridge.

Gillett, S. L. (1988). Orbital planes of visual binary stars are randomly oriented: a statistical demonstration. *Astronom. J.* **96**, 1967–1970.

Jupp, P. E. and Cuypers, J. (1994). Testing the uniformity of ambiguous directional data subject to selection, in preparation.

Jupp, P. E. and Mardia, K. V. (1979). Maximum-likelihood estimators for the matrix von Mises-Fisher and Bingham distributions. *Ann. Statist.* **7**, 599–606.

Jupp, P. E. and Mardia, K. V. (1989). A unified view of the theory of directional statistics, 1975–1988. *Internat. Statist. Rev.* **57**, 261–294.

Jupp, P. E. and Spurr, B. D. (1983). Sobolev tests for symmetry of directional data. *Ann. Statist.* **11**, 1225–1231.

Kendall, D. G. and Young, G. A. (1984). Indirectional statistics and the significance of an asymmetry discovered by Birch. *Monthly Notices Roy. Astronom. Soc.* **207**, 637–647.

Kresák, L. (1982). Comet discoveries, statistics and observational selection. In: *Comets*, pp. 56–82, L. L. Wilkening (Ed.), University of Arizona Press, Tucson.

Magnusson, P. (1988). Pole determinations of asteroids. In: *Asteroids II*, R. P. Binzel *et al.* (Eds), University of Arizona Press, Tucson.

Mardia, K. V. (1972). *Statistics of Directional Data*. Academic Press, New York.

Mardia, K. V. (1975). Statistics of directional data (with discussion). *J. Roy. Statist. Soc. Ser. B* **37**, 349–393.

Marsden, B. and Williams, G. V. (1993). *Catalogue of Cometary Orbits.* 8th Ed. International Astronomical Union and the Minor Planet Center, Harvard.

Pensado, J. and Lahulla, J. F. (1967). Orientación de los planos orbitales y distribución de las órbitas de las estrellas dobles visuales. *Bol. Astr. Obs. Madrid* **9**, 21–35.

Watson, G. S. (1983). *Statistics on Spheres*, University of Arkansas Lecture Notes in the Mathematical Sciences, Vol. 6. Academic Press, New York.

Worley, C. E. and Heintz, W. D. (1983). Fourth catalog of orbits of visual binary stars. *Publ. US Naval Obs.* **XXIV** Part VII, 1–53.

Multivar. Statist., pp. 135–139
E.-M. Tiit *et al.* (Eds)

SIMPLE METHODS FOR FITTING CIRCLES OR POINTS TO SPHERICAL DATA

RICHARD WILLIAM FAREBROTHER
Department of Economic Studies, Victoria University of Manchester, Manchester M13 9PL, United Kingdom

ABSTRACT

In this paper we develop several alternative methods for fitting a point or a circle to a set of independent observations on the surface of the unit sphere. Our discussion includes the methods of Jupp and Mardia (1980) and Fisher (1985) as special cases.

1. INTRODUCTION

In their recent book Fisher, Lewis and Embleton (1987), henceforth FLE, described a single method for fitting a circle of radius $\sin\theta$ to a set of independent point observations on the surface of the unit sphere. In this paper we will suggest several alternative procedures which, in the limiting case $\theta = 0$, may also be used to fit a point to the same data. In particular we obtain the conventional mean direction, Fisher's (1985) spherical median and a related spherical mean as special cases of our general procedures.

2. PRELIMINARIES

Let P_i be the point on the surface of the unit sphere at geographical latitude $90° - \theta_i^*$ $(0° \leqslant \theta_i^* \leqslant 180°)$ and geographical longitude $\varphi_i^* - 180°$ $(0° \leqslant \varphi_i^* \leqslant 360°)$, and let $\theta_i = \pi\theta_i^*/180$ and $\varphi_i = \pi\varphi_i^*/180$, then P_i has cartesian coordinates

$$z_i = (\sin\theta_i \cos\varphi_i,\ \sin\theta_i \sin\varphi_i,\ \cos\theta_i)' \tag{2.1}$$

relative to the standard axes e_1, e_2, and e_3, where $E = [e_1 e_2 e_3]$ is the 3×3 identity matrix (see (FLE, 1987, pp. 18–19)). In particular, if we set $\theta_i = \pi/2$, then P_i is a point on the equator with coordinates

$$x_i = (\cos\varphi_i,\ \sin\varphi_i,\ 0)'. \tag{2.2}$$

Similarly, if we set $\theta_i = 0$, then we obtain the coordinates of the north pole

$$w = e_3 = (0, 0, 1)'. \tag{2.3}$$

Combining equations (2.1), (2.2) and (2.3) we have

$$z_i = x_i \sin\theta_i + w\cos\theta_i \tag{2.4}$$

for the coordinates of a general point relative to the standard axes. Premultiplying this expression by the 3×3 orthonormal matrix $U = [u_1 u_2 u_3]$ we have

$$y_i^* = U(x_i \sin\theta_i + w\cos\theta_i) \tag{2.5}$$

for the coordinates of a general point relative to the orthogonal axes u_1, u_2, and u_3.

3. JUPP AND MARDIA'S PROBLEM

Given n independent observations $y_1, y_2, \ldots, y_n$ on the surface of the unit sphere and the n associated angles $\varphi_1, \varphi_2, \ldots, \varphi_n$ (and hence the n explanatory variables $x_1, x_2, \ldots, x_n$), the finite sample variant of Jupp and Mardia's (1980) problem discussed by FLE (1987, p. 239) is to choose the orthonormal matrix U and the common angle θ to minimise the sum of the squared deviations between y_i and Uz_i, or

$$\sum_{i=1}^{n}(y_i - Uz_i)'(y_i - Uz_i). \tag{3.1}$$

Let Y and Z be the $n \times 3$ matrices with i-th rows y_i' and z_i', respectively, then problem (3.1) may be rewritten in the form of an orthogonal Procrustes problem and we have to choose U and θ to minimise

$$\operatorname{tr}(Y - ZU')'(Y - ZU'), \tag{3.2}$$

or, equivalently, to maximise

$$\operatorname{tr} UB = n^{-1}\operatorname{tr} UZ'Y, \tag{3.3}$$

where

$$B = n^{-1}Z'Y = V\sin\theta + w\bar{y}'\cos\theta, \tag{3.4}$$

$$V = n^{-1}\sum_{i=1}^{n} x_i y_i' \quad \text{and} \quad \bar{y} = n^{-1}\sum_{i=1}^{n} y_i. \tag{3.5}$$

Let P and Q be 3×3 orthonormal matrices and let D be a 3×3 diagonal matrix with non-negative diagonal elements such that B has singular value decomposition $B = PDQ'$. Then it is well known that expression (3.3) is maximised for a fixed nonsingular matrix B, when U is set equal to $\hat{U} = QP' = (B'B)^{-1/2}B'$ (see (Farebrother, 1985; Gower, 1984, p. 762) or (Jupp and Mardia, 1980)). The case of a singular B matrix is also discussed by Farebrother (1985).

We have thus determined an appropriate value for U when θ is given. To determine an appropriate value for θ when U is given we have to maximise expression (3.3) or, equivalently,

$$\operatorname{tr}(UV)\sin\theta + \bar{y}'Uw\cos\theta \tag{3.6}$$

subject to $0 \leqslant \theta \leqslant \pi$. If $\operatorname{tr}(UV) \leqslant 0$ and $\hat{y}'Uw > 0$ then (3.6) is maximised at $\theta = 0$; and if $\operatorname{tr}(UV) \leqslant 0$ and $\bar{y}'Uw < 0$ then it is maximised at $\theta = \pi$. On the other hand if $\operatorname{tr}(UV) > 0$ then there is an interior maximum at $\theta = \hat{\theta}$ where $\hat{\theta}$ is the unique solution of

$$\tan\hat{\theta} = \operatorname{tr}(UV)/\bar{y}'Uw. \tag{3.7}$$

Note that Jupp and Mardia's (1980, p. 171) and FLE's (1987, p. 239) expressions for $\hat{\theta}$, their $\hat{\alpha}$, are incorrect.

Thus corrected this pair of results may be used as the basis of an iterative procedure to determine the values of U and θ which minimise expression (3.1). Unfortunately there does not seem to be a simple algebraic formula for the positive definite square root of the 3×3 positive definite matrix $(B'B)^{-1} = QD^{-2}Q'$ so that it is necessary to find the singular values and right eigenvectors of B at each stage. (In the 2×2 case such a formula is easily derived.)

Finally we note that expression (3.1) is minimised subject to $\theta = 0$ when u_3 is set equal to $\bar{y}(\bar{y}'\bar{y})^{-1/2}$. This expression defines the conventional mean direction, discussed by FLE (1987, pp. 29–31).

4. ORTHOGONAL EUCLIDEAN REGRESSION

Let r_i denote the 3×1 matrix $r_i = y_i - Uz_i$, then the model of the previous section is estimated by choosing U and θ to minimise the sum of the squared Euclidean (or L_2-norm) distances between y_i and Uz_i, or

$$\sum_{i=1}^{n} \left(r_{1i}^2 + r_{2i}^2 + r_{3i}^2\right). \tag{4.1}$$

As alternatives to this criterion we could have chosen U and θ to minimise the sum of the Euclidean distances

$$\sum_{i=1}^{n} \left(r_{1i}^2 + r_{2i}^2 + r_{3i}^2\right)^{1/2} \tag{4.2}$$

or the sum of the L_1-norm distances

$$\sum_{i=1}^{n} \left(|r_{1i}| + |r_{2i}| + |r_{3i}|\right). \tag{4.3}$$

However the model adopted by Jupp and Mardia assumes that an explanatory variable φ_i is associated with each observation y_i. If this is not so, then we may implicitly set

$$d_i = \left[(u_1'y_i)^2 + (u_2'y_i)^2\right]^{1/2}, \quad \cos\varphi_i = u_1'y_i/d_i \quad \text{and} \quad \sin\varphi_i = u_2'y_i/d_i,$$

and choose u_3 and θ to minimise expressions (4.1) or (4.2) which may now be rewritten as

$$\sum_{i=1}^{n} \left[(d_i - \sin\theta)^2 + (u_3'y_i - \cos\theta)^2\right]^{p/2}, \tag{4.4}$$

where $p = 1$ or $p = 2$. Now $\sin\theta$ and $d_i = \sin t_i$ represent the distances from Uz_i and y_i to the u_3 axis while $\cos\theta$ and $u_3'y_i = \cos t_i$ give the corresponding heights above the equatorial plane so that expression (4.4) represents the sum of the p-th powers of the orthogonal Euclidean distances from the observed points $y_1, y_2, \ldots, y_n$ on the surface of the unit sphere to the fitted circle.

5. ORTHOGONAL ARC REGRESSION

In this section we shall replace the Euclidean distances of Section 4 by arc lengths (which are perhaps more appropriate when we are concerned with analysis of spherical data) and will develop an orthogonal regression criterion in this context.

Now it is well known that the angle subtended at the centre of the unit sphere by the points y_0 and y_i on its surface, and hence the arc length between them, is given by $q_i = \cos^{-1}(y_0'y_i)$. Thus the arc length between the pole $y_0 = u_3$ and the point y_i is given by $t_i = \cos^{-1}(u_3'y_i)$ and the arc length between the pole u_3 and any point $y_i^* = Uz_i$ on the fitted circle by $\theta = \cos^{-1}(u_3'y_i^*) = \cos^{-1}(z_{3i})$. Both of these values are non-negative, so that the arc length between y_i and the fitted circle is given by $|t_i - \theta|$ and we have to choose u_3 and θ to minimise the sum of the p-th powers of these lengths, or

$$\sum_{i=1}^{n} |t_i - \theta|^p, \tag{5.1}$$

where $p = 1$ or $p = 2$. In particular, if $\theta = 0$ and $p = 1$ then the value of u_3 which minimises expression (5.1) defines Fisher's (1985) spherical median (see (FLE, 1987, p. 111)). Similarly if $\theta = 0$ and $p = 2$ then expression (5.1) defines a spherical mean which is distinct from the mean direction. All four variants of this problem may be illustrated by extending Farebrother's (1987) mechanical models for orthogonal regression in the plane to the same problem on the sphere. By projecting the forces associated with these mechanical models onto the cone which is tangential to the fitted circle, or onto and orthogonal to the plane of the fitted circle, we have two models for a solution procedure but it seems simpler to follow FLE and apply a standard optimisation procedure to expression (5.1).

Finally we note that expression (4.4) is close to (5.1) if θ and the t_i are all sufficiently close to zero.

6. CONCLUSION

In this paper we have suggested several methods for fitting a point or a circle to a set of independent observations on the surface of the unit sphere. The statistical properties of these estimators are not known but they can readily be explored by means of bootstrap techniques (see (Fisher, 1985)). Further it is clear that several of our estimators are maximum likelihood estimators under appropriate distributional assumptions (see (Jupp and Mardia, 1980)).

REFERENCES

Farebrother, R. W. (1985). Linear unbiased approximators of the disturbance in the standard linear model. *Linear Algebra Appl.* **67**, 259–274.

Farebrother, R. W. (1987). Mechanical representations of the L_1 and L_2 estimation problems. In: *Statistical Data Analysis Based on the L_1 Norm and Related Methods*, pp. 455–464, North-Holland, Amsterdam.

Fisher, N. I. (1985). Spherical medians. *J. Royal Stat. Soc.* (*Series B*) **47**, 342–348.

Fisher, N. I., Lewis, T. and Embleton, B. J. J. (1987). *Statistical Analysis of Spherical Data*. Cambridge University Press, Cambridge.

Gower, J.C. (1984). Multivariate analysis: ordination, multidimensional sealing, and allied topics. In: *Handbook of Applicable Mathematics, Vol. 6*, pp. 727–781, Wiley, Chichester.

Jupp, P. E. and Mardia, K. V. (1980). A general correlation coefficient for directional data and related regression problems. *Biometrika* **67**, 163–173.

Part Four

MULTIVARIATE NONPARAMETRIC MODELS AND THEIR APPLICATIONS

Multivar. Statist., pp. 143–161
E.-M. Tiit *et al.* (Eds)

THE USE OF HELLINGER DISTANCE IN GRAPHICAL DISPLAYS OF CONTINGENCY TABLE DATA*

C. RADHAKRISHNA RAO
Department of Statistics, Pennsylvania State University,
University Park, PA 16802

ABSTRACT

A general theory is developed for representing population profiles characterized by multiple measurements in a low dimensional Euclidean space. The basic input to the problem is a measure of distance between two population profiles. A well known method for representing row or column profiles in a contingency table using a chisquare type distance between profiles is the correspondence analysis. It is suggested that a similar analysis based on Hellinger distance between profiles has some advantages and is better suited for studying the configuration of profiles. Some examples are given to show the differences between the two analyses.

An asymmetric biplot technique which is useful in interpreting differences in row (column) profiles in terms of column (row) categories is developed.

1. CANONICAL COORDINATES

The problem we consider may be stated as follows. Let $X = (X_1, \ldots, X_m)$ be an $s \times m$ data matrix, with the i-th column vector X_i representing measurements of s variables made on the i-th population (individual or unit). The column vector X_i will be referred to as the i-th population profile (PP). The PP's can be represented as m points in an s-dimensional vector space R^s with a specified inner product and the associated norm

$$(x, y) = x'My, \qquad x, y \in R^s, \tag{1.1}$$

$$\|x\| = (x, x)^{1/2}, \qquad x \in R^s, \tag{1.2}$$

where M is a positive definite matrix and x' represents the transpose of x. We may call this the Mahalanobis or M-space (Mahalanobis, 1936). In practical situations, it may be necessary to attach a weight $w_i \geqslant 0$ to the i-th PP, $i = 1, \ldots, m$, such that $w_1 + \cdots + w_m = 1$, the exact use of which will be detailed in the following discussion. We represent the vector $(w_1, \ldots, w_m)'$ by

*Research sponsored by the Army Research Office under Grant DAAH04-93-G-0030.

w and the diagonal matrix with w_i as the i-th diagonal element by W. The M-space with weight as an additional dimension will be referred to as the WM-space. (In our discussion we consider W as a general positive definite matrix to cover more general applications).

The problem is to find a $k \times m$ matrix

$$Y = (Y_1, \ldots, Y_m) \tag{1.3}$$

with $k < s$ for representing the PP's in a k-dimensional Euclidean space (E^k) with the usual inner product ($x'y$ for $x, y \in E^k$) and the k-vector Y_i as the profile of the i-th population, in such a way that the relative positions of the PP's in the M-space (in terms of distances between profiles) are preserved to the extent possible in E^k. For this purpose, we need to have a criterion for measuring the loss of information in reducing the dimension of the profile space, by minimizing which we obtain an optimum solution for (1.3).

The relative positions of the PP's in the M-space can be described by what may called a *configuration matrix*

$$C = (X - \xi 1')' M (X - \xi 1') = \big((X_i - \xi)' M (X_j - \xi)\big) = (c_{ij}) \tag{1.4}$$

where ξ is some chosen reference (profile) vector and the c_{ij}'s represent the distances and angles between profiles as explained in the Figure 1.

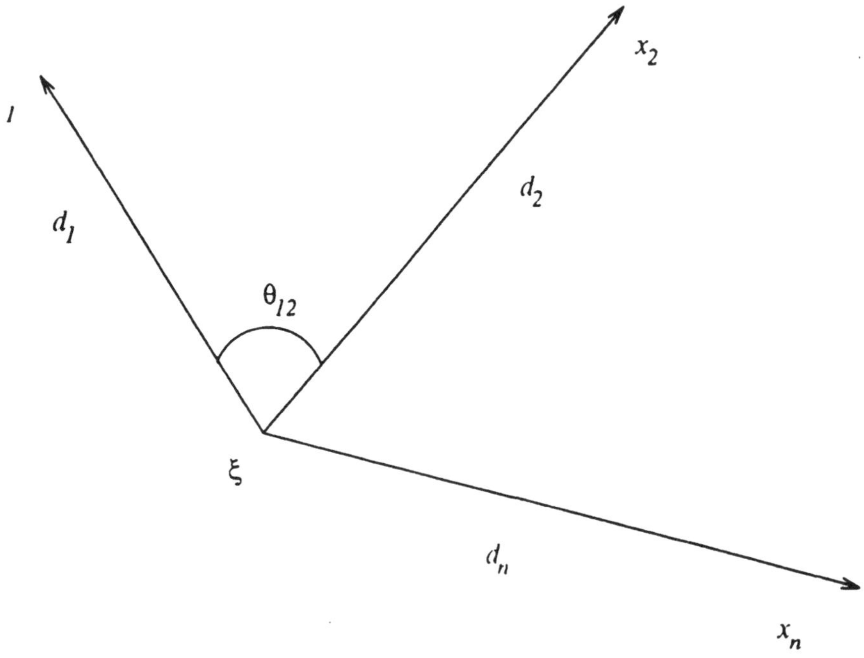

Figure 1. Configuration of the profiles in M-space ($d_i = \sqrt{c_{ii}}$, $d_i d_j \cos\theta_{ij} = c_{ij}$).

The corresponding configuration about the origin in the reduced space E^k is $Y'Y$. The problem then reduces to minimizing

$$\|C - Y'Y\| \tag{1.5}$$

with respect to Y, a $k \times m$ matrix as defined in (1.3), for a suitably chosen matrix norm. The following theorem provides the solution.

THEOREM. *Consider the singular value decomposition (s.v.d.)*

$$M^{1/2}(X - \xi 1')W^{1/2} = \lambda_1 U_1 V_1' + \cdots + \lambda_s U_s V_s', \tag{1.6}$$

where $M^{1/2}$ and $W^{1/2}$ are symmetric square roots of M and W. Then the choice

$$Y = \begin{pmatrix} \lambda_1 V_1' W^{-1/2} \\ \vdots \\ \lambda_k V_k' W^{-1/2} \end{pmatrix} \tag{1.7}$$

minimizes (1.5) *for any* (W, W)*-invariant norm as defined in Note* 1 *below. The matrix Y in* (1.7) *is of order $k \times m$ and the i-th column gives the coordinates for the i-th population in the reduced space E^k. We call these coordinates the canonical coordinates for populations* (*CCP*), *which was introduced in an early paper by the author* (*see* (Rao, 1948)).

A detailed proof of the above theorem and applications can be found in (Rao, 1964, 1979, 1980, 1985).

Note 1. (A, B)-invariant norm of an $m \times n$ matrix is the usual norm (satisfying the postulates of a norm) with the additional condition

$$\|CZD\| = \|Z\| \quad \text{for any} \quad C, D \ni C'AC = A, \quad D'BD = B \tag{1.8}$$

for all $m \times n$ matrices Z, where C is an $m \times m$ matrix, D is an $n \times n$ matrix, and A and B are positive definite matrices of orders m and n respectively. This is a generalization given in (Rao, 1980) of a unitarily invariant norm defined by von Neumann (1937) with A and B as unit matrices.

Note 2. In our applications, we indicate some choices of the reference vector ξ. However, we note that a further minimization of (1.5) with respect to ξ leads to the choice

$$\hat{\xi} = (1'W1)^{-1} XW1 \tag{1.9}$$

where 1 is the column vector of unities.

Note 3. We determined the canonical coordinates by minimizing (1.5), which provided a closed form solution. There are other possibilities leading to better approximations which are computationally complex. For instance if d_{ij} is the distance in the basic space between populations i and j, however it is defined, and δ_{ij} is the corresponding distance in the Euclidean space, we may determine Y by

$$\min \sum_i \sum_j \left(\frac{\delta_{ij}}{f(d_{ij})} - 1 \right)^2 \tag{1.10}$$

where f is a monotonic function, or by

$$\min \max \left| \frac{\delta_{ij}}{f(d_{ij})} - 1 \right|. \tag{1.11}$$

For other methods in this direction, reference may be made to Eslava-Gomez and Marriott (1993).

Note 4. The concept of canonical variates (coordinates) was introduced in an early paper by the author (Rao, 1948) for graphical representation of taxonomical units characterized by multiple measurements. This was, perhaps, the first attempt to reduce high dimensional data to two or three dimensions using an objective criterion for purposes of graphical displays. Since then, graphical representation of multivariate data for visual examination of clusters, outliers and other structures in the data has been an active field of research. Some of the developments are biplots (Gabriel, 1971; Gifi, 1990; Gower, 1993; Greenacre, 1993) multidimensional scaling (Kruskal and Wish, 1978), correspondence analysis (Benzécri, 1992; Greenacre, 1984), Chernoff's faces (Chernoff, 1993) and parallel coordinates (Mahalanobis *et al.*, 1949; Wegman, 1990). Cavalli-Sfroza (1991) uses canonical coordinates (variables) in interpreting the evolution of human populations.

2. LOSS OF INFORMATION

In terms of the s.v.d. (1.6), the configuration matrix (1.4) of the profiles is

$$C = C_s = W^{-1/2}(\lambda_1^2 V_1 V_1' + \cdots + \lambda_s^2 V_s V_s') W^{-1/2} \tag{2.1}$$

while that based on the canonical coordinates (1.7) in E_k is

$$C_k = Y'Y = W^{-1/2}(\lambda_1^2 V_1 V_1' + \cdots + \lambda_k^2 V_k V_k') W^{-1/2} \tag{2.2}$$

and the deficiency is

$$D_1 = C_s - C_k = W^{-1/2}(\lambda_{k+1}^2 V_{k+1} V_{k+1}' + \cdots + \lambda_s^2 V_s V_s') W^{-1/2}. \tag{2.3}$$

An overall measure of loss of information is the ratio

$$\frac{\operatorname{tr} W^{1/2} D_1 W^{1/2}}{\operatorname{tr} W^{1/2} C_s W^{1/2}} = 1 - \frac{\lambda_1^2 + \cdots + \lambda_k^2}{\lambda^2} \tag{2.4}$$

where $\lambda^2 = \lambda_1^2 + \cdots + \lambda_s^2$. It is customary to compute the ratios

$$\frac{\lambda_1^2}{\lambda^2}, \frac{\lambda_2^2}{\lambda^2}, \ldots \tag{2.5}$$

expressed as percentages to indicate the relative importance of the canonical coordinates in different dimensions.

It is more important to assess the distortions in the inter profile squared distances due to reduction of dimensionality. The $m \times m$ matrix of these squared distances denoted by S can be computed from the configuration matrix C by the formula

$$S = c1' + 1c' - 2C = \left(d_{ij}^2\right) \tag{2.6}$$

where c is the vector of the diagonal elements of C. The corresponding matrix in the reduced space is

$$S_{(k)} = c_{(k)}1' + 1c'_{(k)} - 2C_{(k)} \tag{2.7}$$

so that the matrix

$$D_2 = S - S_{(k)} = \left(d_{ij(k)}^2\right) \tag{2.8}$$

measures the deficiencies in the distances due to reduction of dimensionality. An over all measure of deficiency is the ratio

$$\frac{\sum\sum w_i w_j d_{ij(k)}^2}{\sum\sum w_i w_j d_{ij}^2} = \frac{\lambda_{k+1}^2 + \cdots + \lambda_s^2}{\lambda_1^2 + \cdots + \lambda_s^2} \tag{2.9}$$

which is the same as (2.4).

The weighted dispersion matrix between profiles in the original space is

$$B = (X - \xi 1')W(X - \xi 1')' = (b_{ij}) \tag{2.10}$$

while the corresponding matrix in the reduced k-dimensional space is

$$B_{(k)} = M^{-1/2}(\lambda_1^2 U_1 U_1' + \cdots + \lambda_k^2 U_k U_k')M^{-1/2} = (b_{ij(k)}). \tag{2.11}$$

The proportion of the between profile variance in the i-th variable explained by the first k canonical variates (coordinates) is

$$b_{ii(k)}/b_{ii}, \qquad i = 1, \ldots, s. \tag{2.12}$$

For an interpretation of the canonical coordinates in different dimensions it would be useful to compute the proportion of variance in each variable explained by each of the canonical variates, i.e., to obtain a decomposition of (2.12) in terms of canonical variates. For this purpose, we introduce the matrices

$$E_1 = M^{-1/2}(\lambda_1 U_1, \ldots, \lambda_s U_s) = (e_{ij}) \tag{2.13}$$

$$E_2 = \left(e_{ij}/\sqrt{b_{ii}}\right) = (f_{ij}) \tag{2.14}$$

where b_{ii} is as defined in (2.10). Let $E_{i(k)}$ be the matrix obtained by retaining only the first k columns in E_i for i=1, 2. Then it is seen that

$$E_1 E_1' = B, \qquad E_{1(k)} E_{1(k)}' = B_{(k)}. \tag{2.15}$$

Let us consider the matrix $E_{1(k)}$ and define what may be called *canonical coordinates for variables* (CCV) in k dimensions as shown in Table 1.

If we plot the variables as points in E^k using the row coordinates in different dimensions, then the scalar products of the vectors representing the variables are the elements of $B_{(k)}$, the best k-dimensional approximation to B.

There is some advantage in plotting the variables using the standardized coordinates (f_{ij}) defined in (2.14) as shown in Table 2.

The magnitudes in the right hand block of Table 2 indicate the influence of different variables in each dimension (canonical variate) in the reduced space. This may enable us to associate each dimension with certain variables.

Table 1.
Canonical coordinates for variables

Variable	dim 1	dim 2		dim k
1	e_{11}	e_{12}	$\cdots$	e_{1k}
2	e_{21}	e_{22}	$\cdots$	e_{2k}
$\vdots$	$\vdots$	$\vdots$	$\vdots$	$\vdots$
s	e_{s1}	e_{s2}	$\cdots$	e_{sk}

Table 2.
Standardized CCV's and the variance explained by each canonical variate

Variable	Standardized coordinates			Proportion of variance explained			
	dim 1	...	dim k	dim 1	...	dim k	total
1	f_{11}	$\cdots$	f_{1k}	f_{11}^2	$\cdots$	f_{1k}^2	$\sum f_{1i}^2$
$\vdots$	$\vdots$		$\vdots$	$\vdots$		$\vdots$	$\vdots$
s	f_{s1}	$\cdots$	f_{sk}	f_{s1}^2	$\cdots$	f_{sk}^2	$\sum f_{si}^2$

We may plot the variables using the CCV's in the same chart as the canonical coordinates for the profiles. It is seen that all variable points lie inside the unit sphere in E^k, and the variables close to the surface of the sphere have greater influence on the canonical variates.

It may also be mentioned that it is the usual practice in a biplot to represent the i-th variable as a directed line using the direction cosines proportional to the i-th row elements in the matrix

$$E_{1(k)} = M^{-1/2}(U_1, \ldots, U_k) \tag{2.16}$$

in which case the projections of a profile point in these directions are proportional to the approximate coordinates of the profile in the original space (see (Gabriel, 1971; Greenacre, 1993)).

Note 5. The choices of M and W as inputs in the analysis of canonical coordinates need some discussion. The choice of M is related to the measure of distance between two profiles, and as such it depends on the nature of the problem under investigation. Similarly the weight to be attached to each population is at the choice of the investigator. The general theory developed in Sections 1 and 2 is applicable for any choices of M and W.

Note 6. In taxonomical classification M is generally taken to be the inverse of the variance-covariance (dispersion) matrix of the variables within a population, leading to Mahalanobis distance. Also w_i is taken to be n_i/n where n_i is the number of individuals sampled from population i to estimate the mean values defining the i-th population profile, and $n = n_1 + \cdots + n_m$. In such a case when the variables within a population have an s-variate normal distribution, we can use the singular values λ_i in (1.6) to test hypotheses of the type

H_k: the true population profiles can be represented in E^k without any loss of information.

The appropriate test statistic for this purpose is

$$n\left(\lambda_{k+1}^2 + \cdots + \lambda_s^2\right) \tag{2.17}$$

which is asymptotically distributed as chisquare on $(s - k - 1)(m - k - 1)$ degrees of freedom (see (Rao, 1973, pp. 556-560) for details).

Such a test is not available for other choices of M and W.

3. TWO WAY CONTINGENCY TABLES

We consider dichotomous categorical data with s rows and m columns and n_{ij} observations in the (i, j)-th cell. Define

$$N = (n_{ij}), \qquad n_{i\bullet} = \sum_{j=1}^{m} n_{ij}, \qquad n_{\bullet j} = \sum_{i=1}^{s} n_{ij}, \qquad n = \sum_{1}^{s}\sum_{1}^{m} n_{ij},$$

$$R = \text{Diag}\,(n_{1\bullet}/n, \ldots, n_{s\bullet}/n), \qquad C = \text{Diag}\,(n_{\bullet 1}/n, \ldots, n_{\bullet m}/n),$$

$$P = n^{-1}NC^{-1} = \begin{pmatrix} p_{1|1} & \cdots & p_{1|m} \\ \vdots & \vdots & \vdots \\ p_{s|1} & \cdots & p_{s|m} \end{pmatrix}, \qquad \text{column profiles,} \tag{3.1}$$

$$Q = n^{-1}R^{-1}N = \begin{pmatrix} q_{1|1} & \cdots & q_{m|1} \\ \vdots & \vdots & \vdots \\ q_{1|s} & \cdots & q_{m|s} \end{pmatrix}, \qquad \text{row profiles,} \tag{3.2}$$

$$p = R1, \qquad q = C1.$$

The problem is to represent the column (row) profiles as points in E^k, $k < s$, such that the Euclidean distances between points reflect specified affinities between the corresponding column (row) profiles.

The technique developed for this purpose by Benzécri (1992) is known as correspondence analysis (CA) which can be identified as canonical coordinate analysis. For instance, to represent the column profiles, we need to choose

$$X = P, \qquad M = R^{-1}, \qquad W = C \tag{3.3}$$

and apply the analysis of Section 1. Thus one finds the s.v.d. of

$$R^{-1/2}(P - p1')C^{1/2} = \lambda_1 U_1 V_1' + \cdots + \lambda_s U_s V_s' \tag{3.4}$$

giving the coordinates in E_k

$$\begin{pmatrix} \lambda_1 V_1' C^{-1/2} \\ \vdots \\ \lambda_k V_k' C^{-1/2} \end{pmatrix}. \tag{3.5}$$

Implicit in this analysis is the choice of measure of affinity between the i-th and j-th profiles as the squared distance

$$d_{ij}^2 = \frac{(p_{1|i} - p_{1|j})^2}{p_1} + \cdots + \frac{(p_{s|i} - p_{s|j})^2}{p_s} \tag{3.6}$$

which is the chisquare distance. The squared Euclidean distance between the points representing the i-th and j-th profiles in E^k, the reduced space, is an approximation to (3.6). Thus the clusters we see in the Euclidean representation is based on the affinities measured by the chisquare distance (3.6).

Why should one choose the chisquare distance to measure the affinities between profiles? Some of the advantages mentioned by Benzécri and Greenacre are as follows.

1. Note that the expression in (3.4)

$$R^{-1/2}(P - p1')C^{1/2} = R^{1/2}(Q - 1q')C^{-1/2} = T \tag{3.7}$$

so that if we need a representation of the row profiles in E^k, we use the same s.v.d. as in (3.4)

$$R^{1/2}(Q - 1q')C^{-1/2} = \lambda_1 U_1 V_1' + \cdots + \lambda_s U_s V_s' \tag{3.8}$$

leading to the row coordinates

$$\left(\lambda_1 R^{-1/2} U_1, \ldots, \lambda_k R^{-1/2} U_k\right) \tag{3.9}$$

so that no extra computations are needed if we want a representation of the row profiles also.

2. It is easy to see that

$$n\left(\lambda_1^2 + \cdots + \lambda_k^2\right) = n \operatorname{tr} TT' = \sum\sum \frac{(n_{ij} - np_i q_j)^2}{np_i q_j} \tag{3.10}$$

with T as in (3.7), which is the Pearson chisquare statistic for testing independence between the attributes in a contingency table. Thus the computations involved in CA automatically allow us to test for independence, and also test for the dimensionality of the space of profiles using statistics of the type

$$n\left(\lambda_i^2 + \cdots + \lambda_s^2\right), \qquad i = 1, 2, \ldots, \tag{3.11}$$

as discussed in (2.17).

3. CA is only an exploratory data analysis to examine the configuration of row and column profiles in a general way, so that a particular convenient choice of the distance measure can serve the purpose.

On the other hand, there seem to be some drawbacks in using the chisquare distance.

1. The chisquare distance (3.6) is not a function of the i-th and j-th column profiles only. It involves the marginal profile which is a weighted average of the individual column profiles. The weights depend on the observed numbers of individuals in the column categories. These numbers may not have any relevance to the problem under study, especially when the columns represent different populations from each of which some individuals are chosen and classified according to row categories. In such a case the marginal profile depends on the actual sample sizes chosen or realized for different populations. The examples discussed in the next Section 4.1 show that the derived configurations in the reduced space may be sensitive to the sample numbers.

2. The marginal profile depends on the set of populations included in CA. The CA's based on a given set of populations (S_1) and an extended set of populations (S_1, S_2) may provide different configurations to the subset S_1 as shown in the examples of Section 4.1.

3. There is no special advantage in plotting the row and column profiles in the same chart. There is no particular interpretation of the relative positions of the

row profiles in terms of those of column profiles. Indeed one could use different distance measures for column and row profiles and study configurations of the column and row profiles separately.

4. Since the chisquare distance uses the marginal proportions in the denominator, undue emphasis is given to the categories with low frequencies in measuring affinities between profiles.

An alternative to the chisquare distance which has several advantages is the Hellinger distance between the i-th and j-th column profiles defined by

$$d_{ij}^2 = \left(\sqrt{p_{1|i}} - \sqrt{p_{1|j}}\right)^2 + \cdots + \left(\sqrt{p_{s|i}} - \sqrt{p_{s|j}}\right)^2 \tag{3.12}$$

which depends only the i-th and j-th column profiles. In such a case, the Euclidean distance in the reduced space between the i-th and j-th column profiles is an approximation to (3.12). For the derivation of canonical coordinates we choose

$$X = \begin{pmatrix} \sqrt{p_{1|1}} & \cdots & \sqrt{p_{1|m}} \\ \vdots & \vdots & \vdots \\ \sqrt{p_{s|1}} & \cdots & \sqrt{p_{s|m}} \end{pmatrix},$$

$$M = I, \qquad W = C = \text{Diag}\,(n_{\bullet 1}/n, \ldots, n_{\bullet m}/n) \tag{3.13}$$

and consider the s.v.d.

$$(X - \xi 1')C^{1/2} = \lambda_1 U_1 V_1' + \cdots + \lambda_s U_s V_s'. \tag{3.14}$$

We may choose $\xi' = (\xi_1, \ldots, \xi_s)$ as

$$\xi_i = \sqrt{p_i} = \sqrt{n_{i\bullet}/n}, \quad \text{or} \tag{3.15}$$

$$= n^{-1}\left(n_{\bullet 1}\sqrt{p_{i|1}} + \cdots + n_{\bullet m}\sqrt{p_{i|m}}\right). \tag{3.16}$$

The canonical coordinates in E^k for the column profiles choosing ξ as in (3.15) or (3.16) are

$$\begin{pmatrix} \lambda_1 V_1' C^{-1/2} \\ \vdots \\ \lambda_k V_k' C^{-1/2} \end{pmatrix}. \tag{3.17}$$

It can be shown that the statistic

$$4n\left(\lambda_1^2 + \cdots + \lambda_s^2\right) \tag{3.18}$$

is asymptotically distributed as chisquare on $(s-1)(m-1)$ degrees of freedom to test independence in the two way contingency table. Further, hypotheses specifying the dimensions of the subspace in which the profiles can be represented can also be tested in the same way as in (3.11) using the residual singular values.

The advantages in using Hellinger distance between profiles are the following.

1. The distance measure depends only on the profiles of the concerned pair. It is not altered when an extended set of profiles is considered.

2. The distance measure does not depend on the sample sizes on which the profiles are estimated.

3. The choice of ξ as in (3.16) provides a better approximation of the distances in the reduced space. However, if a representation of the row profiles is also needed we take $X = Q'$ where Q is the matrix defined in (3.2) and compute the s.v.d. of

$$(Q' - \eta 1')R^{1/2} = \mu_1 A_1 B_1' + \cdots + \mu_s A_s B_s'$$

leading to the canonical coordinates for row profiles

$$\begin{pmatrix} \mu_1 B_1' R^{-1/2} \\ \vdots \\ \mu_k B_k' R^{-1/2} \end{pmatrix}. \tag{3.19}$$

4. If we choose ξ as in (3.15), then the matrix in (3.14)

$$(X - \xi 1')C^{1/2} = \left(\sqrt{\frac{n_{ij}}{n}} - \sqrt{\frac{n_{i\bullet}}{n}\frac{n_{\bullet j}}{n}}\right)$$

which is symmetric in i and j. Then, the same s.v.d. as in (3.14) could be used for computing the canonical coordinates

$$\begin{pmatrix} \lambda_1 U_1' R^{-1/2} \\ \vdots \\ \lambda_k U_k' R^{-1/2} \end{pmatrix}$$

for the row profiles, as in the case of CA.

5. In (3.13), we used the marginal proportions as weights. With Hellinger distance, there is some flexibility in the choice of weights. We could have closen $W = I$ or in a manner which preserves some particular distances better than the others in the reduced space. Such a flexibility does not exist in correspondence analysis.

4. SOME EXAMPLES

4.1. *Example* 1. Consider for example the data on smoking habits (s_1, s_2, s_3) of different categories of employees (A, B, C, D, E) given in Tables 3 and 4, one of which has data on two additional categories of employees.

Table 3.

	Frequencies				Profiles			
	A	B	C	Total	A	B	C	Average
s_1	5	0	0	5	1	0	0	5/11
s_2	0	5	0	5	0	1	0	5/11
s_3	0	0	1	11	1	1	1	1/11
Total	5	5	1	11	1	1	1	1

Table 4.

	Frequencies						Profiles					
	A	B	C	D	E	Total	A	B	C	D	E	Average
s_1	5	0	0	4	1	10	1	0	0	0.8	0.2	10/21
s_2	0	5	0	1	4	10	0	1	0	0.2	0.8	10/21
s_3	0	0	1	0	0	1	0	0	1	0	0	1/21
Total	5	5	1	5	5	21	1	1	1	1	1	1

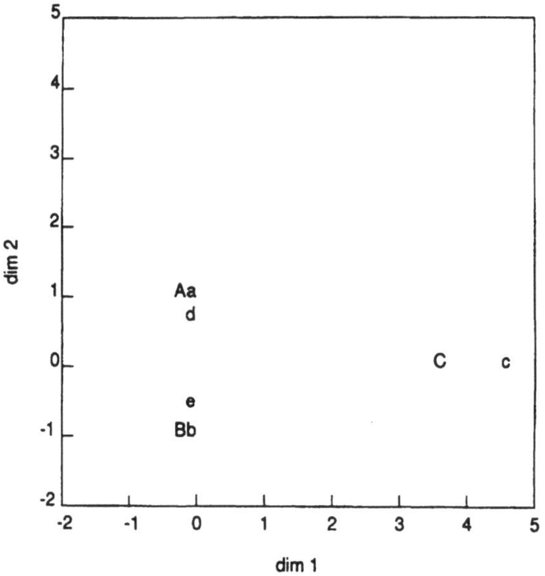

Figure 2. Cofiguration of canonical coordinates based on chisquare distance. (Lower case letters are used when the analysis is based on all the categories).

The two dimensional representations of the employee categories obtained through correspondence analysis using the formula (3.4) in Tables 3 and 4 are given in Figure 2, where the lower case letters are used for the data of Table 4. We make the following observations.

1. The relative positions of A, B and C change when additional employee categories are introduced, although their individual profiles remain the same.

2. The profiles for A, B, C in Table 3 suggest that they are equally distant between each other, and the correspondence analysis would have revealed this if the sample size for category C had been the same as that for A and B.

Thus, with chisquare distances, there is instability of the configuration of the populations with changes in the sample sizes and the addition or deletion of populations.

The two dimensional representations of the employee categories based on Hellinger distance using the formula (3.13) are given in Figures 2 and 3. It is

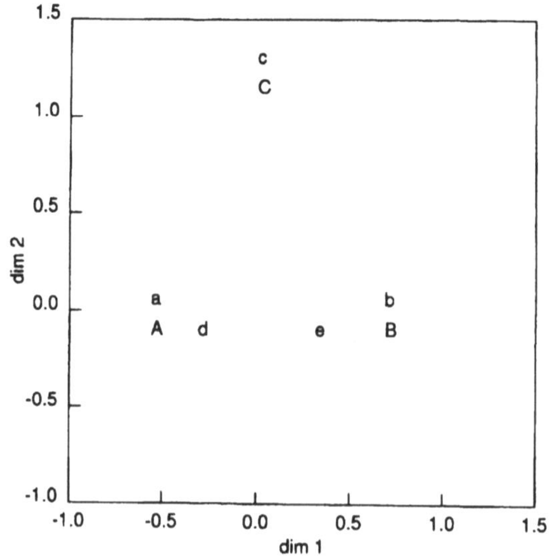

Figure 3. Cofiguration of canonical coordinates based on Hellinger distance. (Lower case letters are used when the analysis is based on all the categories).

seen that the graph based on Table 3 shows that A, B, and C are equidistant from each other and the graph based on Table 4 shows that the positions of A, B, C represented by lower case letters are not changed when the additional categories *d* and *e* are introduced. Thus, the use of Hellinger distance seems to provide a more satisfactory solution.

4.2. *Example* 2. We consider the data (from (Greenacre, 1993b)) on 796 scientific researchers classified according to their scientific discipline and funding category as shown in Table 5.

Table 5.
Scientific disciplines by research funding categories

Scientific discipline		Funding category a	b	c	d	e	Total
Geology	G	3	19	39	14	10	85
Biochemistry	B_1	1	2	13	1	12	29
Chemistry	C	6	25	49	21	29	130
Zoology	Z	3	15	41	35	26	120
Physics	P	10	22	47	9	26	114
Engineering	E	3	11	25	15	34	88
Microbiology	M_1	1	6	14	5	11	37
Botany	B_2	0	12	34	17	23	86
Statistics	S	2	5	11	4	7	29
Mathematics	M_2	2	11	37	8	20	78
Total		31	128	310	129	198	796

The canonical coordinates for the first three dimensions and percentage of variance explained by each are given in Table 6 for the analyses based on the chisquare distance (correspondence analysis) and the Hellinger distance (alternative). The formulas (3.3)–(3.5) are used for the analysis based on chisquare and the formulas (3.13)–(3.17) for that based on Hellinger distance.

Table 6.
Canonical coordinates for the scientific disciplines in the first three dimensions

Subjects	Chisquare Distance			Hellinger Distance		
	Dim 1	Dim 2	Dim 3	Dim 1	Dim 2	Dim 3
G	0.076401	0.302569	−0.087749	−0.031140	0.167408	−0.048245
B_1	0.179892	−0.454996	−0.151716	−0.129374	−0.242174	−0.077614
C	0.037644	0.073353	0.042371	−0.021144	0.040433	0.028254
Z	−0.327365	0.102283	0.064515	0.138850	0.045255	0.056894
P	0.315552	0.026997	0.108688	−0.165340	0.010679	0.023844
E	−0.117495	−0.291712	0.107330	0.049451	−0.129906	0.082901
M_1	0.012766	−0.109656	−0.041435	−0.004913	−0.052588	−0.008439
B_2	−0.178695	−0.038501	−0.129055	0.151404	−0.036559	−0.108025
S	0.124638	0.014162	0.107190	−0.066639	0.011763	0.052571
M_2	0.106751	−0.061316	−0.175688	−0.050307	−0.037572	−0.078006
% var.	47.20	36.66	13.11	45.87	34.10	16.57

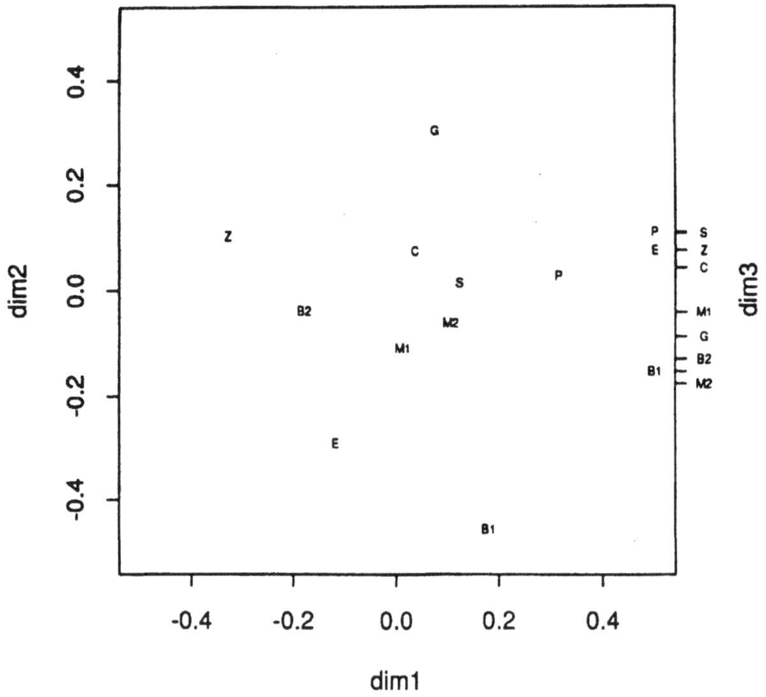

Figure 4. Configuration of scientific disciplines using Chisquare distance. (Correspondence Analysis).

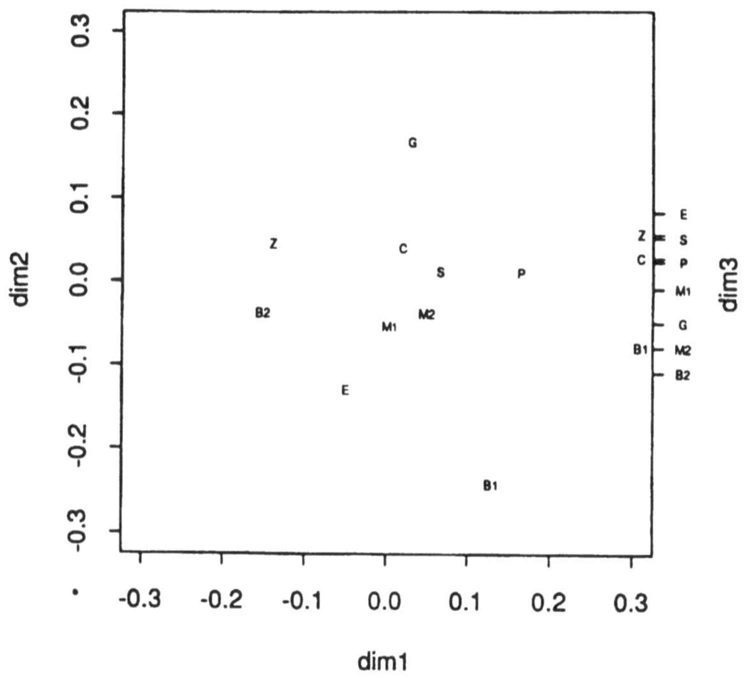

Figure 5. Configuration of scientific disciplines using Hellinger distance. (Alternative to Correspondence Analysis).

The plots of the scientific disciplines (subjects) using the canonical coordinates based on the chisquare and Hellinger distances are given in Figures 4 and 5 respectively. The coordinates in the third dimension are plotted on a line on the right hand side of the two dimensional plot. This will be of help in visualizing the plot in three dimensions and in interpreting the distances in the two dimensional plot. Thus, although B_2 and E appear to be close to each other in the two dimensional chart, they are clearly separated in the third dimension. No additional distances in the third dimension are involved in the case of P, C, S, Z, and E. (It may be noted that the squared distance between any two points in the three dimensional plot is equal to the sum of the squared distances in the two dimensional plot and in the third dimension).

It is of interest to note in this example that the configuration of the scientific disciplines in three dimensions obtained by both the methods are very similar. The percentage variance explained in each dimension is nearly the same for both the methods.

The canonical coordinates for the variables (funding categories) are computed using the formulas (2.13) and (2.14). These are obtained from the same s.v.d. used to compute the canonical coordinates for the scientific disciplines. Table 7 gives the standardized canonical coordinates for the funding categories a, b, c, d, e, using the formula (2.14) for the analyses based on the chisquare and Hellinger distances.

The standardized canonical coordinates for the funding categories are plotted in Figure 6 (for chisquare distance) and in Figure 7 (for Hellinger distance). It may be noted that all the points lie within the unit circle. It is customary to represent the canonical coordinates for the subjects and variables in one chart. We are using separate charts in order to explain the salient features of the configuration of the variables. The following interpretations emerge from the study of Table 7 and Figures 6 and 7.

Table 7.
Standardized canonical coordinates for funding categories in the first three dimensions

Funding category	Chisquare Distance				Hellinger Distance			
	dim 1	dim 2	dim 3	% var	dim 1	dim 2	dim 3	% var
a	0.419	−0.733	−0.206	75.6	0.385	−0.712	−0.332	76.5
b	0.406	−0.905	0.082	99.1	0.261	−0.955	0.111	99.2
c	−0.573	−0.628	−0.094	73.1	−0.613	−0.651	−0.017	80.1
d	−0.993	−0.097	0.054	99.9	−0.995	−0.054	0.065	99.7
e	−0.657	−0.189	−0.718	98.3	−0.742	−0.229	−0.602	96.5

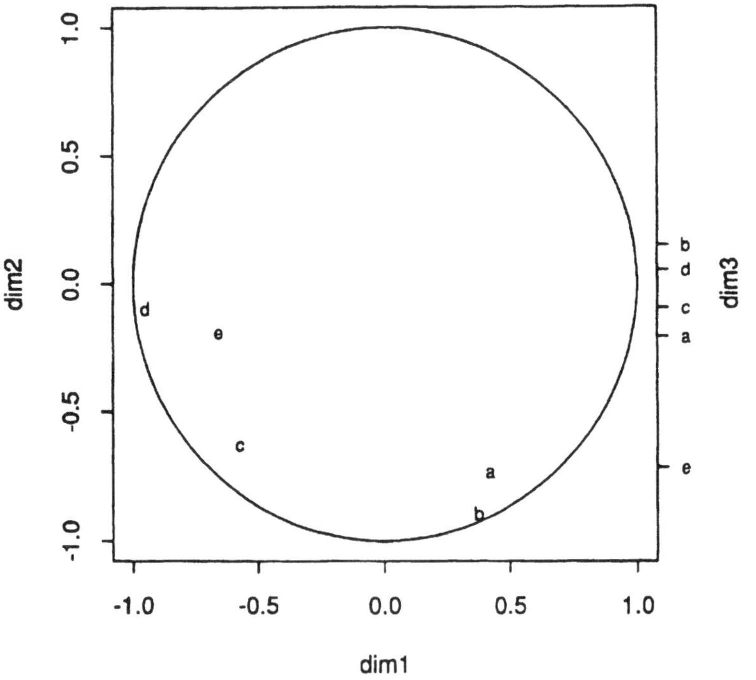

Figure 6. Configuration of funding categories using standardized canonical coordinates based on Chisquare distance.

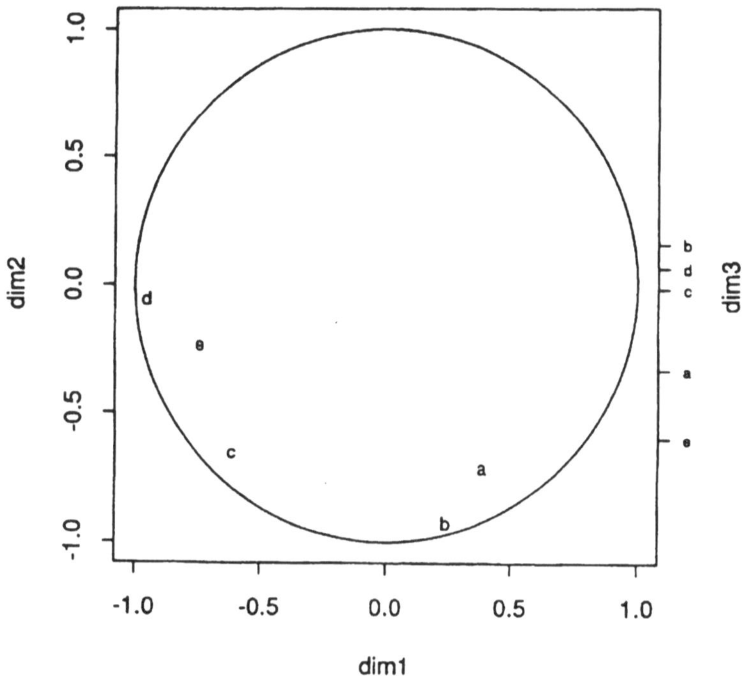

Figure 7. Configuration of funding categories using standardized canonical coordinates based on Hellinger distance.

1. The configurations of the funding categories as exhibited by Figures 6 and 7 obtained by using the chisquare and Hellinger distances are very similar.

2. All most all the variation in the funding categories b, d and e is captured in the first three canonical coordinates of the scientific disciplines. A large percentage of variation in a and c is explained by the first three coordinates.

3. The first dimension is strongly influenced by d, e, c, the second dimension by b, a, c and the third dimension by e.

Thus the use of standardized coordinates for variables enables us to interpret the different dimensions in terms of observed variables. There are other ways of plotting the coordinates of the variables as mentioned in the paragraphs below Table 4. Such biplots having a different interpretation are discussed in (Gabriel, 1971; Gifi, 1990; Gower, 1993; Grenacre, 1993a).

5. CONCLUSION

A general theory is developed for representing high dimensional "population by variable" data in a low dimensional Euclidean space. The first step in the problem is the specification of the basic space in which the populations can be considered as points and the choice of a distance measure between points characterizing the differences between populations, which in a practical problem depends on the nature of the investigation involved. The second step is the transformation of the points (representing the populations) from the basic space to a low dimensional Euclidean space. This has to be done in such a way that the configurations of the points as defined by the specified distances in the basic space and the Euclidean distances in the reduced space are as similar as possible. A closed form solution is obtained when the basic space is a vector space endowed with an inner product and the associated norm.

When we have data in the form of a two way contingency table with the frequency in the (i, j)-th cell representing the number of individuals from population j having category i of an attribute, a well known method for representing the populations in a low dimensional Euclidean space is correspondence analysis. The basic space in this case is a vector space where each population is represented by the vector of relative frequencies of the different categories of an attribute and the distance between vectors is defined by a chisquare type formula. Such a distance function is not an intrinsic measure of difference between two populations as it depends not only on the differences between their relative frequencies, but also on the average relative frequencies computed from the set of populations under study. Thus the configuration of any subset of populations depends on what other populations are included in the analysis, and also on the relative numbers of individuals observed from each population, as shown in the example of Subsection 4.1. An alternative approach of representing a population by the vector of the square roots of relative frequencies and defining distance between two populations by the Hellinger formula does not have the drawbacks associated with the chisquare type formula. In addition, the new analysis has the same advantage of providing tests of significance for homogeneity (or dimensionality) of the populations as in correspondence analysis based on the chisquare formula. Thus the method of dimensionality reduction based on Hellinger distance appears to be a better tool than that on the chisquare distance in exploratory data analysis.

A method of biplots which enables an interpretation of different dimensions in the reduced Euclidean space in terms of the original variables is discussed and illustrated in the example of Subsection 4.2.

It may be noted that Hellinger distance is one possible choice which depends only on the differences in population profiles. There are numerous other choices (see for instance Rao (1982)). In any particular problem, there may be some guidelines for the choice of an appropriate distance. But in a purely exploratory data analysis Hellinger distance is an attractive alternative in terms of stability of the configuration in the reduced space and also in providing appropriate tests of significance for reduction of dimensionality.

In the particular example illustrated in this article, the configurations in the reduced space happen to be similar under correspondence analysis and the alternative by using Hellinger distance. This may be so in many practical situations, but Hellinger distance provides an insurance against distortions due to selection of populations to be studied and the sample sizes of individuals observed for measurements from each population.

REFERENCES

Benzécri, J. P. (1992). *Correspondence Analysis Handbook.* Marcel Dekker, New York.

Cavallis-Sfroza, L. L. (1991). Genes, peoples and languages. *Scie. Amer.* **265**, 104–110.

Chernoff, H. (1973). The use of faces to represent points in k-dimensional space graphically. *J. Amer. Statist. Assoc.* **68**, 361–368.

Eslava-Gomez, G. and Marriott, F. H. C. (1993). Criteria to represent groups in the plane when the grouping is unknown. *Biometrics* **49**, 1088–1098.

Gabriel, K. R. (1971). The biplot graphical display of matrices with applications to principal component analysis. *Biometrika* **58**, 453–467.

Gifi, A. (1990). *Nonlinear Multivariate Analysis.* Wiley, New York.

Greenacre, M. J. (1984). *Theory and Applications of Correspondence Analysis.* Academic Press, London.

Greenacre, M. J. (1993a). Biplots in correspondence analysis. *J. Appl. Statist.* **20**, 251–269.

Greenacre, M. J. (1993b). *Correspondence Analysis in Practice.* Academic Press, San Diego.

Gower, J. C. (1993). Recent advances in biplot methodology. In: *Multivariate Analysis: Future Directions, Vol. 2*, pp. 295–325, C. M. Cuadras and C. R. Rao (Eds), North-Holland, Amsterdam.

Kruskal, J. B. and Wish, M. (1978). *Multidimensional Scaling.* Sage, Beverly Hills.

Mahalanobis, P. C. (1936). On the generalized distance in statistics. *Proc. Nat. Inst. Sci.* **12**, 49–55.

Mahalanobis, P. C., Mazumdar, D. N. and Rao, C. R. (1949). Anthropometric survey of United Provinces, 1941. A statistical study. *Sankhyā* **9**, 90–324.

Von Neumann, J. (1937). Some matrix inequalities and metrization of metric spaces. *Tomsk. Univ. Rev.* **1**, 286–299.

Rao, C. R. (1948). The utilization of multiple measurements in problems of biological classification (with discussion). *J. Roy. Statist. Soc. Ser. B* **10**, 159–193.

Rao, C. R. (1964). The use and interpretation of principal component analysis in applied research. *Sankhyā* **26**, 329–357.

Rao, C. R. (1973). *Linear Statistical Inference and its Applications.* 2nd edn., Wiley, New York.

Rao, C. R. (1979). Separation theorems for singular values of matrices and their applications in multivariate analysis. *J. Multivariate Anal.* **9**, 362–377.

Rao, C. R. (1980). Matrix approximations and reduction of dimensionality in multivariate statistical analysis. In: *Multivariate Analysis V*, pp. 3–22, P. R. Krishnaiah (Ed.), North-Holland, Amsterdam.

Rao, C. R. (1982). Diversity and dissimilarity coefficients: a unified approach. *Theoret. Population Biology* **21**, 24–43.

Rao, C. R. (1985). Tests for dimensionality and interaction of mean vectors under general and reducible covariance structures. *J. Multivariate Anal.* **16**, 173–184.

Wegman, E. J. (1990). Hyperdimensional data analysis using parallel coordinates. *J. Amer. Statist. Assoc.* **85**, 664–675.

Multivar. Statist., pp. 163–180
E.–M. Tiit *et al.* (Eds)

BIVARIATE GENERALIZATIONS OF THE MEDIAN

A. NIINIMAA
Department of Mathematical Sciences, University of Oulu,
SF-90570 Oulu 57, Finland

ABSTRACT

Small (1990) and Farebrother (1992) give good survey articles of multidimensional medians. In this paper some recent contributions in this area are reviewed. The asymptotic efficiency, influence function and breakdown properties as well as computational aspects of different multivariate medians are discussed.

1. INTRODUCTION

The median of a univariate sample $\mathbf{X} = \{x_1, \dots, x_n\}$ is the value $\hat{\theta}(\mathbf{X})$ of θ which minimizes the sum of distances

$$T(\theta, \mathbf{X}) = \sum_{i=1}^{n} |x_i - \theta|.$$

One can also determine the univariate median using the gradient (derivative) of the objective function $T(\theta, \mathbf{X})$ with respect to θ, namely

$$Q(\theta, \mathbf{X}) = -\sum_{i=1}^{n} \operatorname{sign}(x_i - \theta).$$

Then $Q(\hat{\theta}(\mathbf{X}), \mathbf{X}) = 0$. The "centered rank" of θ, $Q(\theta, \mathbf{X})$, is an increasing function of θ, and, if the sample points are ordered by this function, the usual natural ordering is obtained. $T(\theta, \mathbf{X})$ gives an alternative criterion function to order the observations according to their distance from the median. Such an ordering is not often applied in the univariate case but it appears to be useful in higher dimensions where exist no natural basis for ordering the data. If F is a continuous distribution function with median $\theta(F)$ the hypothesis $\theta(F) = \theta_0$ can be tested using the gradient test statistics $Q(\theta_0, \mathbf{X})$. In fact, this gradient test is simply the sign test since

$$Q(\theta_0, \mathbf{X}) = 2\ \#(x_i < \theta_0) - n$$

where # is the count function, i.e., $\#(x_i > 0)$ counts the number of cases for which $x_i > 0$.

The sample median, the sign test and the ordering of the univariate data set were obtained by defining an object function $T(\theta, \mathbf{X})$ which measures the depth of the point θ inside the data set. There are many ways to define the notion of data depth in the multidimensional case. In this paper we discuss only extensions that give direct analogues of univariate median in the sense that they coincide with the usual $\Re^1$ median in the univariate case. The gradient vector of the depth function is analogous to the univariate rank and the gradient vector calculated at the null hypothesis point gives a generalization of the univariate sign test statistics (see (Hettmansperger *et al.*, 1992)). The attention is here restricted to the concept of the multivariate median. Liu (1992) and Hettmansperger *et al.* (1992) gave detailed discussions on the multivariate sign and rank tests.

When a set of n observations is taken from a normal distribution $N(\theta, \sigma^2)$ the most popular estimators of θ are the sample mean and sample median. They are both unbiased, but the efficiency of the univariate median relative to the sample mean is only $2/\pi = 0.637$. The lack of efficiency for the normal data is however compensated by the robustness of the univariate median. It is possible to move the sample mean to infinity by moving just a single observation. For odd n, the sample median equals the order statistics $x_{(n+1)/2}$. In this case the displacement of one observation has either no effect at all on the sample median or the median is shifted to the nearest data point just before or just after the original median. When n is odd, more than half of the observations are needed to cause the breakdown of the sample median.

Hampel (1971) introduced the notion of the breakdown point — the smallest percentage of contaminated points that can make the estimator arbitrarily large. The sample mean has breakdown of 0% and the univariate sample median 50% breakdown. An alternative tool for studying robustness is the Hampel's (1974) influence function. If T is an estimator based on a large number of observations, then the influence function $IF(x)$ measures approximately the effect caused by an additional observation x on the estimator T (see (Hampel *et al.*, 1986)). In the univariate case the median has bounded and the mean unbounded influence function.

Of course, it is possible to make a compromise between robustness and efficiency. In the univariate case the Hodges–Lehman estimate is defined as a median of all pairwise averages of data points. Using the median of pairwise averages instead of the usual sample median causes the breakdown point to drop down from 50% to 39.3%, but the Gaussian efficiency of the Hodges–Lehman estimate is as big as 0.955 (Lehmann, 1975, p. 377). This idea has been generalized also to the multivariate case. See (Brown, 1982; Chaudhuri, 1992).

In this paper the following multivariate medians are discussed: the vector of marginal medians, the spatial median, the Oja's median (Oja, 1983) and the simplicial depth median (Liu, 1990). A brief review of other generalizations of the median is also given in Section 7. The definitions and results are given only in the bivariate case. With a little loss in generality the paper is in this way made more readable. Most ideas can be easily extended to the general multivariate setting.

The terminology used by Small (1990) is adopted with the exception that the term spatial median is used instead of L_1 median. To avoid details it is assumed that the bivariate data set $\mathbf{X} = \{\mathbf{x}_1, \ldots, \mathbf{x}_n\}$ is in the general position, i.e., any 3 points of $\mathbf{X}$ determine a triangle with a nonzero area.

Multivariate medians are not the only robust estimates of location. See for example (Rousseeuw and Leroy, 1987; Hampel *et al.* 1987).

2. REGULAR SETS

Let $\mathbf{X} = \{x_1, \ldots, x_n\}$ be a univariate data set with even number of observations. Choose B_i to be the half space $\{x \mid x - x_i \geqslant 0\}$ or $\{x \mid x - x_i \leqslant 0\}$ which contains the majority of observations in $\mathbf{X}$. The minimization of the sum $\sum |x_i - \theta|$ produces the solution set

$$\bigcap_{i=1}^{n} B_i = [x_{(n/2)}, x_{(n/2+1)}].$$

Thus θ is the median of the data set $\mathbf{X}$ if and only if for each i the point θ is lying in the majority set B_i.

We now make an attempt to generalize this idea into the bivariate case with data set $\mathbf{X} = \{\mathbf{x}_1, \ldots, \mathbf{x}_n\}$ (see also (Singh, 1991; Jeyaratnam, 1992)). The line through points $\mathbf{x}_i$ and $\mathbf{x}_j$ splits the space $\Re^2$ into two parts. Let B_{ij} then be the closed half-space containing the majority of observations, i.e., $\{\mathbf{x} \mid \det(\mathbf{x} - \mathbf{x}_i\ \mathbf{x} - \mathbf{x}_j) \geqslant 0\}$ or $\{\mathbf{x} \mid \det(\mathbf{x} - \mathbf{x}_i\ \mathbf{x} - \mathbf{x}_j) \leqslant 0\}$. The first trial to

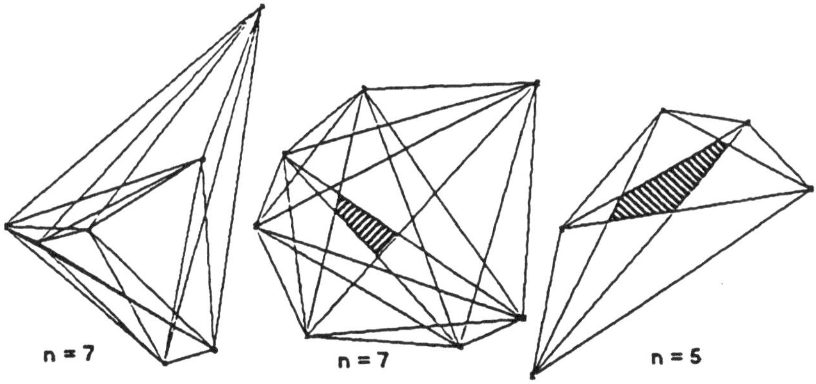

Figure 1. A non-regular set consisting of 7 data points and two regular data sets with 7 and 5 observations. The set $\bigcap_{i<j} B_{ij}$ is shaded.

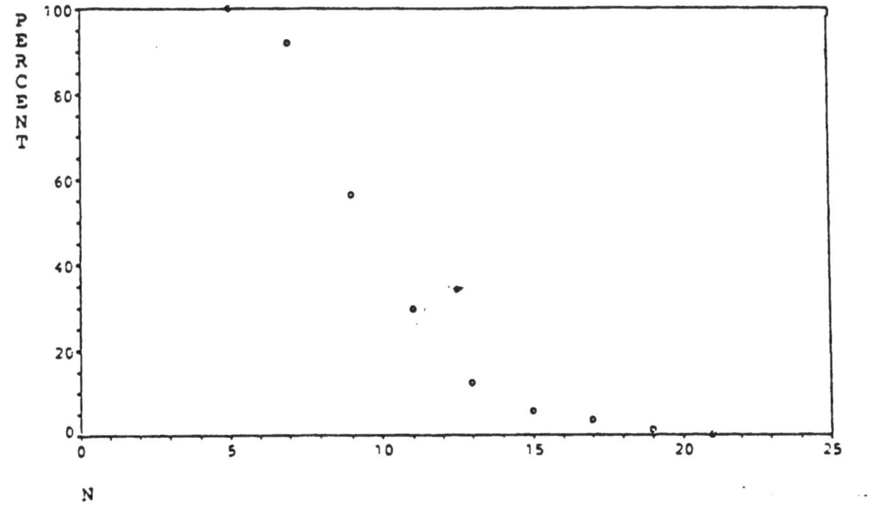

Figure 2. The estimated probability to get a regular data set in the case of bivariate normal data based on a large number of simulations.

define the bivariate median (set) would then be: any point $\theta \in \bigcap_{i<j} B_{ij}$ is the median of the data set $\mathbf{X}$. Unfortunately the set $\bigcap_{i<j} B_{ij}$ is often empty.

DEFINITION 2.1. $\mathbf{X}$ (for odd number of observations) is *regular* if $\bigcap_{i<j} B_{ij} \neq \varnothing$.

For continuous distributions, the probability to get a regular sample decreases rapidly with odd sample size n. This is illustrated in Figure 2 with Monte Carlo experiments in the case of the bivariate normal data.

We will show later that, for regular data sets, most of the bivariate generalizations of the medians coincide.

3. THE VECTOR OF MARGINAL MEDIANS

Let $\mathbf{X} = \{\mathbf{x}_1, \ldots, \mathbf{x}_n\}$ be a bivariate data set. The first criterion function is the sum of componentwise absolute distances (also called Manhattan distances or taxi distances)

$$T(\mathbf{\Theta}, \mathbf{X}) = \sum_{i=1}^{n} \left(|x_{i1} - \theta_1| + |x_{i2} - \theta_2| \right).$$

The criterion function can be minimized componentwise, and so the vector of marginal medians is obtained.

The univariate sample median is asymptotically normally distributed with mean $\theta(F)$ (the median of F) and variance $(4n)^{-1}(f(\theta(F)))^{-2}$. The density function is usually unknown and must be estimated from the data. See for example (Rao, 1988) for references. The component-wise median, of course, has the same univariate 64 percent efficiency when the underlying model is multivariate normal, while, as we will see later, the efficiencies of the other generalized medians increase with the dimension.

Let $F(x_1, x_2)$ be the cumulative distribution function (c.d.f.) of a bivariate random variable $\mathbf{x}$, and let F_1 and F_2 be the corresponding marginal distribution functions. David (1981, p. 25) gives the joint distribution function of the bivariate order statistics in terms of the cumulative distribution functions of $\mathbf{x}$.

$$\begin{aligned} H(u,v) &= P\big(x_{1(r)} \leqslant u, x_{2(s)} \leqslant v\big) \\ &= \sum_{i=r}^{n}\sum_{j=s}^{n}\sum_{k=\max(0,i+j-n)}^{\min(i,j)} \frac{n!}{k!(i-k)!(j-k)!(n-i-j-k)!} \\ &\quad \times \big(F_1(u) - F(u,v)\big)^{i-k}\big(F_2(v) - F(u,v)\big)^{j-k} \\ &\quad \times \big(1 - F_1(u) - F_2(v) + F(u,v)\big)^{n-i-j+k} F(u,v)^k. \end{aligned}$$

If the sample size n is odd, the function $H(u,v)$ gives directly the cumulative distribution function of the sample marginal median vector $(\hat{\theta}_1, \hat{\theta}_2)$, and for even n the distribution function can be easily calculated. Several other methods to generalize the concept of median are reviewed in this paper, but only in this case the exact distributions are known. If the cumulative distribution function $F(x,y)$ is known the covariance matrix of the marginal medians can be calculated numerically, using the function $H(u,v)$. If $F(x_1,x_2)$ is not known, it can be substituted by its sample version F_n. In fact this is a bootstrap technique for a covariance matrix estimation, but without any simulations (Maritz, 1991).

The asymptotic distribution function of $\sqrt{n}(\widehat{\Theta} - \Theta(F))$ is normal with expectation vector 0 and covariance matrix

$$\begin{pmatrix} \dfrac{1}{4f_1^2(\theta_1)} & \dfrac{\rho}{2f_1(\theta_1)f_2(\theta_2)} \\ \dfrac{\rho}{2f_1(\theta_1)f_2(\theta_2)} & \dfrac{1}{4f_2^2(\theta_2)} \end{pmatrix},$$

where $\rho = E\{\operatorname{sign}(x_1 - \theta_1)\operatorname{sign}(x_2 - \theta_2)\}$. See (Babu and Rao, 1988; Rao, 1988).

Let F be a univariate continuous distribution with a density function f and median 0. The influence function of the univariate sample median is (e.g. (Hettmansperger, 1984, p. 22))

$$IF(y) = \begin{cases} -\dfrac{1}{2f(0)}, & y \leqslant 0, \\ \dfrac{1}{2f(0)}, & y > 0. \end{cases}$$

Figure 3. The curve gives the marginal medians corresponding to all choices of the coordinate axes.

The function $IF(y)$ has only two different values and similarly if $(0,0)^T$ is the median of a bivariate distribution the influence function of the component-wise median **IF(y)** maps each quadrant into a single point (Niinimaa and Oja, 1995).

In the case of spatial data, for example, the coordinate axes can be thought to be arbitrary. The vector of marginal medians is not, however, rotation equivariant.

In Figure 3 the coordinate axes are rotated in order to illustrate the coordinate dependency of the vector of marginal medians. In dimensions higher than two it is even possible to construct data sets so that the vector of marginal medians lies outside the convex hull of the points (Rousseeuw and Leroy, 1987).

4. THE SPATIAL MEDIAN

The spatial median was first introduced to define a geographical center of a population in the beginning of this century (see (Small, 1990) for references). Also names "L_1 median", "geometrical median", the "mediancentre" and "Weber point" have been used (Small, 1990).

DEFINITION 4.1. Let $\mathbf{X} = \{\mathbf{x}_1, \ldots, \mathbf{x}_n\}$ be a bivariate data set points. The spatial median for $\mathbf{X}$ is defined as a point in $\Re^2$ which minimizes

$$T(\mathbf{\Theta}, \mathbf{X}) = \sum_{i=1}^{n} |\mathbf{x}_i - \mathbf{\Theta}|$$

where $|\cdot|$ is the Euclidian distance. The spatial median $\mathbf{\Theta}(F)$ of a bivariate distribution F is a point which minimizes

$$\dot{E}_F|\mathbf{X} - \mathbf{\Theta}|.$$

where x has a c.d.f. F.

The spatial median is unique if the data set $\mathbf{X}$ is in the general position (Milasevic and Duchame, 1987). The gradient vector of the objective function $T(\mathbf{\Theta}, \mathbf{X})$ is

$$\mathbf{Q}(\mathbf{\Theta}, \mathbf{X}) = \begin{pmatrix} -\sum_{i=1}^{n} \cos \varphi_i \\ -\sum_{i=1}^{n} \sin \varphi_i \end{pmatrix}$$

where φ_i is the direction angle of $\mathbf{x}_i - \mathbf{\Theta}$, $i = 1, \ldots, n$. Note that the gradient vector does not involve the distances $r_i = |\mathbf{x}_i - \mathbf{\Theta}|$, $i = 1, \ldots, n$, and thus the gradient is not affected by the moving the observation points along the rays connecting $\mathbf{\Theta}$ and $\mathbf{x}_i$. Therefore, it seems plausible that single outliers cannot have a large effect on the values of the spatial median. In fact, the spatial median has breakdown 50 percent (Lopuhaä and Rousseeuw, 1991). For the influence functions, see (Niinimaa and Oja, 1993).

The spatial median is orthogonally equivariant, not scale equivariant, and therefore different solutions are obtained, for example, just by transforming units from pounds and inches to kilograms and centimetres. Rao (1988) gives an affinely equivariant version of the spatial median. His median is obtained as a minimizer of the sum

$$\sum_{i=1}^{n} \left[(\mathbf{x}_i - \mathbf{\Theta})^T \mathbf{S}^{-1} (\mathbf{x}_i - \mathbf{\Theta})\right]^{1/2}$$

where $\mathbf{S}$ is the sample covariance matrix.

Brown (1983) showed that the asymptotic distribution function of $\sqrt{n}(\widehat{\mathbf{\Theta}} - \mathbf{\Theta}(F))$ is multivariate normal with the expectation vector 0 and covariance matrix $\mathbf{U}^{-1}\mathbf{C}\mathbf{U}^{-1}$ where

$$\mathbf{U} = \begin{pmatrix} E(r^{-1} \sin^2 \varphi) & -E(r^{-1} \sin \varphi \cos \varphi) \\ -E(r^{-1} \sin \varphi \cos \varphi) & E(r^{-1} \cos^2 \varphi) \end{pmatrix}$$

and

$$\mathbf{C} = \begin{pmatrix} E(\cos^2 \varphi) & E(\sin \varphi \cos \varphi) \\ E(\sin \varphi \cos \varphi) & E(\sin^2 \varphi) \end{pmatrix},$$

r and φ are now the length and the direction angle of $\mathbf{X} - \mathbf{\Theta}(F)$ where F is the c.d.f. of $\mathbf{x}$. In the case of a circular normal distribution the asymptotic efficiency of the spatial median relative to the sample mean vector, is 0.785, exceeding that of the marginal median vector 0.637. For bivariate spherically normal data the efficiency increases to 0.966 if the spatial median of pairwise averages (Hodges–Lehman estimate) is used (Chaudhuri, 1992). The price to be paid is the loss in robustness.

Jackknife and bootstrap methods have been proposed for the estimation of the dispersion matrix of the spatial median (see (Bose and Chaudhuri, 1991) for references).

Likewise in the univariate case, the depth function gives rise to an ordering of the data points from the center outwards. The companion sign test of the spatial median, the angle test, has been given by Brown (1983).

The spatial median can be calculated by the Newton–Raphson method. The move from the kth iterate $\mathbf{\Theta}_k$ to the next one is taken using the formula

$$\mathbf{\Theta}_{k+1} = \mathbf{\Theta}_k - \boldsymbol{\Gamma}_k^{-1}(\mathbf{\Theta}_k)\mathbf{Q}(\mathbf{\Theta}_k)$$

where n is the sample size and

$$\boldsymbol{\Gamma}_k^{-1}(\mathbf{\Theta}_k) = \begin{pmatrix} -\sum r_i^{-1}\cos^2\varphi_i & \sum r_i^{-1}\cos\varphi_i\sin\varphi_i \\ \sum r_i^{-1}\cos\varphi_i\sin\varphi_i & \sum r_i^{-1}\sin^2\varphi_i \end{pmatrix}$$

is the inverse of the second derivative matrix of the objective function $T(\mathbf{\Theta}, \mathbf{X})$. The term r_i^{-1} in the above formula may cause difficulties when $\mathbf{\Theta}$ is close to the observation points. Brown (1982) proposed a method to avoid this problem. The Fortran algorithm by Bedall and Zimmerman (1979) is also an application of the Newton–Raphson method.

5. OJA'S MEDIAN

Before the actual definition of the median of Oja, it may be useful to consider, the related notions by Wilks (1960). If $A(\mathbf{a}, \mathbf{b}, \mathbf{c})$ is the area of the triangle whose vertices are $\mathbf{a}$, $\mathbf{b}$, and $\mathbf{c}$, the usual mean vector $\bar{\mathbf{x}}$ for the bivariate data set $\mathbf{X} = \{\mathbf{x}_1, \ldots, \mathbf{x}_n\}$ is obtained by minimizing the sum

$$W(\mathbf{\Theta}) = \sum_{i<j} A^2(\mathbf{x}_i, \mathbf{x}_j, \mathbf{\Theta}).$$

Wilks also showed that the minimum value of function $W(\mathbf{\Theta})$ is $4|\mathbf{S}|/(n-1)^2$, where $|\mathbf{S}|$ is the determinant of the sample covariance matrix (the generalized variance of Wilks). The univariate sample variance can be rewritten in the form

$$s_x^2 = \frac{1}{n(n-1)} \sum_{i<j} (x_i - x_j)^2$$

and similarly

$$|\mathbf{S}| = \frac{4}{n(n-1)^2} \sum_{i<j<k} A^2(\mathbf{x}_i, \mathbf{x}_j, \mathbf{x}_k).$$

Analogously with the univariate case, using the sum of areas, instead the sum of squared areas as a measure of scatter, we get the following definition of the bivariate Oja median (Oja, 1983).

DEFINITION 5.1. Let $\mathbf{X} = \{\mathbf{x}_1, \ldots, \mathbf{x}_n\}$ be bivariate points. The bivariate median of Oja is defined as a point in $\Re^2$ which minimizes

$$T(\mathbf{\Theta}, \mathbf{X}) = \sum_{i<j} A(\mathbf{x}_i, \mathbf{x}_j, \mathbf{\Theta})$$

and the Oja theoretical median of bivariate distribution F as a point which minimizes

$$E_F A(\mathbf{x}, \mathbf{y}, \mathbf{\Theta}),$$

where $\mathbf{x}$ and $\mathbf{y}$ are independent random variables with c.d.f. F.

The area of the triangle defined by vectors $\mathbf{x}_1$, $\mathbf{x}_2$, and $\boldsymbol{\Theta}$ is

$$A(\mathbf{x}_1, \mathbf{x}_2, \boldsymbol{\Theta}) = \frac{1}{2!} \operatorname{abs} \begin{vmatrix} 1 & 1 & 1 \\ x_{11} & x_{21} & \theta_1 \\ x_{12} & x_{22} & \theta_2 \end{vmatrix}.$$

The criterion function $T(\boldsymbol{\Theta}, \mathbf{X})$ is differentiable almost everywhere. The gradient vector is

$$\mathbf{Q}(\boldsymbol{\Theta}, \mathbf{X}) = \frac{1}{2!} \sum_{i<j} s_{ij}(\boldsymbol{\Theta}) \begin{pmatrix} x_{i2} - x_{j2} \\ -(x_{i1} - x_{j1}) \end{pmatrix}$$

with

$$s_{ij}(\boldsymbol{\Theta}) = \operatorname{sign} \left(\begin{vmatrix} 1 & 1 & 1 \\ x_{i1} & x_{j1} & \theta_1 \\ x_{i2} & x_{j2} & \theta_2 \end{vmatrix} \right).$$

Let

$$l_i(\boldsymbol{\Theta}) = \sum_{j=1}^{n} I\big(s_{ij}(\boldsymbol{\Theta}) > 0\big), \qquad i = 1, \ldots, n.$$

The line passing through points $\boldsymbol{\Theta}$ and $\mathbf{x}_i$ divides the remaining $n-1$ observations into two parts of sizes (numbers of observations) $l_i(\boldsymbol{\Theta})$ and $n-1-l_i(\boldsymbol{\Theta})$.

Using l_i, the gradient vector can be written in form

$$\mathbf{Q}(\boldsymbol{\Theta}, \mathbf{X}) = \sum_{i=1}^{n} \left(l_i(\boldsymbol{\Theta}) - \frac{n-1}{2} \right) \begin{pmatrix} x_{i2} \\ -x_{i1} \end{pmatrix}.$$

If the number of observations is even and the data set $\mathbf{X}$ comes from a continuous distribution the minimization of objective function $T(\boldsymbol{\Theta}, \mathbf{X})$ gives a unique solution with probability one (Oja and Niinimaa, 1985). For odd n, the gradient is zero if $l_i(\boldsymbol{\Theta}) - (n-1)/2 = 0$ for each i. This is the case when all lines defined by $\boldsymbol{\Theta}$ and $\mathbf{x}_i$ split the remaining $n-1$ sample points into two parts of equal size. Let D be the set of points having this property and suppose that $\mathbf{X}$ is in general position. Using the definition of majority sets from Section 2 it, is easy to see that $\bigcap_{i<j} B_{ij} \subset D$. On the other hand if the majority of the observations and the $\boldsymbol{\Theta}$ do not lie on the same side of the line through points $\mathbf{x}_k$ and $\mathbf{x}_l$, it is not possible to have $l_i(\boldsymbol{\Theta}) = (n-1)/2$ for each $\mathbf{x}_i \in B_{kl}$. Hence $D \subset \bigcap_{i<j} B_{ij}$ and consequently $D = \bigcap_{i<j} B_{ij}$. The probability to get a non-unique solution for odd n decreases rapidly with n (see Section 2). Thus, for odd n, the Oja median is

(i) unique if the set $\mathbf{X} = \{\mathbf{x}_1, \ldots, \mathbf{x}_n\}$ is non-regular,

(ii) non-unique ($\bigcap_{i<j} B_{ij}$) if the set $\mathbf{X}$ is regular.

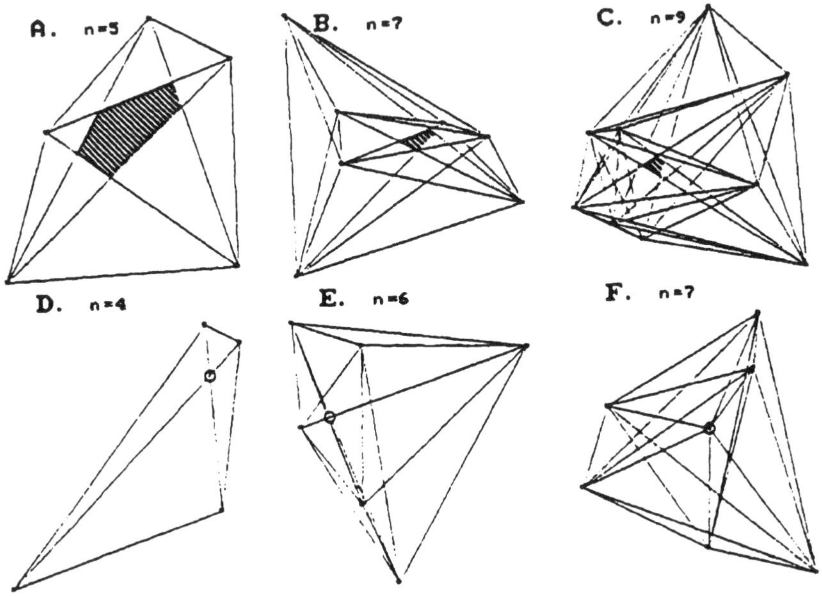

Figure 4. For regular data sets (A, B and C) the minimization of Oja objective function gives a convex set as a solution. The unique solution (o) is obtained (with probability one in the continuous case) if the sample size n is even (D and E), or if the data set is non-regular (F).

Figure 4 gives three data sets (A, B and C) with non-unique Oja's medians and three data sets (D, E and F) with unique Oja's medians.

If $\mathbf{C}$ is a 2×2 nonsingular matrix and $\mathbf{b}$ a 2×1 vector, then

$$A(\mathbf{Cx}_1 + \mathbf{b}, \mathbf{Cx}_2 + \mathbf{b}, \mathbf{Cx}_3 + \mathbf{b}) = \text{abs}\left(|\mathbf{C}|\right) A(\mathbf{x}_1, \mathbf{x}_2, \mathbf{x}_3).$$

The area of the triangle is not affected when the original data is translated or rotated, and the rescaling of the first (or second) components of data vectors $\mathbf{x}_i$ and $\boldsymbol{\Theta}$ by constant c multiplies the area of the triangle by the same constant c. It is then easy to see that the median of Oja is affinely equivariant (Oja and Niinimaa, 1985).

The asymptotic properties of the Oja median are studied in (Oja, 1984; Oja and Niinimaa, 1985; Arcones *et al.*, 1992; Brown *et al.*, 1992; Hettmansperger *et al.*, 1994). If the underlying distribution is multivariate normal the asymptotic efficiency of the Oja multivariate median uniformly outperforms that of the spatial median unless in the circular normal case when the efficiencies are the same (Oja and Niinimaa, 1985). The efficiency of the spatial median and the Oja's median relative to the mean vector increase when the number of the dimensions increases. For example, the spatial median and the Oja median have the same asymptotic variances with circular normal distribution, and their efficiencies increases from 0.637 in the univariate case to 0.905 in 5 dimensions.

Niinimaa and Oja (1995) showed that if F is any circular distribution, i.e., the distribution having the density $f(x_1, x_2) = f({x_1}^2 + {x_2}^2)$, the Oja's median and the spatial median have the same asymptotic variances. In this case the influence functions of the spatial median and the Oja median coincide. When the distribution is spherically symmetric normal the common influence function is

$$\mathbf{IF}\big(r(\cos\phi, \sin\phi)^T\big) = 2\sqrt{\frac{2}{\pi}}\begin{pmatrix}\cos\phi\\ \sin\phi\end{pmatrix}.$$

The bounded influence function for the general elliptic case is also given by Niinimaa and Oja (1995).

The Oja bivariate median breaks down in the presence of just two suitable chosen observations (Niinimaa *et al.*, 1989). We, however, mentioned above that the Oja median has a bounded influence function. This seems to be in a sharp contrast with the 0% breakdown of the Oja median. The explanation is that the breakdown point of the bivariate Oja median depends on the dispersion of the original and corrupted (outlying) observations. If all corrupted observations coincide (dispersion is 0) then the breakdown point is $1/3$. If the dispersion of the corrupted observations is the same as that of the original observations then the breakdown is at least 0.293 (Niinimaa and Oja, 1995).

The companion sign, rank and signed rank tests, and corresponding Hodges–Lehman estimate are studied by Brown and Hettmansperger (1987, 1989) and Hettmansperger *et al.* (1992).

Niinimaa *et al.* (1992) give a Fortran algorithm for the bivariate Oja median. This algorithm can be easily modified for a fast bootstrap estimation. Niinimaa *et al.* (1992) also showed how the general k-variate median can be calculated using L_1 regression algorithms.

6. THE SIMPLICIAL DEPTH MEDIAN

In the univariate case the sample median can also be defined as a point lying in the largest number of intervals constructed from the data pairs (Liu, 1990). The depth function to be maximized is then

$$T(\theta, \mathbf{X}) = \#\big\{\theta \in [x_i, x_j]\big\}.$$

If x_1 and x_2 are two independent observations from univariate continuous distribution F, then the maximization of the criterion function

$$P_F\big(x_{(1)} \leqslant \theta \leqslant x_{(2)}\big) = 2F(\theta)\big(1 - F(\theta)\big)$$

leads to an alternative way to define the usual univariate median. Liu (1990) used triangles instead of line segments and the following definition was then obtained:

DEFINITION 6.1. Let $\mathbf{X} = \{\mathbf{x}_1, \ldots, \mathbf{x}_n\}$ be a bivariate data set. The simplicial depth median is defined as a point $\mathbf{\Theta}$ in $\Re^2$ which maximizes the number of data triangles containing $\mathbf{\Theta}$

$$T(\mathbf{\Theta}, \mathbf{X}) = \sum_{1 \leqslant i < j < k \leqslant n} I\big(\mathbf{\Theta} \in \Delta(\mathbf{x}_i, \mathbf{x}_j, \mathbf{x}_k)\big) = \#\big\{\mathbf{\Theta} \in \Delta(\mathbf{x}_i, \mathbf{x}_j, \mathbf{x}_k)\big\}.$$

The corresponding theoretical median for a c.d.f. F is defined as a point maximizing the probability

$$P_F\big(\mathbf{\Theta} \in \Delta(\mathbf{x}_1, \mathbf{x}_2, \mathbf{x}_3)\big)$$

where $\Delta(\mathbf{a}, \mathbf{b}, \mathbf{c})$ is an open triangle with vertices $\mathbf{a}$, $\mathbf{b}$, and $\mathbf{c}$, and $\mathbf{x}_1, \mathbf{x}_2, \mathbf{x}_3$ a random sample from F.

It is easy to check whether the point $\mathbf{\Theta}$ lies inside the triangle with vertices $\mathbf{x}_1, \mathbf{x}_2$, and $\mathbf{x}_3$. If we choose any two vertices, the vertice not chosen and the point $\mathbf{\Theta}$ must always lie on the same side of the line passing through the two chosen points. Using the notation of Section 5 this means that

$$s_{ij}(\mathbf{\Theta}) = s_{ij}(\mathbf{x}_k), \qquad (i, j, k) \in \big\{(1,2,3), (1,3,2), (2,3,1)\big\}.$$

From n data points one can construct $\binom{n}{3}$ different triangles. $T(\mathbf{\Theta}, \mathbf{X})$ can be computed then by subtracting the number of triangles that do not contain $\mathbf{\Theta}$ from $\binom{n}{3}$, i.e.,

$$\begin{aligned} T(\mathbf{\Theta}, \mathbf{X}) &= \binom{n}{3} - \frac{1}{2}\sum_{i=1}^{n}\Big(\mathop{\#}_{j<k}\big\{s_{ij}(\mathbf{\Theta}) = s_{ik}(\mathbf{\Theta}) = 1\big\} \\ &\quad + \mathop{\#}_{j<k}\big\{s_{ij}(\mathbf{\Theta}) = s_{ik}(\mathbf{\Theta}) = -1\big\}\Big) \\ &= \binom{n}{3} - \frac{1}{2}\sum_{i=1}^{n}\left[\binom{l_i(\mathbf{\Theta})}{2} + \binom{n-1-l_i(\mathbf{\Theta})}{2}\right] \\ &= \frac{n(n-1)(n+1)}{24} - \frac{1}{2}\sum_{i=1}^{n}\left(l_i(\mathbf{\Theta}) - \frac{n-1}{2}\right)^2. \end{aligned}$$

The above formula shows that if the data set $\mathbf{X}$ is regular (Definition 2.1) then the set of points minimizing $T(\mathbf{\Theta}, \mathbf{X})$ is $\bigcap_{i<j} B_{ij}$ and the Liu depth function attains its highest possible value $(n(n-1)(n+1))/24$.

Figure 5 gives a regular data set with 11 observations. If the data set is in the general position, 55 is the highest possible value for Liu's depth function. In this case the median set of Oja, and the closure of the median set defined by Liu coincide.

Liu (1990) used the closed triangles instead of the open ones which favours sample points as median candidates.

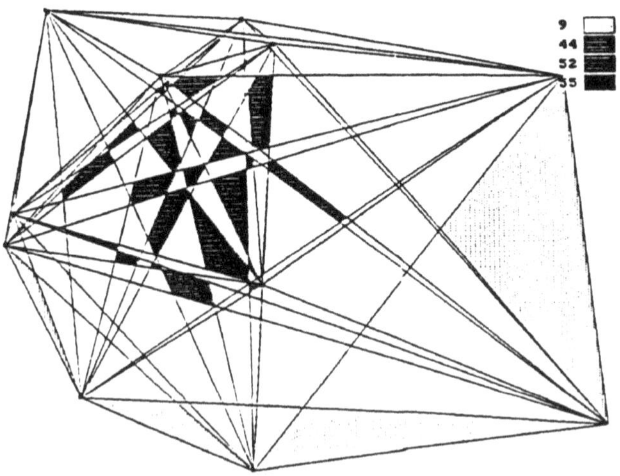

Figure 5. The median region for regular data set is given (the black area). Some values of Liu's depth function $T(\Theta, \mathbf{X})$ are displayed by shading.

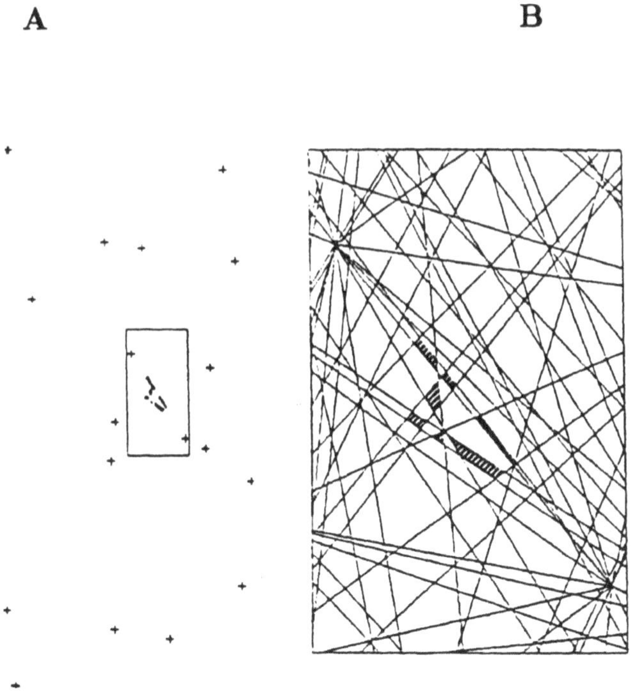

Figure 6. A non-compact Liu median set (shaded area) for 17 observations is given in (A). The rectangle in (A) is enlarged in (B) where the lines connecting the observations are also given.

If the number of observations is large, the solution set may split into several disjoint areas. This is illustrated in Figure 6.

The data depth function $T(\boldsymbol{\Theta}, \mathbf{X})$ can also be used to construct multivariate ranking methods and the corresponding rank tests (see (Liu, 1992)).

The asymptotic properties of Liu's median are studied by Dümbgren (1992). Arcones *et al.* (1992) gave assumptions for proving the asymptotic normality, and also give the formula for the asymptotic covariance matrix of the simplicial depth median. The computations, however, may be quite involved. When the underlying distribution is bivariate normal, the simplicial depth median and the Oja's median have the same asymptotic efficiency.

It is easy to calculate the values of the Liu's depth function $T(\boldsymbol{\Theta}, \mathbf{X})$: let $\mathbf{X} = \{\mathbf{x}_1, \ldots, \mathbf{x}_n\}$ and let L_{ij} be the line passing through the points $\mathbf{x}_i$ and $\mathbf{x}_j$. If one can move from $\boldsymbol{\Theta}_1$ to $\boldsymbol{\Theta}_2$ without crossing any line L_{ij}, then $T(\boldsymbol{\Theta}_1, \mathbf{X}) = T(\boldsymbol{\Theta}_2, \mathbf{X})$. The line L_{ij} can be crossed outside the interval $(\mathbf{x}_i, \mathbf{x}_j)$ without changing the value of the objective function. If one jumps over the line L_{ij} between two points $\mathbf{x}_i$ and $\mathbf{x}_j$ from half space B_{ij} to half space A_{ij}, then the change in the criterion function is $n_2 - n_1$, where n_1 and n_2 are the number of observations in half spaces B_{ij} and A_{ij}, correspondingly. Upon request, the author can provide a C language program utilizing these ideas. Rousseeuw and Ruts (1992) give a Fortran algorithm to calculate the simplicial depth using similar ideas.

If the number of observations is small, the computation is very fast even without a computer, as illustrated in Figure 7.

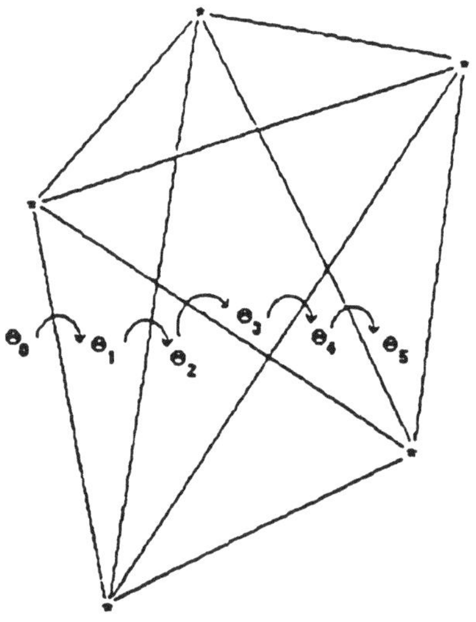

$$T(\Theta_0, X) = 0$$
$$T(\Theta_1, X) = T(\Theta_0, X) + 3 - 0 = 3$$
$$T(\Theta_2, X) = T(\Theta_1, X) + 2 - 1 = 4$$
$$T(\Theta_3, X) = T(\Theta_2, X) + 2 - 1 = 5$$
$$T(\Theta_4, X) = T(\Theta_3, X) + 1 - 2 = 4$$
$$T(\Theta_5, X) = T(\Theta_4, X) + 1 - 2 = 3$$

Figure 7. An example about the calculation of the Liu's simplicial depth function $T(\boldsymbol{\Theta}, \mathbf{X})$. Sample size $n = 5$.

7. OTHER GENERALIZATIONS OF THE MEDIAN

Consider two different characterization of the univariate median (Hotelling, 1929). First, let

$$T(\theta, \mathbf{X}) = \max\left(\#\{i\colon\ x_i \leqslant \theta\}, \#\{i\colon\ x_i \geqslant \theta\}\right).$$

So $T(\theta, \mathbf{X})$ measures how well the θ splits the data $\mathbf{X}$. In the bivariate case, consider then how well the lines going through $\mathbf{\Theta}$ split the data. It is then enough to study lines connecting $\mathbf{\Theta}$, and sample points $\mathbf{x}_i$, $i = 1, \ldots, n$. The sizes of the half spaces are then $l_i(\mathbf{\Theta})$ and $n - 1 - l_i(\mathbf{\Theta})$ (see Section 5). The halfspace median is then defined to be the minimizer of criterion function

$$T(\mathbf{\Theta}, \mathbf{X}) = \max_{1 \leq i \leq n} \left| l_i(\mathbf{\Theta}) - \frac{n-1}{2} \right|.$$

See (Tukey, 1975; Donoho, 1982).

If the bivariate set $\mathbf{X}$ is regular this median, the median of Oja, and the simplicial depth median coincide.

The algorithm of Rousseeuw and Ruts (1992) for simplicial depth calculates also the value of this criterion function.

The univariate sample median can be found successively stripping away the lowest and the highest order statistics. Figure 8 illustrates how the outlying values can be successively deleted in the bivariate case using "convex hull stripping method". This method produces an affinely equivariant median. Donoho and Gasko (1992) showed that the breakdown properties of this median are quite poor. The distribution theory of the convex hull median is difficult (see (Small, 1991)).

The computation of the convex hull median is straightforward. A pair of points $\mathbf{x}_i$ and $\mathbf{x}_j$ are both boundary points if all other points are lying on the same side of the line passing through the points $\mathbf{x}_i$ and $\mathbf{x}_j$. See (Green and Silverman, 1979) for computer programs.

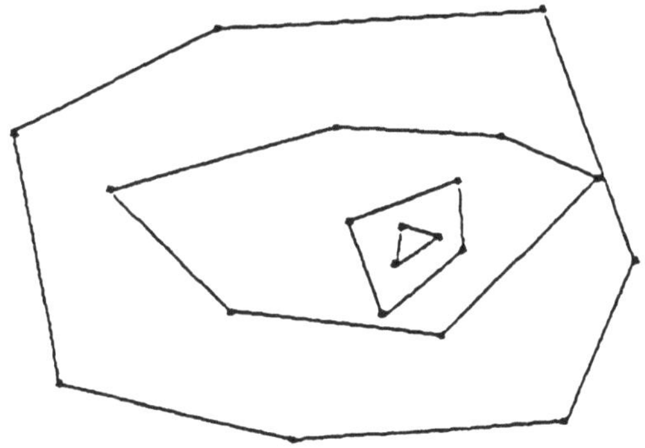

Figure 8. The convex hull stripping method.

There is several other “stripping” or “peeling” methods. Barnett (1976) gives a detailed discussion.

8. DISCUSSION

The vector of marginal medians is a straightforward extension of the univariate median. The concepts of median, sign and rank have usually been linked together with ordering of the data. However, ordering multivariate data is quite problematic (Barnett, 1976). Therefore, it is easy to understand why the classical textbooks by Puri and Sen (1970), Maritz (1981), and Hettmansperger (1984) consider extensions which are just simple combinations of the corresponding univariate statistics. Until recently, multivariate sign and rank tests considered were essentially univariate in nature, and therefore coordinate-dependent. Brown (1983) was the first to show that a genuinely multivariate method (namely the spatial median) may be more efficient than corresponding combined estimate (the vector of marginal medians).

There are still problems hindering the widespread use of the methods presented here. The component-wise median is the only median one can calculate in a reasonably short time for high dimensions and large number of observations. The finite sample distributions of the multivariate medians are not known, good estimation methods for the asymptotic covariance matrices are lacking and further research is therefore needed.

In this paper several different methods to generalize the univariate median are considered. It is hard to make any recommendations. Low efficiency and lack of rotation equivariance make the vector of marginal medians in most cases inferior to other multivariate medians. In many cases the optimization problem determines the median to be used. If a company wishes to find the optimum location for a storage depot to serve n plants in the “Manhattan area”, the median minimizing the sum of componentwise absolute distances is of course the right solution. If the depot-plant distances are measured using the standard L_2-norm, the optimization leads to the spatial median.

Acknowledgement

I have received valuable assistance and guidance for this work from Hannu Oja.

REFERENCES

Arcones, M. A., Chen, Z. and Gine, E. (1992). Estimators related to U-processes with applications to multivariate medians: Asymptotic normality. Technical Report, University of Connecticut.

Babu, G. J. and Rao, C. R. (1988). Joint asymptotic distribution of marginal quantile function in samples from a multivariate population. *J. of Multivariate Anal.* **27**, 15–23.

Barnett, V. (1976). The ordering of multivariate data. *J. Roy. Statist. Soc., Ser. A* **139**, 318–355.

Bose, A. and Chaudhuri, P. (1991). The dispersion of the multivariate median. Technical Report No. 23. Indian Statistical Institute, Calcutta.

Brown, B. M. (1982). Spatial median. In: *Encyclopaedia of Statistical Sciences* **8**, pp. 574–575, N. Johnson and S. Kotz (Eds), Wiley, New York.

Brown, B. M. (1983). Statistical use of the spatial median. *J. Roy. Statist. Soc., Ser. B* **45**, 25–30.

Brown, B. M. and Hettmansperger, T. P. (1987). Affine invariant rank methods in the bivariate location model. *J. Roy. Statist. Soc., Ser. B* **49**, 301–310.

Brown, B. M. and Hettmansperger, T. P. (1987). Invariant tests in bivariate models and the L_1 criterion. 333–344, North Holland, Amsterdam, New York.

Brown, B. M. and Hettmansperger, T. P. (1989). The affine invariant bivariate version of the sign test. *J. Roy. Statist. Soc., Ser. B* **51**, 117–126.

Brown, B. M., Hettmansperger, T. P., Nyblom, J. and Oja, H. (1992). On certain bivariate sign tests and medians. *J. Amer. Statist. Assoc.* **87**, 127–135.

Chaudhuri, P. (1992). Multivariate location estimation using extension of R-estimates through U-statistics type approach. *Ann. Statist.* **20**, 897–916.

David, H. A. (1981). *Order Statistics*. Wiley, New York.

Dietz, E. J. (1982). Bivariate nonparametric tests for the one-sample location problem. *J. Amer. Statist. Assoc.* **77**, 163–169.

Donoho, D. L. (1982). Breakdown properties of multivariate location estimators. Ph.D. Qualifying paper. Harvard University.

Donoho, D. L. and Gasko, M. (1992). The breakdown properties of location estimates based on halfspace depth and projected outlyingness. *Ann. Statist.* **20**, 1803–1827.

Dümbgren, L. (1992). Limit theorems for simplicial depth. *Statist. Probab. Lett.* **14**, 119–128.

Farebrother, R. W. (1992). The geometrical foundations of a class of estimation procedures which minimize sums of Euclidean distances and related quantities. In: *L_1-Statistical Analysis and Related Methods*, pp. 337–349, Y. Dodge (Ed.), North-Holland, Amsterdam.

Green, P. J. and Silverman, B. W. (1979). A comparative study of algorithms for convex hull peeling of bivariate sets. *Comp. J.* **22**, 262–266.

Hampel, F. R. (1971). A general qualitative definition of robustness. *Ann. Math. Statist.* **42**, 1887–1896.

Hampel, F. R. (1974). The influence curve and its role in robust estimation. *J. Amer. Statist. Assoc.* **69**, 375–391.

Hampel, F. R., Rousseeuw, P. J., Ronchetti, E. M. and Stahel, W. A. (1986). *Robust Statistics. The Approach Based on Influence Functions*. Wiley, New York.

Hettmansperger, T. P. (1984). *Statistical Inference Based on Ranks*. Wiley, New York.

Hettmansperger, T. P., Nyblom, J. and Oja, H. (1992). On multivariate notions of sign and rank. In: *L_1-Statistical Analysis and Related Methods*, pp. 267–278, Y. Dodge (Ed.), North-Holland, Amsterdam.

Hettmansperger, T.P., Nyblom, J. and Oja, H. (1994). Affine invariant multivariate one-sample sign tests. *J. Roy. Statist. Soc., Ser. B*, to appear.

Hettmansperger, T. P. and Oja, H. (1994). Affine invariant multivariate multisample sign tests. *J. Roy. Statist. Soc., Ser. B*, to appear.

Hodges, J. L. (1955). A bivariate sign test. *Ann. Math. Statist.* **39**, 523–527.

Hotelling, H. (1929). Stability in competition. *Econom. J.* **39**, 41–47.

Jeyaratnam, S. (1992). A robust affine-equivariant median. *Statist. Probab. Lett.* **11**, 167–168.

Lehmann, E. L. (1975). *Nonparametrics, Statistical Methods Based on Ranks*. McGraw-Hill, New York.

Liu, R. Y. (1990). On a notion of data depth based upon random simplices. *Ann. Statist.* **18**, 405–414.

Liu, R. Y. and Singh, K. (1991). A quality index based on data depth and multivariate rank tests. *J. Amer. Statist. Assoc.*

Liu, R. Y. (1992). Data depth and multivariate rank tests. In: *L_1-Statistical Analysis and Related Methods*, pp. 279–302, Y. Dodge (Ed.), North-Holland, Amsterdam.

Lopuhaä, H. P. and Rousseeuw, P. J. (1991). Breakdown properties of affine equivariant estimators of multivariate location and covariance matrices. *Ann. Statist.* **19**, 229–248.

Maritz, J. S. (1981). *Distribution-Free Statistical Methods*. Chapman and Hall, London.

Maritz, J. S. (1991). Estimating the covariance matrix of bivariate medians. *Statist. Probab. Lett.* **4**, 305–309.

Milasevic, P. and Ducharme, G. R. (1987). Uniqueness of the spatial median. *Ann. Statist.* **15**, 1332–1333.

Niinimaa, A., Oja, H. and Tableman, M. (1990). The finite-sample breakdown point of the Oja bivariate median. *Statist. Probab. Lett.* **10**, 325–328.

Niinimaa, A. and Oja, H. (1995). On the influence functions of certain bivariate medians. *J. Roy. Statist. Soc., Ser. B* **57**, to appear.

Niinimaa, A., Oja, H. and Nyblom, J. (1992). Algorithm AS 277: The Oja bivariate median. *Appl. Statist.* **41**, 611–617.

Oja, H. (1983). Descriptive statistics for multivariate distributions. *Statist. Probab. Lett.* **1**, 327–332.

Oja, H. (1984). Asymptotical properties of estimators based on U-statistics. *Mathematics*. University of Oulu.

Oja, H. and Niinimaa, A. (1985). Asymptotic properties of the generalized median in the case of multivariate normality. *J. Roy. Statist. Soc., Ser. B* **47**, 372–377.

Oja, H. and Nyblom, J. (1989). On bivariate sign tests. *J. Amer. Statist. Assoc.* **84**, 249–259.

Puri, M. L. and Sen, P. K. (1971). *Nonparametric Methods in Multivariate Analysis*. Wiley, New York.

Rao, C. R. (1988). Methodology based on L_1-norm in statistical inference. *Sankhyā, A* **50**, 289–313.

Rousseeuw, P. J. and Leroy, A. M. (1987). *Robust Regression and Outlier Detection*. Wiley, New York.

Rousseeuw, P. J. and Ruts, I. (1992). *Bivariate Simplicial Depth*. Technical Report, Rutgers, University of Antwerpen.

Small, G. (1990). A survey of multidimensional medians. *Internat. Statist. Rev.* **58**, 263–277.

Singh, K. (1991). A notion of majority depth. Technical Report. Rutgers University.

Tukey, J. W. (1975). Mathematics and the picturing of data. In: *Proc. International Congress of Mathematics, Vancover, 1974*, Vol. 2, pp. 523–531.

Wilks, S. S. (1960). Multidimensional statistical scatter. In: *Contributions to Probability and Statistics in Honour of Harold Hotelling*, pp. 486–503, I. Olkin *et al.* (Eds), Stanford University Press.

Multivar. Statist., pp. 181–195
E.–M. Tiit *et al.* (Eds)

DISTANCE-BASED REGRESSION IN SOME HELIOPHYSICAL DATA ANALYSIS

ANNA BARTKOWIAK
Institute of Computer Science, University of Wrocław,
Przesmyckiego 20, 51-151 Wrocław, Poland

ABSTRACT

The paper is an extension of the work of Bartkowiak and Jakimiec (1994, pp. 7–38) and Jakimiec and Bartkowiak (1994, pp. 115–140). They have investigated the functioning of the distance-based regression method proposed by Cuadras and Arenas, (1990, pp. 2261–2279). It appeared that the algorithm used by Bartkowiak and Jakimiec was tested externally on data that had slightly different interdependence structure. Now a kind of sliding procedure is performed — permitting to estimate the activity for small portions of subsequent items.

The main results are the following: the applied distance-based regression — when using Gower or L_1-norm distances — describes quite accurately the dependence of the predicted variable from the explanatory variables when doing this in the training sets, however the predictions made for the small test sets are poor.

The distance-based method builds a regression model from a number of principal coordinates obtained by a method of multidimensional scaling. It is not known how many principal coordinates should be included in the regression model. Two empirical rules indicating significant principal coordinates are compared: the CP coefficient of predictability, and the 'broken stick' rule. The two rules yield similar, although not identical results.

1. INTRODUCTION

We are concerned with predicting the values of a variable Y from p explanatory variables $X_1, \ldots, X_p$. The most standard method to do this is by using a regression of the form

$$Y = b_0 + b_1 X_1 + \cdots + b_p X_p + e$$

with coefficients $b_0, \ldots, b_p$ estimated by the Least Square Error (LSE) method. An interesting alternative — the Distance-Based Regression (DBR) — was introduced by Cuadras and Arenas (1990). Their method works as follows: firstly, a distance matrix is evaluated from the explanatory variables. Next, by methods of MDS (multidimensional scaling, see, e.g. (Mardia *et al.*, 1979)), some principal coordinates denoted in the following by $\tilde{\Gamma}_1, \ldots, \tilde{\Gamma}_m$ are constructed. Finally, an ordinary LSE regression of Y from the constructed principal coordinates is established.

The type of the used distances is a crucial point in DBR. The generality of DBR is established by the fact that the distances can be evaluated in a number of ways. Euclidean (L_2-norm), Absolute value (L_1-norm) and Gower distances are used quite often (for definitions or properties of the mentioned distances see (Gower, 1971, 1985; Cuadras and Arenas, 1990)). It was proven by Cuadras and Arenas, that DBR based on Euclidean distances with p principal coordinates is equivalent to full LSE regression carried out with original data.

Let us introduce the following denotation:

DBR-EUC: distance-based regression constructed from Euclidean distances;

DBR-ABS: distance-based regression constructed from Absolute value distances;

DBR-GOW: distance-based regression constructed from Gower distances.

Bartkowiak and Jakimiec (1994) have used these three distances for predicting solar flare activity. Analysing the results obtained for some real heliophysical data they stated that:

1. The DBR-ABS and the DBR-GOW methods carried out with standardized data gave similar results for the considered data.
2. The DBR-ABS and the DBR-GOW methods when using p principal components and compared to the DBR-EUC method gave higher values of $R^2(\mathbf{y}; \tilde{\Gamma}_1, \ldots, \tilde{\Gamma}_p)$ (denoting the squared multiple correlation coefficient between $\mathbf{y}$, the vector of the gathered values of the variable Y, and the constructed principal coordinates $\tilde{\Gamma}_1, \ldots, \tilde{\Gamma}_p$).
3. The DBR-ABS and the DBR-GOW methods yielded up to $n-1$ PC's which give jointly a multiple correlation coefficient $\approx$ 1.0.

However the 3-rd statement was only true when the algorithm was verified by making predictions on the same data set from which the principal coordinates were constructed. The validation of the DBR-ABS and DBR-GOW methods carried out in external (foreign) data sets with $m > p$ gave worse results as the DBR-EUC method.

Considering the 3-rd statement in more detail it was observed that — when using the same number of PC's — the DBR-ABS and the DBR-GOW methods provided a more accurate prediction of Y then the DBR-EUC method. This was only true when predicting Y in the same data set from which the PC's were constructed. The increasing of the value of R^2 with increasing number of principal coordinates was due to overfitting of the model. When making predictions in a foreign data set (solar flare activity data from the next year), the predictions were systematically biased, and the prediction error for the DBR-ABS and DBR-GOW methods was not smaller than the analogous error obtained from the DBR-EUC method. The prediction quality has been generally deteriorating when taking more than four PC's for the prediction.

The bad quality of the prediction for the heliophysical data — when making predictions in a foreign data set — could be caused by a change of the interrelation structure of the variables characterizing the analysed sunspot groups. Jakimiec and Bartkowiak (1994) considered additionally two more explanatory variables and stated that really, there was some change in the interrelation structure between some variables.

The aim of this paper is to consider once more the quality of predictions made for foreign data sets. We are using the same data, however the analysis is done in a different way. We will be concerned with constructing predictions from smaller data sets and for shorter periods. In Section 2 we describe the elaborated data in more detail. In Section 3 we consider the problem of reduction of the number of PC's taken for prediction; in particular we show two methods of choosing the relevant PC's. In Section 4 we present the results of the evaluations and crossvalidated predictions. In Section 5 we discuss generally the method and the results and give further references.

2. A GENERAL CHARACTERIZATION OF THE DATA AND FORMER RESULTS

The data considered in the following is in principle the same as described in the paper by Bartkowiak and Jakimiec (1994) and Jakimiec and Bartkowiak (1994). Their papers will be hereafter referred to as $B\&J$. We will consider only the first group of these data, i.e., the data items for the year 1988. The data were extracted by M. Jakimiec from the annual publication SGD (1988). The heliophysical problem is, shortly speaking, that having some characteristics of active region on sun disk the given day — the forecasters should make a prediction of the active region activity for the next day. The variables observed the given day describe generally the active region (McIntosh class, sunspot area, magnetic class, and characteristics of the flaring). The predicted variables are mainly describing the intensity of the flaring the next day. The data to be analysed now comprise $n = 130$ items (data vectors) containing values of $p = 8$ continuous predictor variables $X_1, \ldots, X_8$ that are exactly the same as in $B\&J$, and two predicted variables $Y_1 = Fs$, $Y_2 = Fh$, which denote the amount of the 'soft' (Fs) and 'hard' (Fh) X-ray flux the next day. $B\&J$ have considered $Y = MvXY$ and $Y = Fs$ as predicted variables ($MvXY$ denotes maximum value of the intensity of 'soft' component of flaring — with a very irregular distribution). Thus one of the predicted variables considered here is exactly the same as in $B\&J$, while the other is different.

Some features of the considered data are presented in Figures 1–5. In Figure 1 we present histograms for both predicted variables. One can see that the distribution of these variables is not normal, so the basic assumption underlying the classical least squares error (LSE) regression is not met. It is advised in such cases to bring the data nearer to normality by applying to them some

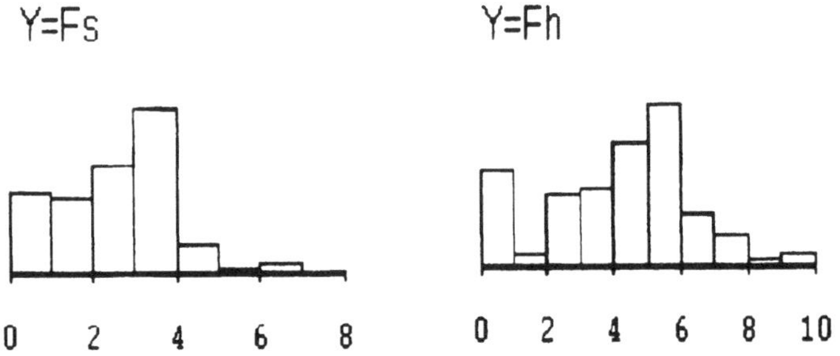

Figure 1. Histograms of the variables $y = Fs$ and $Y = Fh$.

transformations, e.g., the Box–Cox power transformation. We did it — with the results described below.

First of all we have calculated the ordinary LSE regression with untransformed variables. We have used for this purpose the module *make-reg* developed by Cook and Weisberg (1989, 1991) working in the environment of *XLispStat* (Tierney, 1990). The module *make-reg* can choose a 'good' parameter λ for the Box–Cox (power) transformation.

Suppose we deal with the values $x_1, x_2, \ldots, x_n$. The power transformation with power = λ was carried out as follows: firstly we evaluated some auxiliary variables $a_1, \ldots, a_n$ as

$$a_i = \begin{cases} \log(x_i), & |\lambda| < 0.001, \\ (x_i^\lambda)\ /\lambda, & |\lambda| \geqslant 0.001. \end{cases}$$

Let min and max denote the minimum and maximum of the computed auxiliary values $a_1, a_2, \ldots, a_n$.

The transformed values y_i, $i = 1, \ldots, n$, were then defined as

$$y_i = \frac{a_i - \min}{\max - \min}.$$

It came out that the values $\lambda_1 = 0.30$ and $\lambda_2 = 0.40$ were the 'best' for the considered variables Y_1 and Y_2 ('best' means that the Studentized residuals are the closest to the normal distribution — when judged from the normal probability plot).

So we have evaluated the regression with each Y-variable twice: first for the untransformed and then for the transformed predictor variable.

It came out that the applied transformation has only slightly affected R^2, the squared multiple correlation coefficient, which was $R^2 = 0.4846$ for $Y_1 = Fs$ and $R^2 = 0.4804$ for $Y_2 = Fh$ when considering the original, untransformed data. These coefficients have changed less then 0.02 when considering the transformed data.

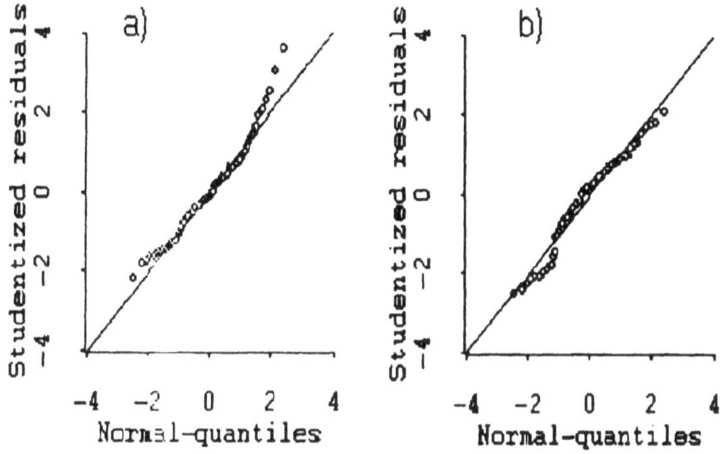

Figure 2. QQ-plots of studentized residuals from untransformed a) and transformed b) variable $y = Fs$.

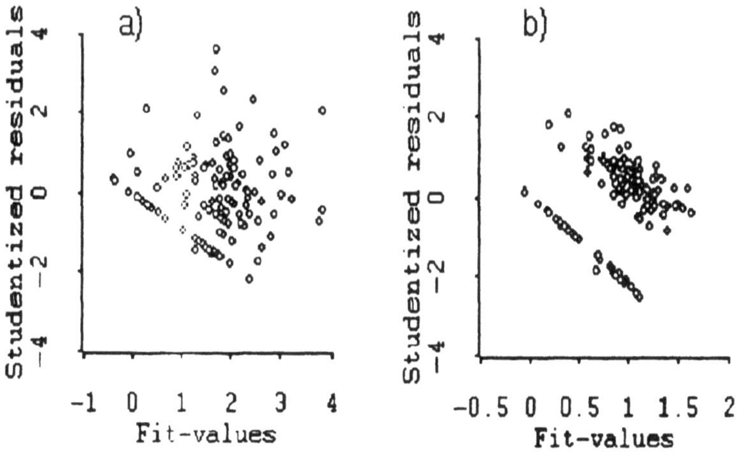

Figure 3. Studentized residuals vs fitted values obtained for untransformed a) and transformed b) variable $y = Fs$.

In Figure 2 we show QQ-plots of internally Studentized residuals obtained from the untransformed (plot a)) and the transformed (plot b)) variable $Y_1 = Fs$.

In Figure 3 we show the same residuals plotted against the fitted values of the variable Y_1. Again, the plot a) exhibits the scatterplot from original data, and plot b) shows the correspondent scatterplot obtained from the regression with the transformed variable Y_1.

Analogous plots constructed for the variable $Y_2 = Fh$ are shown in Figure 4 (QQ-plot of studentized residuals) and Figure 5 (studentized residuals plotted against fitted values). All these plots were obtained from figures drawn initially by the module *make-reg* and next transferred to the *paintbrush* in MS Windows 3.1, wherefrom the PCX file of the figure and also the output was obtained.

Looking at these plots one might be doubtful whether the applied Box–Cox transformation is of any help. The residuals — as plotted against the fitted values — show a very singular pattern. Clearly, the ordinary LSE model is not ideal for our data; hence one should try another model. One alternative might be to carry out the distance-based regression, a non-parametric method that does not need clearly set assumptions on the distribution of the considered variables.

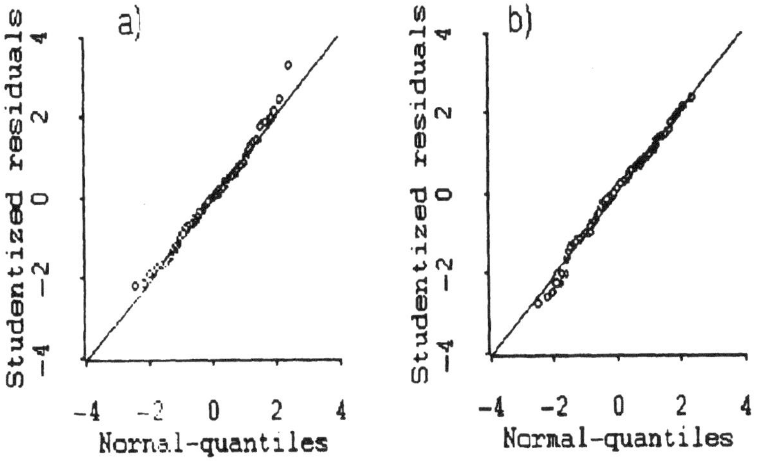

Figure 4. QQ-plots of studentized residuals from untransformed a) and transformed b) variable $y = Fh$.

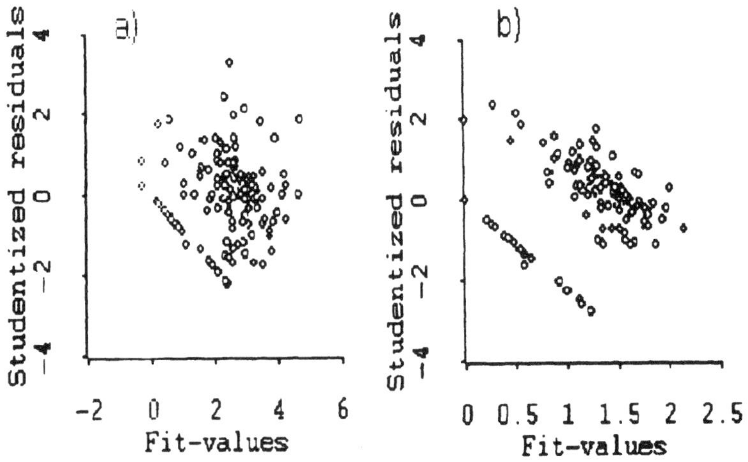

Figure 5. Studentized residuals vs fitted values obtained for untransformed a) and transformed b) variable $y = Fh$.

Therefore the next step was to consider the DB regression. The results of the analysis are reported in detail in the mentioned papers of *B&J*. Their results are that the prediction algorithm evaluated on the basis of data gathered in 1988 ($n = 130$) is not very suitable for making predictions for 1989 ($n = 117$) — because the process of flaring was not stationary that time.

Let us return still for a while to the plots shown in Figure 2 and Figure 4 exhibiting the effects of the Box–Cox transformation. It was observed by an anonymous referee, that looking at these figures one gets the impression that the analysed data sets consists of two quite different subsets. As a result of the transformation the difference between these subsets became more clear.

We have checked which sunspot items are contained in the lower area of the QQ-plots appearing as separated sets. These are items with exactly the same *id* numbers: $91, 106, 13, 121, 42, 55, 102$ for the Figure 2b) and Figure 4b). Also the uppermost Studentized residual is obtained in both figures for the same *id* number: 95. Probably these items have in the variables $X_1, \ldots, X_8$ an interdependency structure different from other items. We do not pursue this topic further here.

In this paper we want to make an evaluation (cross-validation) of the DBR algorithm for nearer in time items for which we hope that the stationarity of the covariance structure might still be preserved. This will be done in Section 4.

3. HOW MANY PRINCIPAL COORDINATES ARE RELEVANT

The distance-based regression gives prediction of Y on the base of the regression equation

$$Y = \beta_0 + \beta_1 \widetilde{\Gamma}_1 + \cdots + \beta_p \widetilde{\Gamma}_m + e. \tag{1}$$

For various reasons only some of the principal coordinates appearing in right side of equation (1) are relevant for prediction of Y. Therefore the predicting equation is usually based on a smaller number k $(< n - 1)$ of principal coordinates:

$$Y = \beta_0 + \beta_1 \widetilde{\Gamma}_{i_1} + \cdots + \beta_k \widetilde{\Gamma}_{i_k} + e_{(k)}. \tag{2}$$

The subscript (k) in the error term $e_{(k)}$ above is meant to indicate that there are k predictors in the equation.

Usually we introduce into the predicting equation those principal coordinates that yield the highest correlation coefficient with the predicted variable Y. In particular the principal coordinate denoted in equation (2) as $\widetilde{\Gamma}_{i_1}$ is that one which shows the highest squared correlation coefficient r^2 with all the constructed principal coordinates $\widetilde{\Gamma}_1, \ldots, \widetilde{\Gamma}_m$:

$$i_1\colon \quad \operatorname{corr}(Y; \widetilde{\Gamma}_{i_1}) = \max_{j=1,\ldots,m} \operatorname{corr}(Y; \widetilde{\Gamma}_j).$$

The problem — how many principal coordinates from the set $\tilde{\Gamma}_1, \ldots, \tilde{\Gamma}_m$ are relevant for the considered data — remains still open. We investigate this problem by making some heuristic evaluations using the CP coefficient and the 'broken stick' rule.

3.1. *The CP coefficient of predictability*

Let **B** denote the positive semi-definite $n \times n$ inner product matrix derived from the distance matrix **D** computed from the training data (formulae for computing **B** are given in (Cuadras and Arenas, 1990; Cuadras *et al.*, 1993; Bartkowiak and Jakimiec, 1994)).

The matrix **B** can be decomposed (spectral decomposition) as

$$\mathbf{B} = \sum_{j=1}^{m} \lambda_j \Gamma_j \Gamma_j^T, \qquad m = \operatorname{rank}(\mathbf{B})$$

with $\lambda_1, \ldots, \lambda_m$ and $\Gamma_1, \ldots, \Gamma_m$ being the eigenvalues and eigenvectors of **B**.

Let $\mathbf{B}_{(i)}$ denote the sum of the first i spectral terms

$$\mathbf{B}_{(i)} = \sum_{j=1}^{i} \lambda_j \Gamma_j \Gamma_j^T.$$

The i-th principal coordinate is defined as $\tilde{\Gamma}_j = \lambda^{1/2} \Gamma_j$.

Cuadras *et al.* (1993) have introduced the CP coefficient of predictability as the following quadratic form:

$$CP(i) = \frac{\mathbf{y}^T \mathbf{B}_{(i)} \mathbf{y}}{\mathbf{y}^T \mathbf{B} \mathbf{y}}, \qquad i = 1, \ldots, m, \tag{3}$$

where $\mathbf{y}$ denotes the $n \times 1$ vector of observed values of the Y variable.

Cuadras *et al.* (1993) have shown that the $CP(i)$ coefficient can also be computed by the formula

$$CP(i) = \frac{\sum_{j=1}^{i} r_j^2 \lambda_j}{\sum_{j=1}^{m} r_j^2 \lambda_j}, \tag{4}$$

where $r_j = \operatorname{corr}(\mathbf{y}; \tilde{\Gamma}_j)$ and $\lambda_1, \ldots, \lambda_m$ are the eigenvalues of **B**.

Obviously $0 < CP(i) \leqslant 1$, $CP(m) = 1$ and $CP(i) \leqslant CP(i+1) \quad \forall i \in \{1, \ldots, m-1\}$.

The quantity $1 - CP(i)$ shows the amount of the denominator in (3) that is not explained yet.

In Table 1 we show the values $1 - CP(i)$ for $i = 1, \ldots, 14$ computed for the whole group of the considered heliophysical data containing $n = 130$ data

Table 1.
Values of $1 - CP(i)$ indicating for unexplained part of $\mathbf{y}^T\mathbf{B}\mathbf{y}$

Predicted: $Y = Fs$				Predicted: $Y = Fh$			
i	EUC	ABS	GOW	i	EUC	ABS	GOW
1	0.0122	0.0269	0.0301	1	0.0138	0.0296	0.0328
2	0.0114	0.0260	0.0253	2	0.0124	0.0288	0.0282
3	0.0104	0.0260	0.0252	3	0.0118	0.0288	0.0279
4	0.0041	0.0248	0.0240	4	0.0069	0.0260	0.0252
5	0.0001	0.0248	0.0237	5	0.0003	0.0260	0.0251
6	0.0001	0.0103	0.0104	6	0.0003	0.0128	0.0126
7	0.0000	0.0098	0.0097	7	0.0000	0.0122	0.0120
8	0.0000	0.0098	0.0097	8	0.0000	0.0122	0.0119
9		0.0073	0.0078	9		0.0070	0.0073
10		0.0073	0.0075	10		0.0070	0.0071
11		0.0072	0.0068	11		0.0068	0.0062
12		0.0055	0.0057	12		0.0054	0.0054
13		0.0055	0.0057	13		0.0053	0.0054
14		0.0054	0.0056	14		0.0051	0.0052
15		0.0053	0.0055	15		0.0050	0.0051

vectors. The principal components that reduce the value $1 - CP(i)$ the most, are underlined.

It is striking that most of the predictive power (about 97% or even more) contains in the first principal coordinate. When considering ABS and GOW distances, then also the 6-th principal coordinate seems to have some, though very small, predictive power (it reduces the remaining $1-CP(i)$ by about 1%).

When comparing distances — it appears that the first principal component obtained when using the EUC distance is a little more predictive as the first principal coordinate obtained when using the other two distances.

3.2. *The 'broken stick' rule*

This 'broken stick' rule is very appealing by its simplicity. It relays on some probability concepts which are also simple. We suppose that we have a stick of unit length, which is broken, at random, into p segments. Then it can be shown that the expected length l_i of the i-th longest segment is (Jolliffe 1986, p. 95):

$$l_i = \frac{1}{p} \sum_{j=i}^{p} \frac{1}{j}. \tag{5}$$

The standard deviation of the estimated expected length can be obtained by simulation.

Now suppose (H_0) that the predicted variable Y is uncorrelated with the predictors (principal coordinates used in the regression equation), and all the correlations are spurious. However, when using L_1-norm or Gower distances we obtain a complete description of Y by the constructed principal coordinates (see statement 3 in the introduction), and the sum of the squared correlations equals 1.0:

$$\sum_{j=1}^{p} r_j^2 = 1 \qquad \text{with} \qquad r_j^2 = r^2(\tilde{\Gamma}_j, Y).$$

If none of the principal components is correlated with Y, but — due to overfitting — the sum of the squared correlations with Y equals 1, then the the subdivision of the unit sum into components r_j^2 is purely random — as in the case when breaking the stick of unit length into p parts. If the length of some parts of the broken stick (or some squared correlations r_j^2) do not correspond with the distribution of the segments obtained under H_0 then we are justified to say that the stick was not broken at random (the correlations are not spurious, i.e., are not obtained purely at random).

We have applied this rule to our data when considering the squared correlations r_i^2 $(i = 1, \ldots, m)$ between the vector $\mathbf{y}$ and the constructed principal coordinates $\tilde{\Gamma}_1, \ldots, \tilde{\Gamma}_m$. When considering the Euclidean distances (EUC) we got $m = 8$ PC's, when considering Absolute value distances (ABS) and Gower distances (GOW) we got $m = 129$ PC's. When considering Euclidean distances, the length of the stick was adjusted to the total sum of all the squared correlation coefficients, i.e., to the value 0.4846 for $Y = Fs$ and 0.4804 for $Y = Fh$.

In Table 2 and 3 we show the highest squared correlation coefficients $r_{(i)}^2$ with the lower and upper confidence intervals (LOW, UPP) obtained working with the EUC distance (Table 2) and the ABS and GOW distances (Table 3). The bounds were obtained by simulation of $r = 2000$ randomly (with uniform distribution) broken sticks. From these simulations the respective standard deviations of the i-th segment were calculated; the bounds were then obtained as the average length of the i-th segment $\pm$ one standard deviation. Values

Table 2.
Values of $r_{(i)}^2$, squared correlation coefficients shown in descending order, and limits (LOW, UPP) from 'broken stick' rule when using Euclidean (EUC) distances; h denotes the initial id number of the principal coordinate yielding the $r_{(i)}^2$

Predicted: $Y = Fs$, EUC					Predicted: $Y = Fh$, EUC				
(i)	$r_{(i)}^2$	h	LOW	UPP	(i)	$r_{(i)}^2$	h	LOW	UPP
1	*0.4180	1	0.1207	0.2084	1	*0.3974	1	0.1195	0.2067
2	0.0325	5	0.0830	0.1265	2	0.0517	5	0.0802	0.1259
3	0.0257	4	0.0570	0.0914	3	0.0192	4	0.0565	0.0908
4	0.0030	3	0.0388	0.0686	4	0.0079	7	0.0384	0.0680
5	0.0030	7	0.0251	0.0518	5	0.0018	2	0.0249	0.0510
6	0.0012	8	0.0145	0.0371	6	0.0016	3	0.0150	0.0372
7	0.0011	2	0.0065	0.0252	7	0.0009	8	0.0068	0.0253
8	0.0001	6	0.0009	0.0136	8	0.0000	6	0.0008	0.0142

Table 3.
Values of $r^2_{(i)}$, squared correlation coefficients shown in descending order, and confidence limits (LOW, UPP) from 'broken stick' rule when using ABS and GOW distances; h denotes the initial *id* number of the principal coordinate yielding the $r^2_{(i)}$

	ABS distances				GOW distances				Conf. limits	
(i)	$Y = Fs$	h	$Y = Fh$	h	$Y = Fs$	h	$Y = Fh$	h	LOW	UPP
1	*0.4159	1	*0.3992	1	*0.4130	1	*0.3967	1	0.0334	0.0511
2	*0.0539	6	*0.0473	6	*0.0516	6	*0.0466	6	0.0292	0.0399
3	0.0240	80	*0.0379	9	0.0238	22	*0.0346	9	0.0268	0.0345
4	0.0220	43	*0.0342	80	0.0224	46	0.0218	57	0.0250	0.0312
5	0.0218	62	0.0194	21	0.0206	84	0.0212	46	0.0235	0.0288
6	0.0202	12	0.0182	43	0.0200	61	0.0203	80	0.0222	0.0269
7	0.0190	9	0.0181	62	0.0186	57	0.0200	81	0.0212	0.0255
8	0.0183	51	0.0165	111	0.0184	36	0.0199	84	0.0203	0.0242
9	0.0180	36	0.0165	12	0.0180	81	0.0180	22	0.0194	0.0230
10	0.0176	24	0.0158	83	0.0146	9	0.0168	61	0.0187	0.0220
11	0.0167	84	0.0151	36	0.0145	124	0.0163	124	0.0180	0.0211
12	0.0167	21	0.0150	51	0.0138	111	0.0151	111	0.0174	0.0203
13	0.0158	29	0.0149	57	0.0133	12	0.0145	36	0.0168	0.0195
14	0.0145	111	0.0143	84	0.0132	80	0.0134	100	0.0163	0.0188
15	0.0143	46	0.0137	125	0.0130	29	0.0128	77	0.0158	0.0182

which exceed the upper confidence bound are supposed to be significant and are marked in Table 2 and Table 3 by an asterisk '*'.

One can see again, that r_1^2 representing the linear Pearsonian correlation coefficient between the first principal coordinate and the Y variable, is dominating for all distances. When considering the EUC distance, this is the only one significant principal coordinate. When considering ABS or GOW distances, we got some more principal coordinates indicated as significant: for both Y variables and for both distances the 6-th principal coordinate yields a correlation coefficient equal ≈ 0.05.

Then, for $Y = Fh$ only, the 9-th principal coordinate yields $r^2_{(3)} = 0.0379$ when using the ABS distance, and $r_{(3)} = 0.346$ when using the GOW distance.

Somehow also the 80-th principal coordinate exhibiting the value $r^2_{(4)} = 0.0342$ — being above the established limits [0.0250, 0.0312] — is indicated as significant for the variable $Y = Fh$.

3.3. *A comparison of the results yielded by the CP coefficient and the 'broken stick' rule*

The primary results indicating for the 1-st and the 6-th principal coordinates as significant are yielded by both methods. The 'broken stick' method indicates for some additional weekly significant principal coordinates exhibiting rather small correlation coefficients of magnitude 0.0379–0.0342. So one is justified to say that both methods — as carried out for the analyzed heliophysical data — indicate for the same principal coordinates as the most significant ones.

From computational point of view the CP method is much easier to carry out: we can proceed step-wise, computing the subsequent principal coordinates step-by-step and checking for their importance. When using the broken stick method we need to compute all the principal coordinates — what might be cumbersome when the distance matrix **D** and the inner product matrix **B** are large.

4. QUALITY OF THE PREDICTIONS FROM THE SLIDING ALGORITHM

Now we will compute the DB regression again for the considered heliophysical data, $n = 130$ items, gathered in 1988; the analysis will be carried out in a special layout, though. The former experience of $B\&J$ was that the DB regression performed exceedingly well in the training data sets, i.e., when making validation of the constructed prediction algorithm in the same data set, from which the principal coordinates appearing in the prediction algorithm were constructed. $B\&J$ considered two groups of data gathered during 1988 (1-st group) and 1989 (2-nd group). They obtained, that the events in 1989 cannot be satisfactorily predicted from the DB regression constructed by the 1988 data; et vice-versa. Jakimiec and Bartkowiak (1994) have shown that the interrelation structure of the considered variables has changed during the period from 1988 to 1989 — therefore the prediction algorithm elaborated for the year 1988 was not appropriate for the year 1989.

Having the above-mentioned facts in mind we proceeded now in a different way. We wanted the validation data to have a structure as similar as possible to those in the training data. To achieve this goal we took advantage of the fact that the data items gathered in our base file were ordered according to their appearance on the sun disk. Our hope was, that after subdividing the whole group into smaller subgroups containing observations adjacent in time and then making predictions by these small groups for very near (in time) subsequent items, we might get a better prediction quality.

We performed a kind of sliding procedure. Firstly we fixed $nn = 30$ as the size of one subgroup. We fixed also k, the number of the principal coordinates to be included into the regression equation. Next, we proceeded the following way: for the first 30 items we evaluated the respective DBR and used it for prediction the values of Y_1 (and Y_2) for the items no. $31, \ldots, 35$. The predicted values for these items will be denoted as $\hat{y}_{31}, \ldots, \hat{y}_{35}$.

Next, we considered the items no. $6, 7, \ldots, 35$ as the training group and the items no. $36, \ldots, 40$ as the test group and we obtained the predicted values $\hat{y}_{35}, \ldots, \hat{y}_{40}$ for the items contained in this small test group. The procedure was repeated until the original group with $n = 130$ items was exhausted.

In this way we got the cross validated predictions for the items no. $31, \ldots, 130$. Let us state clearly that each prediction was made by a distance based regression constructed by a small group of independent items being close in time.

The predictions were made using DBR-EUC and DBR-ABS methods only.

Analogous evaluations were carried out fixing $nn = 50$. The results are very similar to those obtained for $nn = 30$ and hence will not be reported here.

To evaluate the predicted values we have used the computational algorithm described by Cuadras and Arenas (1990), also by Bartkowiak and Jakimiec (1994). The estimation of Y was done on the basis of k principal coordinates, with k chosen somehow arbitrarily. How many principal coordinates must be taken for predictions? From results presented in Section 3 one might say that the first principal coordinate is the most relevant and dominating. The second most significant, although of much lesser importance, could be for our data the 6-th one.

To investigate this problem from practical point of view we have done the evaluations with $k = 1, 2, 4$ and 8 principal coordinates. With each k the expected values $\hat{y}_i^{(k)}$, $i = 31, \ldots, 130$, were evaluated using the sliding algorithm described above. Then we calculated the residuals $e_i^{(k)} = y_i(\mathit{observed}) - \hat{y}_i^{(k)}(\mathit{estimated})$. Let us say clearly that these are external residuals, calculated from the 'test' samples.

In Figure 6 we show the residuals for $Y = Fs$ evaluated with $k = 1$ (res1), $k = 2$ (res2), $k = 4$ (res4) and $k = 8$ (res8) principal coordinates. For comparative purposes we show also the boxplot for the differences $y_i - \bar{y}$. Figure 7 shows the same for the variable $Y = Fh$.

Looking at the Figures 6 and 7 one can see that the quality of the predictions is poor. Certainly the DBR-ABS method does not yield better predictions, i.e., those with lower error. Only the first principal coordinate, if any, seems to be relevant. The distribution of the residuals seems to be fairly symmetric. When taking seemingly irrelevant principal coordinates we do not get nonsense values, and that is somehow optimistic. As compared with the results of $B\&J$, the expected (i.e., predicted) values are not biased systematically — so in this respect the present analysis yielded a better result than the former one carried out by $B\&J$.

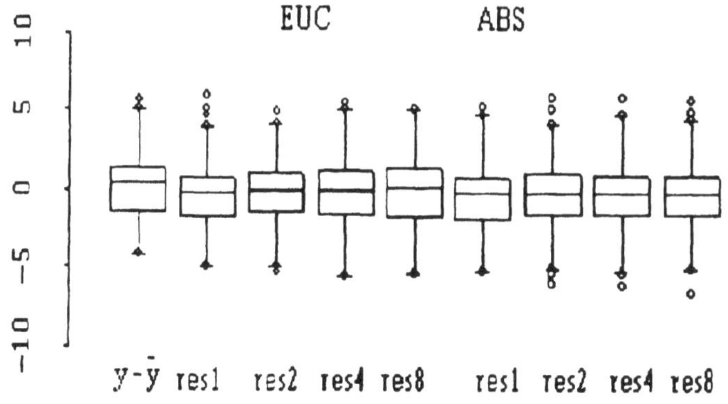

Figure 6. Boxplots for residuals from $y = Fs$ obtained by distance-based regression using Euclidean or Absolute value distances.

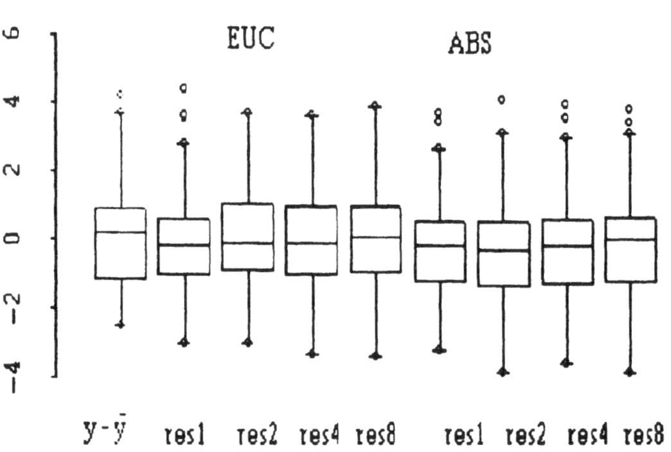

Figure 7. Boxplots for residuals from $y = Fh$ obtained by distance-based regression using Euclidean or Absolute value distances.

5. DISCUSSION

The process of sunspot flaring has an 11-year cycle and is essentially time dependent. It was stated in $B\&J$ that external predictions made from data gathered during time interval of one year are systematically biased. Now we got the result that, when taking smaller subsamples of the data and predicting from them the sun spot activities for a small number of other subsequent items, the bias is no more present. However, the prediction quality remains still poor, and the applied DBR-ABS method does not yield better results as the ordinary LSE method.

It seems that the data does not contain the information needed for successful prediction. May be we could obtain better predictions when adding some new variables, in particular some mixed type variables describing additional characteristics of sunspot group evolution, like magnetic or velocity fields and radio burst.

Further discussion on the topic 'How many principal coordinates are relevant' can be found in (Bartkowiak, 1994).

An example of better predictions when considering mixed type predictors and using Gower distances is shown in (Bartkowiak and Liebhart, 1994).

Acknowledgements

The author wishes to express her thanks to an anonymous referee for reading thoroughly the manuscript, indicating for unprecise statements, and specially, for giving the comments on Figures 4–6.

REFERENCES

Bartkowiak, A. and Jakimiec M. (1994). Distance-based regression in prediction of solar flare activity. *Qüestiió* **18**, 7–38.

Bartkowiak, A. (1994). Practice of computing distance-based regression — how many PC's are relevant?. In: *Proceedings of the Statistical Computing Section.* Toronto, 1994, submited.

Bartkowiak, A. and Liebhart, J. (1995). Estimation of the spirometric residual volume (RV) by a regression built from Gower distances. *Biometrical J.* to appear.

Cook, R. D. and Weisberg, S. (1989). Regression diagnostics with dynamic graphics. With discussion by D. B. Carr, D. Pregibon, A. C. Atkinson, L. Tierney, R. L. Welsh. *Technometrics* **31**, 277–311.

Cook, R. D. and Weisberg, S. (1991). Dynamic graphics and regression diagnostics using XLISP-STAT. Technical Report No. 565, University of Minnesota, Department of Statistics.

Cuadras, C. M. and Arenas, C. (1990). A distance based regression model for prediction with mixed data. *Comm. Statist. Theory Methods* **19**, 2261–2279.

Cuadras, C. M., Arenas, C. and Fortiana, J. G. (1993). Further aspects of a distance based model for prediction including non linear regression. Paper presented at *LINSTAT'93*, Poznań.

Gower, J. C. (1971). A general coefficient of similarity and some its properties. *Biometrics* **27**, 857–874.

Gower, J. C. (1985). Measures of similarity, dissimilarity, and distance. In: *Encyclopedia of Statistical Sciences, Vol. 5*, pp. 397–405, N. Johnson and S. Kotz (Eds), Wiley, New York.

Jakimiec, M. and Bartkowiak, A. (1994). Short-term solar flare predictions by distance-based regression I. Bearalert Regions in 1988 and 1989 — Continuous Predictors. *Acta Astronomica* **44**, 115–140.

Jolliffe, I. T. (1986). *Principal Component Analysis.* Springer, New York, Berlin.

Mardia, K. V., Kent, J. T. and Bibby, J. M. (1979). *Multivariate Analysis.* Academic Press, London, New York.

SGD — *Solar Geophysical Data, Comprehensive Results.* NOAA Boulder, USA. 1988, 1989.

Tierney, L. (1990). *Lisp-Stat, an Object-Oriented Environment for Statistical Computing and Dynamic Graphics.* Wiley, New York.

Multivar. Statist., pp. 197–210
E.–M. Tiit *et al.* (Eds)

SOME APPLICATIONS OF MULTIVARIATE ANALYSIS IN ENVIRONMENTAL RESEARCH*

M. GREENACRE[1] and R. FIELER[2]
[1]Facultat de Ciències Econòmiques i Impresarials, Universitat Pompeu Fabra, Balmes 132, 08008 Barcelona, Spain

[2]Akvaplan-niva, Strandtorget 2B, N-9001 Tromsø, Norway

ABSTRACT

An important problem in environmental research is the relationship between the biological composition at a sampling point and concomitant environmental information at the same location. Both the biological and environmental descriptions of an an area involve complex, multidimensional observations, of the order of hundreds of possible species and tens of possible environmental indicators. Dimension-reducing techniques of multivariate analysis can be used to simplify the species-environment relationship. Two different approaches to this problem are described and compared, one using multidimensional scaling and nonparametric correlation measures, and the other a combination of correspondence analysis and regression called canonical correspondence analysis.

1. INTRODUCTION

This work forms part of a larger project commissioned by Philips Petroleum Company (Norway) and also supported by the Norwegian Council of Science to look at ways of improving the statistical design and data analysis of monitoring surveys of the North Sea oil fields. In 1989 the Paris Commission published guidelines for environmental monitoring near North Sea oil platforms[3]. The company Akvaplan-niva, based in Tromsø, Norway, is one of the companies commissioned to perform annual surveys of the oil fields according to the Paris Commission guidelines. After collecting a very large database of environmental data from 1990 to 1992, the present project was set up to look at these data beyond the prescribed methodology and to make recommendations about future data exploitation. While most of our work centred on the survey design, for

*This research was supported by Philips Petroleum Company, Norway, the Norwegian Council of Science and Akvaplan-niva; the first author also thanks the University of South Africa (UNISA) in Pretoria, Universitat Pompeu Fabra (UPF) in Barcelona and the Advanced Research Centre (ARC) of Sant Fruitòs de Bages for their support.

[3]See *Guidelines for Monitoring Methods to be Used in the Vicinity of Platforms in the North Sea*, Paris Commission Report prepared by the Norwegian State Pollution Authority (SFT).

example on the sampling scheme and the detectability of changes, we also spent a significant part of our research looking at the relationship between biological and environmental variables, which is the subject of the present paper.

In this paper we review and compare two different ways of analysing environmental survey data using multivariate techniques. We are specifically interested in methods which identify environmental variables which are highly associated with the biological differences found in the marine sampling. Both of these approaches rely on multivariate methods to reduce the dimensionality of the data in the hope of simplifying the quest for the environmental-biological relationship. The first is a combination of multidimensional scaling and nonparametric methods while the second is a constrained version of correspondence analysis called 'canonical correspondence analysis'.

2. DATA

The samples are obtained by 'grab sampling' of the ocean floor at fixed locations called 'stations' situated in the vicinity of an oil drilling platform in the North Sea. There are two sets of variables of interest at each station. One is a very large set of biological variables, specifically abundance data for hundreds of species which may be identified in a sample. The second is a smaller set of environmental variables which includes chemical measurements of heavy metal deposits such as barium and cadmium, measurements of organic material and hydrocarbons, and sedimentological measurements such as percentage silt, clay and pelite (or mud).

The particular data set which we use as an example is the 1992 data set from the Ekofisk Centre oilfield, consisting of 13 stations. A total of 248 species were identified in samples from these stations, so the biological data set is a 13×248 matrix $\mathbf{B}$ of abundances, where b_{ij} = the abundance of species j at station i. For each station we also have a large set of environmental measurements. For purposes of illustration in this paper and to simplify our task, we have chosen a reduced set of five of these measurements, namely pelite (one of the most important sedimentological parameters, measured as a percentage), total hydrocarbon content (g/T), total organic material (%), and two heavy metals barium and cadmium (g/T). To observe how a totally unrelated variable behaves in our analyses, we added a sixth variable, an artificial variable generated randomly from a lognormal distribution. We denote the 13×6 matrix of environmental data $\mathbf{E}$. A schematic diagram of the data we use is shown in Figure 1, along with part of the numerical data. Although this is not directly relevant to our methodology in this paper, we point out that each biological observation is established from separate identifications of five replicate samples at the same station, while each environmental observation (except the sediment variables such as pelite) is the mean of three replicate samples. This replication increases the reliability of each of the biological and environmental measurements.

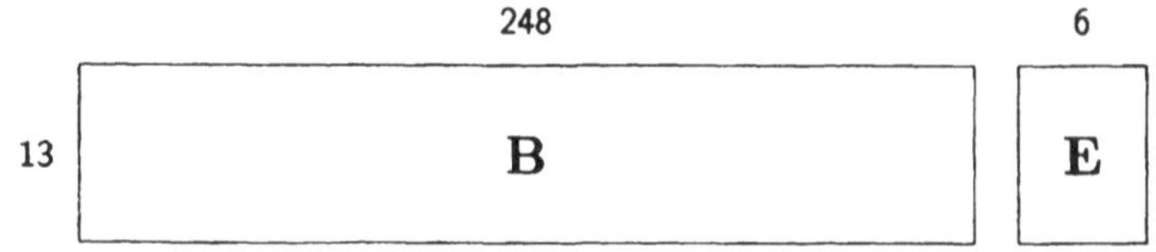

	BIOLOGICAL SPECIES					ENVIRONMENTAL			
	Myr.ocu	Cha.set	Mys.bid	Amp.fal	. . .	Pelite	THC	. . .	LN
Stn 4	193	34	30	49	. . .	2.9	8	. . .	1.05
Stn 8	79	4	11	58	. . .	14.9	9	. . .	4.20
Stn 9	150	247	36	66	. . .	3.8	103	. . .	0.09
Stn 12	72	19	65	47	. . .	5.3	6	. . .	2.48
.	.	.	.	.	. . .	.	.	. . .	.
.	.	.	.	.	. . .	.	.	. . .	.
.	.	.	.	.	. . .	.	.	. . .	.

Figure 1. Format of matrices of biological data (**B**) and environmental data (**E**). Below a small example of the numerical data is shown: the first four stations as rows, the first four species as columns, then columns showing the first two environmental variables and the lognormal random variable LN.

3. METHODOLOGY

The problem is to relate the large corpus of species abundance data to the environmental data. We shall demonstrate two approaches to this problem, both of which use multivariate analysis to reduce the dimensionality of the two sets of data. Both have in common the application of multidimensional scaling techniques to the biological data, and both also share the objective of choosing subsets of the environmental variables which are 'most related' to the "biological distances" computed between stations. Both methods have found favour in different sectors of the environmental research community, but we are not aware of objective comparisons of their respective merits and demerits. We now outline each approach in turn before illustrating their application.

3.1. *The 'non-metric' approach of the Plymouth school*

To our knowledge this approach is described only in journals of marine biology, notably in the papers of (Field *et al.*, 1982; Clarke and Ainsworth, 1993). Gray *et al.* (1990) describe applications to earlier data from the same oil fields of interest in this paper. Clarke (1993) reviews the merits of this approach and introduces improvements such as permutation tests to obtain measures of statistical significance.

The approach, summarized schematically in Figure 2, is as follows. Separate distance or dissimilarity measures are computed between the stations based on the biological data and environmental data respectively. In the case

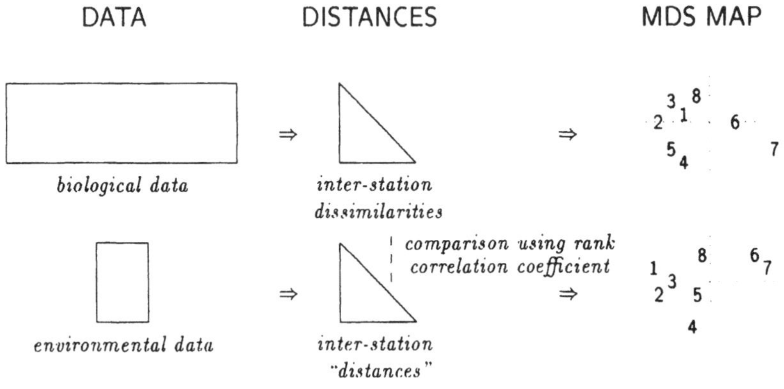

Figure 2. Schematic description of non-metric approach.

of the biological abundance data the measure of choice is the Bray–Curtis dissimilarity index, computed as follows:

$$\delta_{ik} = \frac{\sum_j |x_{ij} - x_{kj}|}{\sum_j (x_{ij} + x_{kj})} \tag{1}$$

where the values x_{ij} are some transformation b_{ij}^{α} of the biological abundance values, the most popular choice being the double-square-root transformation, i.e., $\alpha = 0.25$. This transformation is necessary to give the rarer species a higher contribution in the index, which is otherwise dominated by the more frequent species. An advantage of the Bray–Curtis index is that it has an upper bound of 1, when there are no species in common at two stations. Thus a natural inverse measure of similarity between stations is $1 - \delta_{ik}$, which is often multiplied by 100 and expressed as a percentage. The disadvantages of this index are first, the subjective choice of power transformation applied to the original abundances, and second, the inherent assumption that each grab sample is exactly the same volume, which is not always the case. Clearly, however, the Bray–Curtis index is an easy measure for biologists to understand and this is of great importance when reducing the complex biological differences between stations to a single index.

Concerning the environmental data, each variable is suitably transformed and standardized, the invariable choice being the log-transformation because of the long-tailed distributions of the variables. Distances between stations may then be computed using an ordinary Euclidean distance function:

$$d_{ik} = \left(\sum_j (y_{ij}^* - y_{kj}^*)^2 \right)^{1/2}, \tag{2}$$

where the values y_{ij}^* are the transformed values of the original environmental observations y_{ij}. Such an environmental distance matrix $\mathbf{D}$ can then be compared to the inter-station Bray–Curtis dissimilarity matrix Δ to assess their agreement, by a procedure which we shall describe shortly. The matrix Δ is regarded as a fixed 'target' for this exercise, while a large number of different matrices $\mathbf{D}$ can be computed for different subsets of environmental variables. The subset of variables which gives a set of inter-station distances which matches the Bray–Curtis dissimilarities the closest is regarded as giving the most successful 'explanation' of community structure.

Let us denote by $\mathbf{D}_E$ an environmental distance matrix computed using (2) on a set of environmental variables E. The target matrix Δ and the candidate matrix $\mathbf{D}_E$ can be compared using a variation of the Spearman rank-correlation coefficient:

$$\rho_{ik} = 1 - \frac{6}{N(N-1)} \sum_j \frac{(\gamma_{ij} - g_{kj})^2}{(\gamma_{ij} + g_{kj})},$$

where γ and g are the ranks of the indices in Δ and distances in $\mathbf{D}$ respectively. Low distances, i.e., high similarities, correspond to low ranks. Compared to the usual form of the Spearman coefficient, which does not include a factor $(\gamma_{ij} + g_{kj})$ in the denominator, the above coefficient downweights the contribution of the large number of terms involving distant pairs of stations (large distances or low similarities), and upweights the contribution of the small distances (high similarities). In the context of using the rank correlation coefficient to assess the agreement between the elements of two triangular distance matrices, such an adaptation to favour local matching of distances is considered highly desirable by Clarke and Ainsworth (1993).

Finally, visualization of the inter-station Bray–Curtis indices and of the inter-station environmental distances can be achieved using either metric or nonmetric multidimensional scaling (MDS) algorithms.

3.2. *Correspondence analysis approach*

Correspondence analysis (CA) is a method for visualizing contingency tables, and has found widespread use amongst ecologists for displaying tables of abundance data from ecological surveys. In the ecological context the underlying assumption of CA is that the species follow a unimodal response model along 'composite gradients'. A composite gradient is a theoretical variable which coincides with the concept of a factor in exploratory factor analysis. The use of CA in the context of ecology is explained in greater detail by Ter Braak (1985, 1986) and Jongman *et al.* (1987).

Inherent in CA is the use of the so-called *chi-square distance*. In our present example, the chi-square distance between stations is:

$$\chi_{ik} = \left(\sum_j \left(\frac{b_{ij}}{b_{i+}} - \frac{b_{kj}}{b_{k+}} \right)^2 \bigg/ \frac{b_{+j}}{b_{++}} \right)^{1/2}, \qquad (3)$$

where b_{ij} are the abundance values, with station (row) and species (column) totals b_{i+} and b_{+j} respectively, and total abundance b_{++} (see, for example (Greenacre, 1984, 1993)). Notice first that CA measures distance between relative abundance measures b_{ij}/b_{i+}, not the original values b_{ij}. Second, the presence of the factor b_{+j}/b_{++} in the denominator of each squared term in (3) serves to downweight the larger differences between the more abundant species and upweight the smaller differences between the rarer species (cf. the double square-root transformation of the abundances before calculating the Bray–Curtis index, which had the same objective). This type of normalisation of frequency data is often justified by assuming a Poisson distribution of the frequencies of each species, in which case the chi-square distance implies a standardisation of the variance of each species.

An advantage of CA of biological abundance data is that joint maps of the stations and species are possible in order to interpret the relative positions of the stations. A similar strategy is possible when analysing the environmental data by principal component analysis, which is in effect a metric scaling of the inter-station Euclidean distances. Each environmental variable can be displayed as a vector in the station map to interpret the stations' positions.

Having performed a CA of the biological data, we can relate the environmental data to the identified gradients (principal axes) using the supplementary point idea in CA (see, for example (Greenacre, 1993, Chapter 12)), or by simply regressing the gradients against the environmental variables. This is often referred to as 'indirect gradient analysis' because the environmental information is related to the gradient *a posteriori* and plays no direct role in the gradient determination. It frequently occurs that the principal gradients revealed by CA are not strongly related to the environmental variables. Then a new question arises whether there is a subspace of the full space of the species data which not only contains a maximum percentage of the species inertia, but which satisfies the additional condition that it is *directly* related to the environmental data. This is called 'direct gradient analysis' and a constrained version of correspondence analysis, called *canonical correspondence analysis* (CCA), is designed specifically to satisfy this double objective.

CA and CCA thus both aim to explain a maximum amount of the total inertia in a low-dimensional map. In CA we have the simple decomposition of inertia:

$$\text{Total Inertia} = \text{Inertia Displayed in Map} + \text{Residual Inertia}.$$

In CCA we have a more complex decomposition of inertia:

$$\text{Total Inertia} = \text{Inertia in Constrained Subspace} + \text{Remaining Inertia},$$

where

$$\begin{aligned}&\text{Inertia in Constrained Subspace}\\&\quad = \text{Inertia Displayed in Map} + \text{Residual Inertia}.\end{aligned}$$

By constraining the solution we automatically exclude a part of the inertia from the map (called 'Remaining Inertia' above).

4. RESULTS

4.1. *Non-metric approach*

Using the first approach we first compute the Bray–Curtis indices between the 13 stations. Using metric MDS we obtain the two-dimensional display shown in Figure 3.

The contrast between the 11 stations in the polluted neighbourhood of the oil platform and the two reference stations is clearly displayed. Amongst the polluted stations on the left of the map we see a gradient from stations 8, 12, and 18 at the top down to stations 24, 14, and 15 at the bottom, where the latter stations are known to be more polluted.

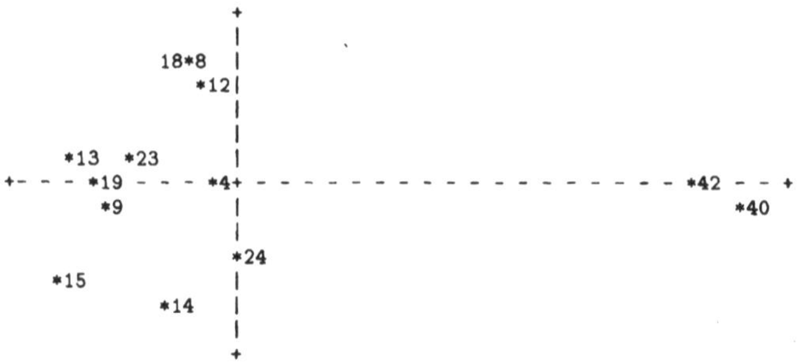

Figure 3. Two-dimensional MDS map of Bray–Curtis indices.

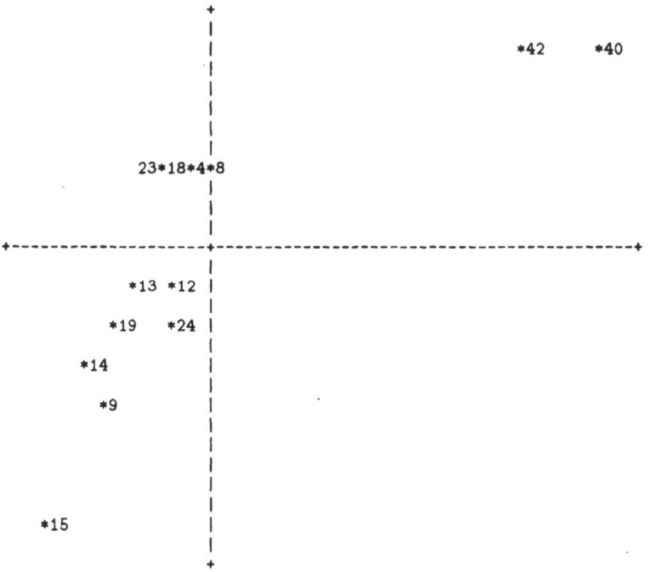

Figure 4. Scatterplot of stations according to log-barium (x-axis) and log-cadmium (y-axis).

We then performed 63 consecutive analyses of the environmental data, using all possible subsets of the six variables. In each case we computed the inter-station distances and then matched these to the Bray–Curtis indices using the adapted rank-correlation coefficient described earlier. Table 1 shows the top 12 and bottom 12 subsets and their respective rank correlations.

The most important subset of variables is the single variable 4 (barium), generating distances which have a correlation of 0.654 with the Bray–Curtis indices. This is followed by the pair of variables 4 and 5 (barium and cadmium), with a correlation of 0.621. The dummy lognormal random variable first appears in the twelfth position, along with the important pair of variables 4 and 5. The last positions are almost all dominated by variable 6 and variable 3 also appears frequently in these positions. Thus it appears that the third variable is quite unimportant, similar to the lognormal variable which was intentionally included as a variable uncorrelated with the data at hand.

Figure 4 shows a scatterplot of the stations according to the log-transformed values of barium and cadmium. The inter-station distances in this plot thus have

Table 1.

Rank correlation coefficients ρ between environmental distances and biological indices, for different subsets of variables

ρ	subset	
0.654	4	variable 1 = pelite (Pel)
0.621	4 5	variable 2 = total hydrocarbon content (THC)
0.590	2 4 5	variable 3 = total organic material (TOM)
0.571	2 4	variable 4 = barium (Ba)
0.545	1 2 4 5	variable 5 = cadmium (Cd)
0.544	2 3 4 5	variable 6 = random lognormal (LN)
0.543	1 4 5	
0.537	1 2 4	
0.533	3 4	
0.532	3 4 5	
0.528	2 3 4	
0.525	4 5 6	
⋮	⋮	
0.148	2 3	
0.131	1	
0.124	1 2 6	
0.105	1 2 3 6	
0.097	2 6	
0.081	1 3	
0.069	2 3 6	
−0.024	1 3 6	
−0.027	1 6	
−0.027	3	
−0.148	3 6	
−0.204	6	

a correlation of 0.621 with the Bray–Curtis distances which were approximately displayed in Figure 3.

4.2. *Correspondence analysis approach*

The correspondence analysis solution in two dimensions is shown in Figure 5.

An important difference between the CA map above and the map of Figure 3 based on Bray–Curtis indices is the separation of station 24. Otherwise the display also shows the gradient from non-polluted stations 40 and 42 on the right curving down to the left to the polluted stations 14, 9 and 15. An advantage of the CA map is that we can get a corresponding map of the species themselves in order to identify why station 24 is separate, or for example, what species characterize the polluted stations at the bottom of the map. In Figure 6 we show the species map, and indicate some species abbreviations. Thus we can see that the species *Ech.sp*, *Myi.ocu*, and *Cyl.cyl* are those that are particularly associated with station 24, whereas a species such as *Amp.fal* characterizes non-polluted areas (which means that it is absent in polluted areas) and *Chae.set* and *Eum.ock* are found relatively frequently in polluted areas. In the context of ecological data CA is frequently criticized for being too influenced by rare species, a criticism which is in most cases unfounded. It is true that some of the rare species do have unusual sets of relative frequencies (called *profiles* in CA), and this gives them outlying positions in the map. This is the case with species *Ech.sp* and *Eum.ock* in Figure 6. However, a closer inspection of the contributions which these points make to the axes shows that they contribute very little owing to their low mass, the mass being proportional to their total abundance. It is important in CA to interpret the

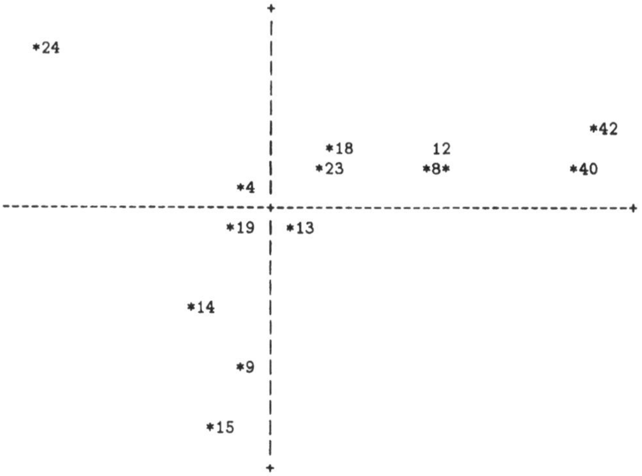

Figure 5. Correspondence analysis of biological abundance data showing stations (rows); Inertia, first axis: 0.2457 (31.4%), second axis: 0.2044 (26.1%).

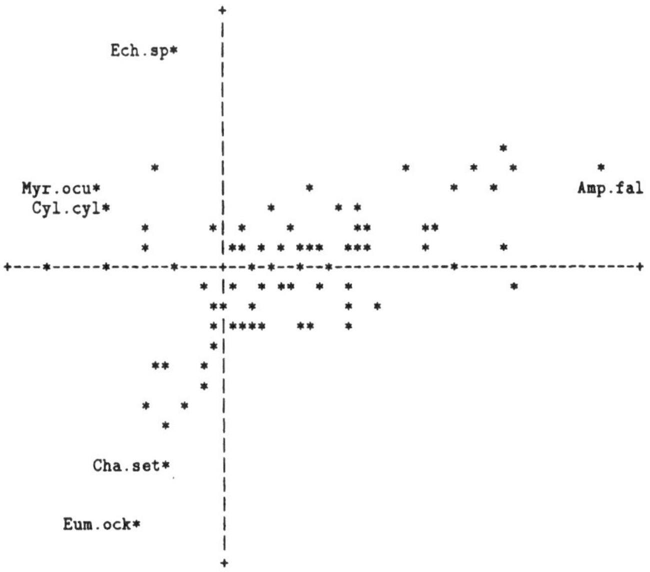

Figure 6. Correspondence analysis of biological abundance data showing selected species.

map in conjunction with the numerical contributions of the points to obtain a clear idea of which points really determine the map.

Coming now to the environmental data, which are measurements, these are more readily visualized using a biplot resulting from a principal component analysis (Gabriel, 1971; Gabriel and Odoroff, 1990; Greenacre, 1991). Figure 7 shows the display of the stations as points and the variables as vectors.

Finally, CCA leads to a visualization of the relationship between the biological and environmental variables, thanks to their joint observation at each station. Figure 8 shows the two-dimensional map of the species, environmental variables and stations, split into two maps for convenience. As described in Section 3.2, the decomposition of inertia is slightly more complicated in CCA. In ordinary CA (Figures 5 and 6) 68% of the total inertia of the data matrix **B** was explained by the two-dimensional map. In the CCA we find that when the species data **B** is constrained by the environmental data **E**, then only 45% of the inertia of **B** is accounted for. In other words, 55% of the biological inertia can not be explained by the environmental variables. Of the 45% of the inertia that can be explained, the map in Figure 8 accounts for 70%.

In Figure 8 we can again see the separation of the reference stations 40 and 42 at the bottom right of the map and a group of five species that also separates out in this quadrant. The difference between this map and the one given in Figures 5 and 6 is that the axes in Figure 8 are linear functions of the six environmental variables. Thus, by construction, the axes of dispersion in the CCA map can be directly interpreted in terms of the environmental variables.

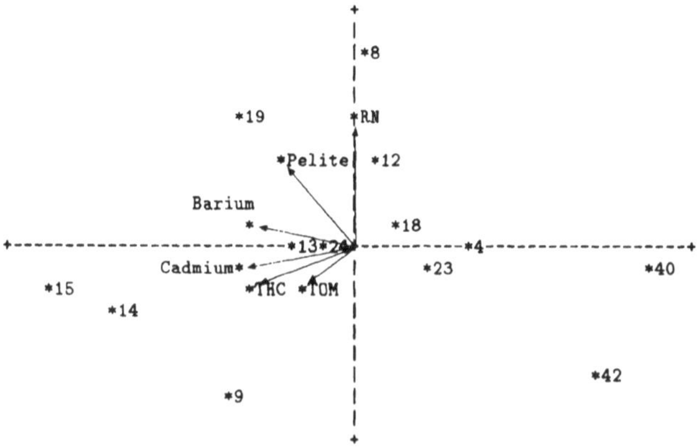

Figure 7. Biplot of environmental data showing stations as points and variables as vectors.

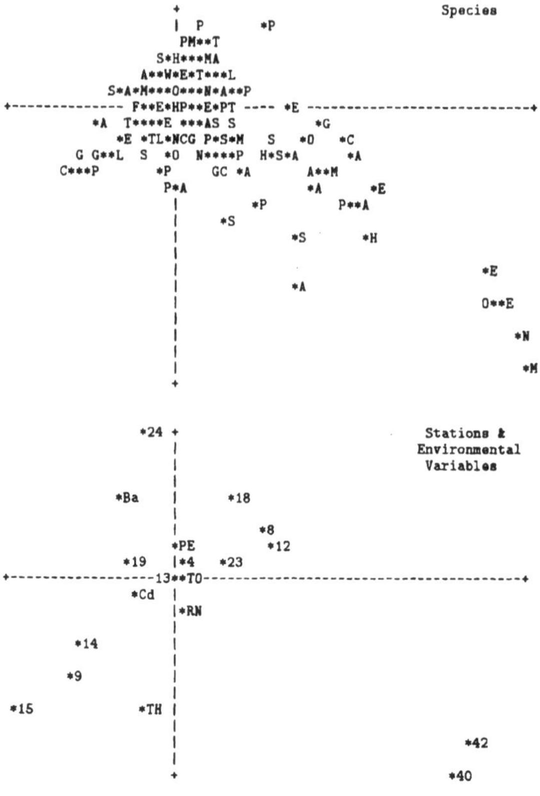

Figure 8. CCA of biological and environmental data, showing species separately and then stations and environmental variables; only single letters are given for species. Inertia in constrained space = 45%, of which 70% explained in this map.

Table 2.

Part of the numerical listing of row (species) contributions in the correspondence analysis of the matrix **B**, showing first three dimensions. Notice the large contributions to the principal axes (in columns labelled CTR) and the small contributions made by the species *Ech.sp* (row 41) and *Eum.ock* (row 67), even though these latter species are outlying in Figure 6. The listing is from the output of program SimCA (Greenacre, 1986, 1988)

I	NAME	QLT	MAS	INR	$k = 1$	COR	CTR	$k = 2$	COR	CTR	$k = 3$	COR	CTR
1	Myr	998	285	184	−582	670	393	404	324	228	50	5	6
2	Cha	982	118	171	−274	66	36	−1010	901	590	128	14	15
3	Amp	811	79	19	−20	2	0	−312	501	37	−244	307	37
4	Mys	505	55	61	603	416	81	130	19	5	249	71	27
5	Amp	947	39	62	955	732	144	251	50	12	453	165	63
⋮	⋮		⋮			⋮			⋮			⋮	
40	Pho	621	1	4	659	180	2	86	3	0	−1027	438	10
41	Ech	829	1	6	−232	14	0	1200	377	8	1295	438	15
42	Cir	599	1	5	1361	497	8	535	77	1	308	25	1
⋮	⋮		⋮			⋮			⋮			⋮	
66	Sca	239	0	2	956	173	1	128	3	0	575	63	1
67	Eum	839	0	1	−397	62	0	−1370	740	3	305	37	0
68	Mod	402	0	2	624	90	0	329	25	0	−1117	287	3
⋮	⋮		⋮			⋮			⋮			⋮	

QLT – quality of point's display in this 3-dimensional solution; MAS – mass of the point; INR – inertia of the point; $k = 1$ – principal coordinate on first dimension; COR – contribution of axis to the inertia of the point; CRD – contribution of point to the inertia of the axis; $k = 2, k = 3$, etc. – same diagnostics for principal axes 2 and 3.

5. DISCUSSION

Our objective has been to explain how the *system* of biological variables is accounted for by the *system* of environmental variables. The stations, or sampling units, form the link between these two systems. Both approaches discussed in this paper use multivariate analysis to try to reduce these systems to simpler entities. In the MDS approach, the simpler entities are distances between the stations, while in the CCA approach the simpler entities are new dimensions, often called 'gradients' in this context, along which the systems, particularly the biological system, vary maximally.

A possible criticism of the MDS approach is that by reducing the two systems separately to two distance matrices and then studying the correlation of these two sets of distances can lead to ignoring a non-monotonic relationship between biological and environmental variables, for example the common unimodal response curve. Figure 9 shows two highly simplified illustrations of five stations plotted according to one biological and one environmental variable, with the first diagram showing a monotonic decreasing relationship and the second a non-monotonic unimodal relationship. On each axis we show the positions of the stations in the separate biological and environmental 'spaces',

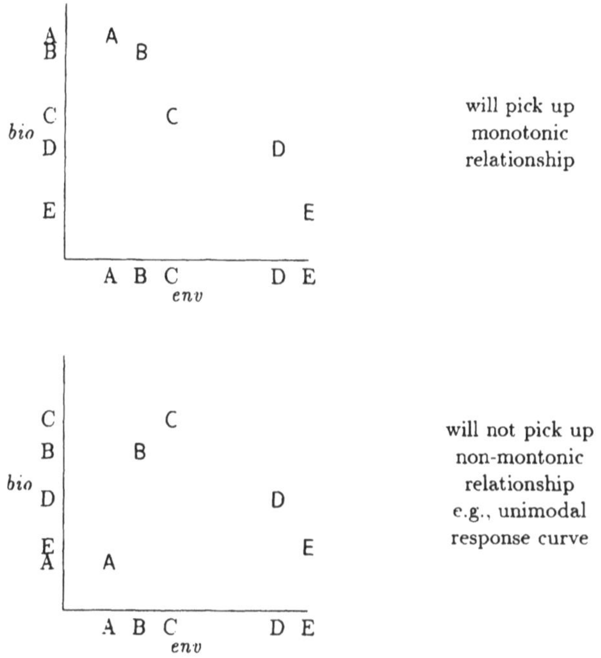

Figure 9. CCA of biological and environmental data, showing species separately and then stations and environmental variables; only single letters are given for species; Inertia in constrained space = 45%, of which 70% explained in this map.

which are uni-dimensional in this simple case. Clearly, the first relationship would be picked up by the MDS approach, since the inter-station distances in the biological and environmental spaces will be highly correlated. The non-monotonic relationship, however, will not be easily identified by the MDS approach because the two sets of distances are no longer highly correlated.

Another point of criticism of both methods is that, being basically exploratory or descriptive methods, neither gets to the crux of the functional relationship between the biological and environmental variables. We could resort to standard methods to model the frequency of one species as a function of a subset of the environmental variables, but what we want to achieve here is an explanation of the whole system of species data. Here the CCA approach seems to have the advantage because it reduces the biological data to a few dimensions and each of these dimensions can be interpreted and explained in terms of the environmental predictors. In the implementation of CCA by ter Braak, there are permutation tests which allow selection of the variables which significantly account for the biological dimensions. But there still seems to be a gap between the descriptive graphical results, along with their numerical diagnostics, and the more concrete objective of quantitative modelling of biological responses in terms of environmental changes.

REFERENCES

Clarke, K. R. (1993). Non-parametric multivariate analyses of changes in community structure. *Austral. J. Ecology.*

Clarke, K. R. and Ainsworth, M. (1993). A method of linking multivariate community structure to environmental variables. *Marine Ecology – Progress Series* **92**, 37–52.

Field, J. G., Clarke, K. R. and Warwick, R. M. (1982). A practical strategy for analysing multispecies distribution patterns. *Marine Ecology – Progress Series* **8**, 37–52.

Gabriel, K. R. (1971). The biplot graphical display of matrices with applications in principal component analysis. *Biometrika* **58**, 453–467.

Gabriel, K. R. and Odoroff, C. L. (1990). Biplots in biomedical research. *Statistics in Medicine* **9**, 469–485.

Gray, J. S., Clarke, K. R., Warwick, R. M. and Hobbs, G. (1990). Detection of initial effects of pollution on marine benthos: an example from the Ekofisk and Eldfisk oilfields, North Sea. *Marine Ecology – Progress Series* **66**, 285–299.

Greenacre, M. J. (1984). *Theory and Applications of Correspondence Analysis.* Academic Press, London.

Greenacre, M. J. (1986). SimCA: a program to perform simple correspondence analysis. *Amer. Statist.* **40**, 230–231.

Greenacre, M. J. (1988). Correspondence analysis on a personal computer. *Chemometrics and Intelligent Laboratory Systems* **2**, 233–234.

Greenacre, M. J. (1991). Biplots in correspondence analysis. *J. Appl. Statist.*

Greenacre, M. J. (1993). *Correspondence Analysis in Practice.* Academic Press, London.

Ter Braak, C. J. F. (1985). Correspondence analysis of incidence and abundance data: properties in terms of a unimodal response model. *Biometrics.*

Ter Braak, C. J. F. (1986). Canonical correspondence analysis: a new eigenvector technique for multivariate direct gradient analysis. *Ecology.*

Jongman, R. H. G., Ter Braak, C. J. F. and Van Tongeren, O. F. R. (1987). *Data Analysis in Community and Landscape Ecology.* Pudoc, Wageningen.

Multivar. Statist., pp. 211–227
E.-M. Tiit *et al.* (Eds)

METHODS FOR LONGITUDINAL SOCIAL NETWORK DATA: REVIEW AND MARKOV PROCESS MODELS

TOM A. B. SNIJDERS

Department of Statistics and Measurement Theory, University of Groningen, Grote Kruisstraat 2/1, 9712 TS Groningen, The Netherlands

ABSTRACT

Social network data pertain to social relations between individuals, or between other actors (countries, firms, etc.). The fact that each relation links two individuals, and each individual can be related with multiple others, results in complicated stochastic dependence structures. Longitudinal studies concerning the change and development of social networks in time are important for progress in sociology and related disciplines, but the literature does not contain enough satisfactory statistical models for longitudinal social network data. The paucity of satisfactory models is related to the difficulty of reconciling mathematical tractability with the expression of substantively interesting processes.

A review is given of models for longitudinal network data. Interesting developments are taking place in applying continuous-time Markov processes to model network change. Such models can incorporate sociologically important effects. They are in many aspects similar to computer simulation models that are common now in sociological theory, but contain a component of random change as an extra ingredient. Difficulties of mathematical tractability can be resolved by using stochastic approximation methods to obtain estimators and tests. This is done, however, at the expense of introducing some new problems: computer intense computations and questions about statistical efficiency. An example of this approach is given using a time series of personal preference relations in a closed group.

1. STATISTICAL MODELS FOR SOCIAL NETWORKS

Social networks, as studied in sociology and other social sciences, are the structures consisting of dyadic relations between individuals or other units (organisations, countries, etc.); a commonly used terms for these units is *actors*, stressing their active role in the constitution of the networks. This paper is concerned with so-called *entire networks*, where all relationships within a given set of n actors are considered. Such a network can be represented by an $n \times n$ matrix $X = (X_{ij})$, where X_{ij} is a number or a vector representing the relation directed from actor i to actor j $(i, j = 1, \ldots, n)$. Usually, self-relations are not considered, so that the diagonal values X_{ii} are non-defined. An often used convention is $X_{ii} = 0$.

Traditionally, much attention has been paid to binary relations. In this case, X is a 0-1 matrix that is naturally seen as the adjacency matrix of a graph or directed graph. Pioneering work in the development of statistical methods for graphs was done by Frank (1971). The substantive meaning of the relationship can be, e.g., friendship, acquaintance, cooperation, opposition, etc.

Special problems that must be dealt with in the statistical analysis of social networks are constituted by the dependence structure of the elements of X:

- the dependence between the two halves X_{ij} and X_{ji} of the *dyad* (X_{ij}, X_{ji});
- the dependence between the elements of each row, corresponding to relations of the same *sender* i, and likewise among the elements of each column, referring to one *receiver* j;
- more complicated dependence structures, such as *transitivity*: "a friend of my friend is also my friend", and *group formation*.

These types of dependence have been modelled by stochastic dependence but also by using common parameters, e.g., row parameters and column parameters referring to sender and receiver characteristics. An important concept for the dyadic dependence is *symmetry* or *reciprocity*, indicating the tendency for X_{ij} and X_{ji} to be similar. In the most simple case of binary (0-1) data, this has led to the classification of dyads as *mutual*, $(x_{ij}, x_{ji}) = (1,1)$, *asymmetric*, (0,1) or (1,0), and *null*, (0,0).

Another problem for the statistical analysis of social networks is

- the highly unbalanced nature of social network data: social structure does not have a crystalline regularity at all, so that there is usually not sufficient reason to assume simple balanced or symmetric structures.

In spite of the rather large amount of work that has been done on statistical and other data-analytic modelling of social network data, in many cases the currently available methods are not completely adequate to model the complexity of social relation patterns.

Longitudinal data is in principle much more informative about the studied phenomena in social networks than cross-sectional data; but they create additional problems for statistical analysis. A review of methods, including exploratory methods, for analysing change in networks was given by Frank (1991). The present paper is restricted to models that specify the joint probability distribution and discuss statistical procedures for time-series $X(t)$, $t \in \mathcal{T}$, of social networks for a constant set of actors.

2. SOME MODELS FOR LONGITUDINAL SOCIAL NETWORKS

2.1. *Discrete Markov chains*

It seems that the first statistical paper on longitudinal social networks was written by Katz and Proctor (1959). They propose a Markov chain model for binary relation networks, in which the dyads are assumed to be independent, each following a Markov chain for the 4 possible outcomes in the outcome space $\{0,1\}^2$, where the transition probabilities for the two asymmetric outcomes (0,1) and (1,0) are obviously required to be identical. This assumption of *dyad independence* has haunted the social network statistical literature, because the models become so complicated without it.

A recent paper that also proposes a Markov chain model is (Carley and Banks, 1995). They consider time series of directed graphs in which changes occur randomly, while the change rate parameter may change (e.g., decay) over time. The assumption of random changes is quite restrictive, and should be tested when these methods are applied.

2.2. *Loglinear models*

Sparked off by the work of Holland and Leinhardt (1981), many papers have appeared about loglinear models for categorical (binary or more complicated) networks. The basic model, called by Holland and Leinhardt the p_1 model, has probability function

$$K(\alpha,\beta,\rho)\exp\left(\sum_i \alpha_i x_{i+} + \sum_i \beta_i x_{+i} + \rho\sum_{ij} x_{ij}x_{ji}\right),$$

where an index replaced by a + indicates summation over that index. Some papers elaborating this approach are (Fienberg and Wasserman, 1981; Fienberg *et al.*, 1985; Wasserman and Weaver, 1985). Instead of using the incidental α_i and β_i parameters, these papers focus mostly on models where the senders and receivers are characterized by observed categorical actor-dependent variables. An extensive review of the literature on the p_1 and other models is given by Wasserman and Faust (1994).

The loglinear approach has been extended to longitudinal data by Wasserman (1987) for 2 time points, and by Wasserman and Iacobucci (1988) for multiple time points. The latter paper distinguishes between associative models, where quite complicated association patterns between the time-dependent values for each dyad (i,j) are modeled, and predictive models, that specify the conditional distribution of a dyad at each given time points, assuming its history is given. These models all assume independence between the dyads, however.

2.3. *Conditionally uniform models*

Another line of statistical models for social networks, mainly represented by directed graphs, is based on conditionally uniform models, i.e., models where the probability distribution is uniform, conditional on a certain set of statistics. Such a model reflects the notion that the conditioning statistics contain something that is relevant in the studied phenomena, and the rest is caused by random factors. Holland and Leinhardt (1975) initiated the study and application of this type of model, emphasizing the uniform model conditional on the dyad count, i.e., the numbers of mutual, asymmetric, and null dyads.

Conditionally uniform models become very complicated, however, when more informative conditioning statistics are considered. For example, conditioning on out-degrees and in-degrees is very relevant in sociological applications. When the relation can be interpreted as friendship or acquaintance, the in-degrees reflect the level of activity of the individual actors, and the out-degrees their popularity. These more complicated conditionally uniform models arise also when exact tests are required for the loglinear models. For example, to obtain an exact test for the hypothesis that the reciprocity parameter ρ in the p_1 model is equal to 0, one needs to condition on the sufficient statistics for the null hypothesis, which is just the vector of in-degrees and out-degrees; this leads to the uniform model, conditional on in- and out-degrees. This conditionally uniform model is difficult to handle statistically because these degrees can be regarded as incidental parameters, so that it is impossible to apply the usual asymptotic arguments. Algorithms for exact tests for this type of model were proposed by Snijders (1991).

Because of these complications, conditionally uniform models have been developed only for relatively simple conditioning statistics, and they are used more as a 'straw man' null hypothesis models than as models that pretend to represent reality. A conditional uniform model for longitudinal binary network data at 2 time points, conditional on the entire digraph at time 1, and on the numbers of newly formed and of disappeared relations for each actor, was developed by Snijders (1990).

2.4. *Continuous-time Markov chain models*

The most promising type of model is the continuous-time Markov model, where it is assumed that the social network, although observed only at discrete time points, follows a Markov process in continuous time. Holland and Leinhardt (1977a, 1977b); Wasserman (1977, 1979, 1980); Mayer (1984) proposed these models and developed estimation techniques for a number of special cases. Crucial for these estimation techniques is that either the dyads are assumed to be independent, so that a reduction by sufficiency leads to an outcome space with few elements, or other symmetrizing restrictions are made. An example of the latter is Wasserman's (1980) popularity model, where the rows of the adjacency matrix are assumed to be independent, and the change rate of each X_{ij} depends only on the column sum (in-degree) X_{+j} which is an index for actor j's popularity, and where the assumption of stationarity is made. Leenders (1995) extended the possibilities for modeling longitudinal binary social network data as a continuous-time Markov processes, by dropping the assumption of stationarity and by allowing change rates for every dyad to be dependent on arbitrary covariates, provided that these remain constant between the moments of observation. In the following sections we present another elaboration of models that are also continuous-time Markov chains.

3. THE NEED FOR MODELS THAT REPRESENT SUBSTANTIVE THEORY

Most of the network models reviewed above are almost trivial from the point of view of sociological theory. The reason is the need to keep these models mathematically tractable in order to apply conventional statistical methods. For example, the assumption of independence between dyads excludes *a priori* almost all sociologically interesting interactions. Conditionally uniform models can serve at best as null models against which to test sociological theories. The continuous-time Markov models for networks offer more scope for expressing sociological theories, but they have not yet found much application because of the statistical difficulties associated with their practical use. The latter models can be seen as statistical models, not explicitly derived from a sociological theory, but in other aspects closely related to the models proposed below.

The remainder of this paper is devoted to statistical estimation of a class of models for evolution of the relation network and of individual behavior in a (non-changing) set of actors that are derived from sociological theories. A more extensive discussion of these models and associated methods for parameter estimation is given in (Snijders, 1995). The models treated are based on the assumption that each actor has his or her own goals which he/she tries to advance in accordance to his/her constraints and possibilities. The approach will be in the spirit of so-called methodological individualism: the driving force behind the network dynamics is constituted by the actors' actions, where each actor takes actions in order to further his own goals, which actions must be in the domain of his own behavior or of the directed relationships from him to others. More specifically, the models are in the rational choice tradition in sociology, as expounded, e.g., by Coleman (1990).

These stochastic models are so complicated that the development for them of statistical methods along classical lines (e.g., maximum likelihood estimators, minimum variance unbiased estimators, likelihood ratio tests) is difficult and seems to border on the impossible. On the other hand, computer simulation of these models is often quite feasible if only the models can be expressed in well-defined probability distributions. It will be shown below that computer simulation opens the possibility of (computer-intensive) methods of statistical inference, although these methods are not necessarily statistically efficient, or optimal in other statistical senses.

4. ELEMENTS OF STOCHASTIC ACTOR-ORIENTED NETWORK MODELS

The outcome space for our dynamic network model has the following basic components.

- The *set of actors*. In this paper a fixed and finite set of actors will be considered; addition of new actors, or exit from the network, will not be taken into consideration. The number of actors is denoted by n.
- The *network of relations* between the actors. All relations considered are directed relations (e.g., liking, esteem or power) because of the approach of methodological individualism. Relations commonly considered as undirected relations, such as friendship or cooperation, will be regarded as mutual directed relationships. Relations may be single, but it is more interesting to consider multiple and/or valued relations. Relationships between actors may (but need not) change in time, and may be determined by the social structure (e.g., hierarchy or kinship relations), but it is more interesting when the relationships can be purposely changed by the authors.

Time is considered in this model as a continuous parameter. The set of actors is fixed, but the relation network may change in time. The *state of the model* is the time-dependent value of the network. The probability distribution of the evolution of the model is assumed to be a continuous-time Markov process with finite outcome space. At random times (following a Poisson process of which the parameter is to be estimated), a random actor may make a change in the relations from him (or her) to the other actors. The transition 'mechanism' for this Markov process is described using the following ingredients.

- The *state of each actor* is a function of the state of the model on which depends his evaluation of this state in terms of his well-being. For example, in a model of friendship networks, the state of an actor could be the number of his friends, this concept is not necessary for the construction of the model, but it is often convenient.
- *Preference or utility functions* for each actor, defined as a function of the state of the actor. Utility functions can be the same for all actors, but in more complicated models they may differ between actors.
- The *actions* that an actor can take. Actions refer to changeable relations between the actors and other actors. For example, an action can be a friendship invitation to another actor, or the acceptance of another actor's invitation to a power contest. It is quite common that in the social structure there exist constraints to the actions that an author can take.
- The *choice* made by an actor to perform a certain action (or to refrain from doing so when the opportunity is offered) depends on the actor's

expectation of the utility of his state after the action. Ideally, the actor chooses the action that offers him the highest expected utility. Two limitations to the principle of utility maximization will be taken into account, however. First, the modeled utility functions will not be a perfect representation of the actors' utilities. Therefore, the utilities that propel the actors' choices may also contain random, i.e., unexplained, elements. (Random utility models are commonly used in econometric modeling (see, e.g. (Maddala, 1983; Pudney, 1989).) Second, an actor's future state may depend also on future actions of others, or on other things unknown to him; moreover, the actor's capacity for strategic foresight and general calculations is bounded. Therefore, instead of strictly maximizing his expected utility, each actor may use a *heuristic way* to approximate the expected utility associated to each of the alternatives for action available to him at a given moment. The choice of action by an actor will therefore be modeled as follows: for each alternative action available to the actor, the approximate, or perceived, expected utility is calculated; the actor does not necessarily choose the alternative with the highest perceived utility, but chooses among the alternatives with given probabilities which are a monotone increasing function of the perceived utility.

Analogous to linear regression modeling, these longitudinal models have to account for a degree of unexplained, or *random change*; the various theoretical effects introduced must push back this random aspect and explain the observed change as much as possible. In the process of model building, a sequence of increasingly complicated models can be fitted to the data, starting with a *null model* of random change.

5. ESTIMATION

A more extensive account of the estimation method proposed in this section, and of the example, is given in (Snijders, 1995).

The models considered in the remainder of this paper are Markov processes $(X(t))$ in continuous time of which the probability distribution is parametrized by a k-dimensional parameter θ. It is not assumed that the distribution of $X(t)$ is stationary. For a discrete set of time points $t = \tau_1, \ldots, \tau_M$, with $M \geqslant 2$, observations on $X(t)$ are available. For the type of models we have in mind, the likelihood function is too complicated to calculate, but Monte Carlo computer simulation of $X(t)$ is possible with the time parameter t starting at an arbitrary value τ_0 with a given outcome $x(\tau_0)$. In other words, a random drawing can be simulated from the conditional probability distribution of $X(t)_{t \geqslant \tau_0}$, given $X(\tau_0) = x(\tau_0)$. Because of the intractable likelihood function, estimation principles such as maximum likelihood are inapplicable. Therefore we propose an unconventional estimation method: the method of moments implemented with Monte Carlo simulation. A related approach to estimation, also based on simulated expected values, was proposed by McFadden (1989) and Pakes and Pollard (1989). In this paper a different procedure is proposed, using stochastic approximation to solve the moment equations.

5.1. *Method of moments*

The method of moments proposed here is based on the conditional distributions of $X(\tau_{m+1})$ given $X(\tau_m)$, because we concentrate on the development dynamics of the network. We consider situations where observations at $M = 2$ time points are available; or where no stationarity assumption is made, so that separate estimations ar performed for each time interval $[\tau_{m-1}, \tau_m]$. Parameter θ can be estimated by a conditional moment estimator based on a k-dimensional statistic of the form

$$S\big(X(\tau_{m-1}), X(\tau_m)\big). \tag{1}$$

The function S must be chosen such that its expectation, conditional on $X(\tau_{m-1})$, is a coordinate-wise increasing function of θ (although this monotonicity is not always easy to prove; we often may have to rely on the intuitive plausibility of this monotonicity). For given data $x(\tau_{m-1})$, $x(\tau_m)$ the estimate $\hat{\theta}$ is defined to be the solution of

$$\begin{aligned} &\mathrm{E}_{\hat{\theta}}\big\{S(X(\tau_{m-1}), X(\tau_m)) \mid X(\tau_{m-1}) = x(\tau_{m-1})\big\} \\ &\quad = S(x(\tau_{m-1}), x(\tau_m)). \end{aligned} \tag{2}$$

The delta method can be used to derive an approximate covariance matrix for $\hat{\theta}$. Denote

$$\Sigma_\theta = \mathrm{Cov}_\theta\big\{S(X(\tau_{m-1}), X(\tau_m)) \mid X(\tau_{m-1}) = x(\tau_{m-1})\big\}, \tag{3}$$

$$D_\theta = \frac{\partial}{\partial\theta}\,\mathrm{E}_\theta\big\{S(X(\tau_{m-1}), X(\tau_m)) \mid X(\tau_{m-1}) = x(\tau_{m-1})\big\}; \tag{4}$$

then the approximate covariance matrix of $\hat{\theta}$ is

$$\mathrm{Cov}\,(\hat{\theta}) \approx D_\theta{}^{-1}\Sigma_\theta D'_\theta{}^{-1}. \tag{5}$$

5.2. *Stochastic approximation*

We are in a situation where we wish to solve the equation (2), while we cannot evaluate the left-hand side explicitly, but we do have a possibility to generate random variables with the desired distribution. Stochastic approximation methods, in particular adaptive variants of the Robbins–Monro procedure, can be used to obtain approximate solutions. For an introduction to stochastic approximation and the Robbins–Monro procedure, and results about the consistency of adaptive Robbins–Monro procedures, we refer to (Ruppert, 1991).

The proposed procedure is represented here in abbreviated notation as a recursive procedure to find the value of θ that solves

$$\mathrm{E}_\theta Z = 0 \tag{6}$$

for some random variable Z with probability distribution $F(\theta)$, where Z and θ are k-dimensional. The random variable Z represents the statistic (1) and its distribution is induced by the conditional distribution of $X(\tau_m)$, given $X(\tau_{m-1}) = x(\tau_{m-1})$.

The basic recursion formula for the Robbins–Monro (1951) procedure with step-size $1/N$ is

$$\theta_{N+1} = \theta_N - \frac{1}{N}\, D_N^{-1} Z_N(\theta_N), \tag{7}$$

where $Z_N(\theta)$ is a random variable with expected value $\mathrm{E}_\theta Z$. These steps are made for $N = 1, 2, \ldots$ up to a suitable total number $N_{\max}$. The optimal value of D_N is the derivative matrix $(\partial \mathrm{E}_\theta Z / \partial \theta)$. In adaptive Robbins–Monro procedures (Venter, 1967), this matrix is estimated during the approximation process as an average of difference quotients. For the stable estimation of these difference quotients, it is essential to use common random numbers (see also (Ruppert, 1991, Section 4.3)). The common random numbers technique is a well-known technique in Monte Carlo simulation, used to reduce the variance of differences between generated random variables. This technique operates by generating two or more random variables using the same stream of random numbers, obtained by employing the same initialisation of the random number generator.

In order to estimate D_θ and Σ_θ, we employ the following procedure. Since the parameters θ_j, $j = 1, \ldots, k$, may have different 'natural scales', positive numbers r_j are used that indicate these scales; they should be in the order of magnitude of the standard errors of the estimates of θ_j, i.e., the square roots of the diagonal elements of (5). These can be determined from earlier experience, or by trial and error. Define e_j as the scaled j-th unit vector ($e_{jj} = r_j$, $e_{jh} = 0$ for $h \neq j$) and let c_N denote a non-increasing sequence of small positive numbers. Generate random variables

$$\begin{aligned} Z_{N0} &\sim F\big(\hat\theta_N\big), \\ Z_{Nj} &\sim F\big(\hat\theta_N + c_N e_j\big) \qquad (j = 1, \ldots, k) \end{aligned} \tag{8}$$

using common random numbers. In order to obtain sufficient stability for the resulting process (7), it is advisable that the derivative matrices D_N be estimated with a precision that is not too low, even in the first steps of the process. For this reason, n_0 simulations are carried out before making any steps of the type (7), just to have a sufficiently stable value for D_N. Specifically, values for (8) are simulated for $N = 1 - n_0, \ldots, 0$, with $\hat{\theta}_N = \hat{\theta}_1$, the initial value. The derivatives $\partial \mathrm{E}_\theta Z_i / \partial \theta_j$ are estimated by

$$D_{Nij} = \frac{1}{N + n_0} \sum_{n=1-n_0}^{N} \frac{Z_{nji} - Z_{n0i}}{c_n r_j}, \tag{9}$$

and the recursion process (7) is carried out for $N \geqslant 1$. We used $n_0 = 10$ and $N_{\max} = 200$ or 400.

An estimate of the covariance matrix Σ_θ can also be obtained from the random variables generated in the recursion process. If $\hat{\theta}_N$ is close to its limiting value $\hat{\theta}$, Z_{N0} generated according to (8) will have approximately the covariance matrix Σ_θ. The expected value $\mathrm{E}Z_{N0} - \mathrm{E}_\theta Z$ is approximately $D_\theta(\hat{\theta}_N - \theta)$. As a consequence, Σ_θ can be estimated by

$$\widehat{\Sigma}_\theta = \frac{1}{N} \sum_{n=1}^{N} H_{Nn} H'_{Nn}, \tag{10}$$

where

$$H_{Nn} = \big(Z_{n0} - \bar{Z}_{(N)0} - D_N(\hat{\theta}_N - \bar{\theta}_{(N)})\big),$$

$$\bar{Z}_{(N)0} = \frac{1}{N} \sum_{n=1}^{N} Z_{n0}, \qquad \bar{\theta}_{(N)} = \frac{1}{N} \sum_{n=1}^{N} \hat{\theta}_n.$$

The resulting estimator for the covariance matrix (5) of $\hat{\theta}$ is

$$\widehat{\mathrm{Cov}}\,(\hat{\theta}) = D_N^{-1} \widehat{\Sigma}_\theta D_N'^{\,-1}. \tag{11}$$

All nice properties of adaptive Robbins–Monro procedures that have been mathematically proved, are asymptotic for $N_{\max} \to \infty$. A good starting value for the recursions is important; from a poor starting value, it will take a very large number of recursion steps (7) to reach the solution of (6). An interactive procedure is used, where the drift in the values for θ_N is monitored, and where the value of N can be reset to a lower value if the drift is still appreciable after a relatively large number of steps.

6. NEWCOMB'S FRATERNITY

The book by Newcomb (1961), reports on an extensive longitudinal study of two groups of students living together in a student fraternity house. In this section a longitudinal model is proposed to express some of the theoretical mechanisms that, according to Newcomb's analysis, govern the development of the friendship network in these groups. This model is intended to give an example of the way of modeling proposed in the preceding section, and as a reconstruction of a part of Newcomb's theory.

6.1. *A model incorporating reciprocity and balance*

The set of actors is the set of $n = 17$ men living in the house in year II; the data used are those reported in the UCINET program (Borgatti, *et al.*, 1992). The relational data are given, for each moment where they are available, by sociometric rankings made by each man of the 16 others. We shall interpret this relation as liking. The ranking matrices are available for 15 almost consecutive weeks. (Data for week 9 are missing.) The ranking of actor j by actor i is denoted r_{ij}, where the value 1 indicates highest preference. The vector

$$r_{i*} = (r_{ij})_{j=1,\ldots,n;\ j \neq i}$$

thus is the permutation of the numbers 1 to $n-1 = 16$ indicating the preference ordering of actor i. The entire preference matrix is denoted by $\mathbf{r}$. The diagonal of this matrix is meaningless, and will be conventionally defined as 0. The weeks are indicated by the time parameter $t = 1, \ldots, 16$ ($t \neq 9$). Ranks r_{ij} or matrices $\mathbf{r}$ referring to a specific time point t are denoted $r_{ij}(t)$ or $\mathbf{r}(t)$, respectively.

The state of the actor, mentioned in Section 4, can be taken to consist of the complete preference matrix. (For this simple model, the concepts of the actor's state is not separately needed; it is mentioned here only for the sake of formal completeness.)

The preference function is the crucial part of the model, and must express some principal parts of the sociological theory developed and used by Newcomb. It will be convenient to work with a *tension function* rather than a utility function; a tension function is a function which the actors wish to minimize, and it may be considered equivalent to a constant minus the utility function. When starting with a bounded utility function, the tension function can be defined as the maximum of the utility function minus its present value. (Hoede (1990) and Zeggelink (1994) use tension functions in a similar way.) The principal effects proposed by Newcomb are *reciprocity* and *balance*, and we shall assume that each actor i wishes to minimize a tension function $p_i(\mathbf{r})$ which is the weighted sum of a reciprocity and a balance component. The reciprocity effect means that the actor prefers that others like him to the same degree as he likes them. The corresponding component of the tension function is defined as

$$p_i^{(1)}(\mathbf{r}) = \sum_{j=1,\ j \neq i}^{n} (r_{ij} - r_{ji})^2. \tag{12}$$

The balance effect means that the actor prefers that others to whom he is close view 'the world' in the same way as he views it himself. The group of other persons in the house is considered as a significant part of the world that determines an important part of the balance effect. Accordingly, the balance effect is understood more restrictively as the actor's preference that his friends in the fraternity house have the same preference order for the various other persons in the house as he has himself. For defining the balance component, we use a non-increasing function $\phi(k)$ defined for $k = 1, \ldots, n-1$ which measures the closeness to i of the actor whom he accords rank $r_{ij} = k$. Assuming, somewhat arbitrarily, that the opinions of actor i's 5 closest friends in the house are especially important to him, this function is defined as

$$\phi(k) = \begin{cases} (6-k)/5 & \text{for} \quad k = 1, \ldots, 5; \\ 0 & \text{for} \quad k > 5. \end{cases} \tag{13}$$

The difference between two actors' views of their housemates is measured by the sum of squared differences of rankings,

$$\sum_{h=1,\ h \neq i,j}^{n} (r_{ih} - r_{jh})^2.$$

The balance component of the tension function is defined as

$$p_i^{(2)}(\mathbf{r}) = \sum_{j=1,\ j \neq i}^{n} \phi(r_{ij}) \sum_{h=1,\ h \neq i,j}^{n} (r_{ih} - r_{jh})^2. \tag{14}$$

The entire tension function is now

$$p_i(\mathbf{r}) = \alpha_1 p_i^{(1)}(\mathbf{r}) + \alpha_2 p_i^{(2)}(\mathbf{r}). \tag{15}$$

The parameters α_1 and α_2 indicate the importance of balance and reciprocity, respectively. We assume that the actor has no strategic foresight, but is completely myopic: he chooses (probabilistically) between the various possible actions on the basis of the value that his tension function will have immediately after the action has been taken.

We now come to the actions that can be taken by the actor. These actions are changes in his preference ordering. It is assumed that the changes in the actors' preferences occur slowly, more or less continuously, and that each actor is immediately aware of the changes in the others' preferences. These frequent but small changes are modeled as follows.

The time unit is a week, but time is regarded as a continuous parameter within weeks. Each actor has opportunities for action, i.e., for changing his preference order, at random time points in the week. These opportunities arise independently for the different actors, and follow, for each actor, a Poisson process with common intensity parameter λ. This means that, at every point on the continuous time axis and for every actor, the time that the actor has to wait until his next opportunity for action has a negative exponential distribution with mean $1/\lambda$; for each actor, the number of opportunities for action per week is a random variable having a Poisson distribution with expected value λ. The actions that each actor may take on these moments are *interchanges of preferences*: when actor i has an opportunity for action and if he has adjacent preferences for j and h, e.g., $r_{ij} = k$, $r_{ih} = k+1$ for some number k, $1 \leqslant k \leqslant n-1$, then he may interchange these preferences leading to $r_{ij} = k+1$, $r_{ih} = k$. When the actor has an opportunity for action, he may also leave his preference order the same. Including this last possibility, the total number of possible actions therefore is n. A sequence of many small changes of this kind can change any ordering into any other ordering.

Given the preference orders of all the actors, the n alternative actions available to actor i at a given moment can lead to n different values for the preference matrix $\mathbf{r}$; indicate these values by $\mathbf{r}^{(1)}$ to $\mathbf{r}^{(n)}$. Their tension values for actor i are $p_i(\mathbf{r}^{(1)})$ to $p_i(\mathbf{r}^{(n)})$. One of these values, associated with the 'no change' alternative, is equal to his present tension. It is assumed that the probability of taking an action is a logistic function of the tension values:

$$P\left\{\text{change to } \mathbf{r}^{(k)}\right\} = \frac{\exp\left(-p_i\left(\mathbf{r}^{(k)}\right)\right)}{\sum_{l=1}^{n} \exp\left(-p_i\left(\mathbf{r}^{(l)}\right)\right)}. \tag{16}$$

This function can be motivated by a random utility argument as in (Maddala, 1983). This functional dependence of choice probabilities on utilities is often called the *multinomial logit model.* Note that the exponential function used in (16) has a unit (and therefore non-variable) scale parameter; this accords with the use of the two free scale parameters α_1 and α_2 in the definition of the tension function p_i.

The three parameters in this model are λ, α_1, and α_2. A higher value of λ leads to more rapid change. A higher value of α_1 will tend to push r_{ij} toward r_{ji}. A higher value of α_2 will have two effects: friendship choices (lower values of r_{ij}) of i are encouraged to those j who have similar preferences as i; and changes of i's preferences are encouraged into the direction of the preferences of his closer friends.

The probabilistic model for friendship development in the fraternity is now complete. Mathematically speaking, it is a continuous-time Markov chain for the discrete stochastic matrix $\mathbf{r}$. Special sub-models are:

- $\alpha_1 = \alpha_2 = 0$: purely random change;
- $\alpha_2 = 0$: change on the basis of reciprocity only.

The parameter λ cannot be set to 0, because that would imply that no change occurs at all. It is possible to consider the model where $\alpha_1 = 0$, $\alpha_2 > 0$, where changes occurs on the basis of balance only while reciprocity plays no role. This sub-model seems, however, very implausible theoretically, so we will not pay attention to this possibility.

6.2. *Statistics for moment estimation*

In order to apply the method of moments, we need to choose statistics that capture the effects of the three parameters in the model. The effects of the parameters were indicated above: λ determines the rate of change, α_1 reciprocity, and α_2 balance. A statistic that is relevant for the amount of change from time t, with matrix $\mathbf{r}(t)$, to time $t+1$, with matrix $\mathbf{r}(t+1)$, is the normed sum of squared differences

$$\operatorname{Dist}(t, t+1) = \frac{6}{n^2(n-1)(n-2)} \sum_{1 \leqslant i,j \leqslant n;\ i \neq j} \big(r_{ij}(t+1) - r_{ij}(t)\big)^2 . \tag{17}$$

The norming factor follows from asymptotic considerations under the null model where $\alpha_1 = \alpha_2 = 0$. Statistics that are relevant for the parameters α_1 and α_2 are the totals for reciprocity and balance over the set of all actors:

$$\operatorname{Rec}(t+1) = \frac{2}{n(n-1)} \sum_{1 \leqslant i < j \leqslant n} \big(r_{ij}(t+1) - r_{ji}(t+1)\big)^2 , \tag{18}$$

$$\operatorname{Bal}(t+1) = \frac{1}{c} \sum_{\substack{1 \leqslant i,j \leqslant n \\ i \neq j}} \phi(r_{ij}(t+1)) \sum_{h \neq i,j} \big(r_{ih}(t+1) - r_{jh}(t+1)\big)^2 , \tag{19}$$

where c is a norming constant,

$$c = n(n-2) \sum_{k=1}^{n-1} \phi(k) .$$

The statistic that has the role of (1) in the estimation process, is defined as

$$S(\mathbf{r}(t+1), \mathbf{r}(t)) = \big(\operatorname{Dist}(t, t+1),\ \operatorname{Rec}(t+1),\ \operatorname{Bal}(t+1)\big). \tag{20}$$

The statistics $\operatorname{Rec}(t+1)$ and $\operatorname{Bal}(t+1)$ depend only on $\mathbf{r}(t+1)$, not on $\mathbf{r}(t)$. However, their probability distribution as it is used in the simulations for the method of moments does depend on $\mathbf{r}(t)$, as this is the starting value for the simulations in the time interval $(t, t+1)$.

6.3. *Parameter estimates*

It is not *a priori* clear whether the parameters λ, α_1 and α_2 may be assumed to be constant during the whole observation period. Therefore, results are presented of estimation of parameters separately for all weeks. The estimation has been performed for all transitions from week t to week $t+1$: $(t = 1, \ldots, 15)$, where weeks 8 and 9 have been taken together because data for the beginning of week 9 are missing. The outcomes of statistics (20) are presented in Table 1.

The parameter estimates are presented in Table 2.

The estimates of λ quickly decrease from $\hat{\lambda} = 192$ for the first period to values around 40 for the later periods. This means that larger changes in preferences occurred in the beginning (the persons living in the house did not know each other before moving in) than later. The estimates for α_1 and α_2

Table 1.
Outcomes of distance, reciprocity and balance statistics

Period t	Dist(t, $t+1$)	Rec($t+1$)	Bal($t+1$)
1	0.3538	26.43	29.89
2	0.1934	24.34	28.83
3	0.1500	25.88	26.57
4	0.1597	27.23	22.11
5	0.1199	29.69	19.91
6	0.0872	30.69	21.25
7	0.0810	26.68	20.16
8–9	0.0960	27.53	20.39
10	0.1067	28.28	18.02
11	0.1123	30.22	20.64
12	0.1062	29.51	20.59
13	0.0787	31.03	21.49
14	0.0948	29.84	18.58
15	0.1012	30.94	17.39

Table 2.
Reciprocity with Balance Model: Robbins–Monro Moment Estimates

Period t	$\hat{\lambda}$	S.E.($\hat{\lambda}$)	$\hat{\alpha}_1$	S.E.($\hat{\alpha}_1$)	$\hat{\alpha}_2$	S.E.($\hat{\alpha}_2$)
1	192.1	32.1	0.0071	0.0015	0.0035	0.0008
2	93.8	15.7	0.0104	0.0042	0.0041	0.0023
3	67.0	11.4	0.0058	0.0023	0.0046	0.0009
4	72.6	14.4	0.0058	0.0024	0.0070	0.0011
5	53.0	9.9	0.0039	0.0037	0.0073	0.0013
6	37.0	4.7	0.0034	0.0053	0.0042	0.0017
7	31.5	3.6	0.0247	0.0062	0.0084	0.0016
8–9	42.5	5.9	0.0096	0.0035	0.0061	0.0019
10	45.0	7.6	0.0076	0.0060	0.0088	0.0027
11	50.3	6.9	0.0035	0.0032	0.0035	0.0016
12	46.7	6.7	0.0086	0.0029	0.0057	0.0011
13	32.9	4.1	0.0028	0.0055	0.0044	0.0018
14	38.7	10.5	0.0081	0.0044	0.0092	0.0041
15	44.3	5.7	0.0056	0.0083	0.0075	0.0054

are quite variable; period 7 stands out with a high value for $\hat{\alpha}_1$. The average estimates are 0.0076 for α_1 and 0.0060 for α_2.

Taken individually, many of the parameter estimates for α_1 and α_2 exceed twice their standard error, which suggests that they are significantly larger than 0. To obtain tests for the significance of the α_1 and α_2 parameters in which the 14 periods are combined, the average parameter estimates for α_1 and α_2 were divided by their standard errors. The resulting t-values were 6.2 for α_1 and 9.5 for α_2, indicating strongly significant reciprocity and balance effects.

REFERENCES

Borgatti, S., Everett, M. G. and Freeman, L. C. (1992). *UCINET IV Version 1.0 Reference Manual*. Analytic Technologies, Columbia.

Carley, K. and Banks, D. (1995). Models for fixed-node networks: Model fitting and model testing. *Social Networks* (to appear).

Coleman, J. S. (1990). *Foundations of Social Theory*. Harvard University Press, Cambridge, London.

Fienberg, S. and Wasserman, S. (1981). Categorical data analysis of directed graphs: model for a single generator. In: *Sociological Methodology 1981*, S. Leinhardt (Ed.), Jossey-Bass, San Francisco.

Fienberg, S. E., Meyer, M. M. and Wasserman, S. (1985). Statistical analysis of multiple sociometric relations. *JASA* **80**, 51–67.

Frank, O. (1971). *Statistical Inference in Graphs*. FOA Repro, Stockholm.

Frank, O. (1991). Statistical analysis of change in networks. *Statistica Neerlandica* **45**, 283–293.

Hoede, C. (1990). Social atoms: kinetics. In: *Social Networks Through Time*, pp. 45–63, J. Weesie and H. D. Flap (Eds), ISOR, Utrecht.

Holland, P. and Leinhardt, S. (1975). Local structure in social networks. In: *Sociological Methodology 1976*, D. R. Heise (Ed.), Jossey-Bass, San Francisco.

Holland, P. and Leinhardt, S. (1977a). A dynamic model for social networks. *J. Math. Sociology* **5**, 5–20.

Holland, P. and Leinhardt, S. (1977b). Social structure as a network process. *Zeitschrift für Soziologie* **6**, 386–402.

Holland, P. and Leinhardt, S. (1981). An exponential family of probability distributions for directed graphs. *JASA* **76**, 33–50.

Katz, L. and Proctor, C. H. (1959). The concept of configuration of interpersonal relations in a group as a time-dependent stochastic process. *Psychometrika* **24**, 317–327.

Leenders, R. Th. A. J. (1995). Models for network dynamics: a Markovian framework. *J. Math. Sociology*, to appear.

Maddala, G. S. (1983). *Limited-Dependent and Qualitative Variables in Econometrics*. Cambridge University Press, Cambridge.

Mayer, T. (1984). Parties and networks: stochastic models for relationship networks. *J. Math. Sociology* **10**, 51–103.

McFadden, D. (1989). A method of simulated moments for estimation of discrete response models without numerical integration. *Econometrica* **57**, 995–1026.

Newcomb, T. (1961). *The Acquaintance Process*. Holt, Rinehart & Winston, New York.

Pakes, A. and Pollard, D. (1991). The asymptotic distribution of simulation experiments. *Econometrica* **57**, 1027–1057.

Pudney, S. (1989). *Modelling Individual Choice*. Basil Blackwell, Oxford.

Robbins, H. and Monro, S. (1951). A stochastic approximation method. *Ann. Math. Statist.* **22**, 400–407.

Ruppert, D. (1991). Stochastic approximation. In: *Handbook of Sequential Analysis*, B. K. Gosh and P. K. Sen (Eds), Marcel Dekker, New York.

Snijders, T. A. B. (1990). Testing for change in a digraph at two time points. *Social Networks* **12**, 359–373.

Snijders, T. A. B. (1991). Enumeration and simulation methods for 0-1 matrics with given marginals. *Psychometrika* **56**, 397–417.

Snijders, T. A. B. (1995). Stochastic actor-oriented dynamic network analysis. *J. Math. Sociology*, to appear.

Venter, J. H. (1967). An extension of the Robbins–Monro procedure. *Ann. Math. Statist.* **38**, 181–190.

Wasserman, S. (1977). Stochastic models for directed graphs. Ph.D. dissertation. University of Harvard, Dept. of Statistics.

Wasserman, S. (1979). A stochastic model for directed graphs with transtion rates determined by reciprocity. In: *Sociological Methodology 1980*, K. F. Schuessler (Ed.), Jossey-Bass, San Francisco.

Wasserman, S. (1980). Analyzing social networks as stochastic processes. *JASA* **75**, 280–294.

Wasserman, S. (1987). The conformity of two sociometric relations. *Psychometrika* **52**, 3–18.

Wasserman, S. and Faust, K. (1994). *Social Network Analysis: Methods and Applications*. Cambridge University Press, New York, Cambridge.

Wasserman, S. and Iacobucci, D. (1988). Sequential social network data. *Psychometrika* **53**, 261–282.

Wasserman, W. and Weaver, S. (1985). Statistical analysis of binary relational data: parameter estimation. *J. Math. Psychology* **29**, 406–427.

Zeggelink, E. P. H. (1994). Dynamics of structure: an individual oriented approach. *Social Networks* **16**, 295–333.

Multivar. Statist., pp. 229–234
E.–M. Tiit *et al.* (Eds)

OPTIMAL INDUSTRIAL CLASSIFICATION

J. S. CHIPMAN[1] and P. WINKER[2,*]
[1]Department of Economics, University of Minnesota,
1035 Management and Economics,
271 19th Avenue South, Minneapolis, MN 55455, U.S.A.

[2]Fakultät für Wirtschaftswissenschaften und Statistik,
Universität Konstanz, Postfach 5560, 78434 Konstanz, Germany

1. INTRODUCTION

Economic data on industrial production, commodity prices, and other similar variables are necessarily presented by government statistical agencies in grouped form, for a number of reasons: (1) the number of different industries and commodities they produce is so huge that it would be too difficult and expensive for statistical agencies to present and for economists to digest and use them in ungrouped form; (2) in most countries, presentation of data on groups of fewer than three industries is prohibited by law. The question then arises as to the optimal classification of industries into groups.

Compilers of commodity and industry classification systems appear to follow (as well as to combine) two alternative criteria: (1) the 'horizontal' method of classification according to which crude raw materials such as metal ores, quarrying products, petroleum, timber, live animals, hides and skins, cotton and wool, grains, etc., are classified together as 'basic materials'; semi-finished goods such as iron and steel, cement, chemicals, wood, paper, textile fabrics, flour, etc., are grouped together as 'semi-finishes'; and ready-to-use products such as automobiles, plastics, utensils, furniture, books, apparel, food, etc. are combined as 'finished products'. On the other hand, the 'vertical' approach groups commodities according to the materials of which they are composed, so that lumber, wood, paper, furniture, and books are in one group, hides and skins, cotton and wool, fabrics, and apparel in another, etc.

In this paper we adopt an objective criterion of classification with regard to price movements along the following lines. We suppose, as economic theory

*Research was supported by the Deutsche Forschungsgemeinschaft, Sonderforschungsbereich 178, 'Internationalisierung der Wirtschaft' at the University of Konstanz; National Science Foundation grant SES-8607652 at the University of Minnesota; and a Senior U. S. Scientist Award from the Humboldt Foundation.

predicts, that within a country following fairly liberal trading policies, domestic price movements will closely follow movements in world prices. A linear homogeneous multivariate multiple-regression model is postulated with the average import and export prices of some 30 to 40 groups of commodities as exogenous (independent) variables and the average domestic prices of these same groups of commodities as endogenous (dependent) variables. The objective is to partition these industries into 6 groups. The number of 6 groups is given by the 'official' mode of aggregation for German industrial production prices. Of course, the method presented in the sequel could be applied to any other given number of groups.

As the classification criterion presented in the following section leads to a problem of high computational complexity, an optimization heuristic is used in order to obtain a 'very good' classification for German price data. The basic idea of the Threshold Accepting algorithm is outlined in Section 3 followed by a comparison of 'official' and 'optimized' groupings of the German data in Section 4.

2. THEORETICAL BACKGROUND

For the German data, the 6 domestic prices could be predicted either by first performing the regression of the 30 to 40 groups and then aggregating the predicted domestic prices into six groups; or by first aggregating the 30 to 40 groups to six, then performing the regression on the grouped data. Following Chipman (1975) the problem can be formulated in terms of the multivariate multiple-regression model

$$Y = XB + E, \tag{1}$$

where Y is the $n \times m$ matrix of n observations on the (endogenous) domestic prices of m commodities, X is the $n \times 2m$ matrix of n observations on exogenous import and export prices, B is a $2m \times m$ matrix of unknown regression coefficients to be estimated, and E is a random $n \times m$ matrix of error terms with zero mean and covariance

$$\mathcal{E}\{(\operatorname{col} E)(\operatorname{col} E)'\} = \Sigma \otimes V, \tag{2}$$

where 'col E' denotes the column vector of successive columns of E, Σ is the $m \times m$ simultaneous covariance matrix and V the $n \times n$ sample covariance matrix. $\mathcal{E}$ denotes the expectation operator. We shall assume that V is positive definite.*

Letting G and H respectively denote $2m \times 2m^*$ and $m \times m^*$ (proper) *grouping matrices*, i.e., matrices with exactly one nonzero (in fact, positive) element in

*The more general case rank $V \leqslant n$ is treated in (Chipman, 1975). The computational results presented in Section 4 were obtained for $V \equiv I_n$.

each row and at least one nonzero element in each column,* it is customary to deal with an aggregative model

$$Y^* = X^* B^* + E^* \tag{3}$$

mimicking the true one, where

$$X^* = XG \quad \text{and} \quad Y^* = YH$$

are $n \times 2m^*$ and $n \times m^*$ matrices of observations on $2m^*$ and m^* aggregative exogenous and endogenous variables respectively.

Following the arguments of Chipman (1975, 1976) a parameter called the 'aggregation bias' can be derived as a mean-square error of forecast by using the grouped regression rather than the detailed regression (Chipman and Winker, 1992, pp. 4–8). Therefore, one may choose as criterion function

$$\alpha = \operatorname{tr} H'B'(I - GG^{\#})'X'V^{-1}X(I - GG^{\#})BHW^* , \tag{4}$$

where $G^{\#}$ is any matrix satisfying

$$X'V^{-1}XGG^{\#}G = X'V^{-1}XG \quad \text{and} \quad X'V^{-1}XGG^{\#} = (X'V^{-1}XGG^{\#})' ,$$

and W^* is some symmetric positive-definite matrix. One may choose the Euclidean metric $W^* = I_{m^*}$; alternatively one may use the 'Mahalanobis distance' defined by the choice

$$W^* = (H'\Sigma H)^{-1}. \tag{5}$$

The main advantage of the Mahalanobis metric is that the measure of aggregation bias is independent of units of measurement.

3. OPTIMIZATION

The problem for the application to price indices is to find a partition of the industries into six groups that minimizes the aggregation bias. This problem falls under the heading of integer programming problems, and with regard to its computational complexity is similar to problems such as the classic travelling salesman problem. In fact, it falls under the heading of so-called NP-complete problems (Winker, 1992) that is, optimization problems for which there exists no algorithm giving an exact solution with a reasonable amount of time and financial resources.

---

*See (Chipman, 1975, p. 135). For the application considered in this paper the prices are already multiplied by their weights. Hence, the nonzero elements in the grouping matrices are restricted to ones. Furthermore, the grouping of the endogenous and exogenous variables was assumed to be identical. Consequently, we find $G \equiv H \otimes I_2$.

We pursue this problem by employing a heuristic integer optimization algorithm known as Threshold Accepting (Dueck and Scheuer, 1991; Dueck and Winker, 1992; Winker, 1994a, 1994b). This is a refined local-search algorithm.

The algorithm starts with an arbitrary partition. The concept of Hamming distance (Hamming, 1950) is used to define a local structure on the set of all possible grouping matrices by neighborhoods; it is defined as the number of differing entries in two grouping matrices of the same order. In each iteration, an element of the sphere of radius 4 around the current grouping matrix is randomly selected as the new current solution. This solution is accepted if it is *not much worse* than the previous one in terms of aggregation bias. It is necessary to admit a temporary worsening of the objective function during the optimization procedure in order to escape from local minima. The criterion for 'not much worse' is determined by a threshold sequence chosen in advance. In our application, the threshold sequence decreases from about 4 percent to 0 percent during the optimization.

Several concepts for defining a local structure by neighborhoods were tried out. Starting from the trivial neighborhood including the whole set of proper grouping matrices smaller neighborhoods were achieved by the use of the Hamming distance with different radii. A reduction in the size of the neighborhoods leads to more stable local structures while at the same time the number of possible exchange steps is reduced for each iteration. The choice of a Hamming distance of 4 proved to give the best results for our application. Further experiments were undertaken for the threshold sequence.

In general, the Threshold Accepting algorithm converges to the global optimum for a 'good choice' of neighborhoods and threshold sequence as the number of iterations goes to infinity (Althöfer and Koschnik, 1991).

4. RESULTS FOR THE APPLICATION TO GERMAN PRICE INDICES

The algorithm presented in the preceeding section has been applied to a model of the 'generalized Stolper–Samuelson mapping' in which domestic prices in a country are functions — via the minimum-unit-cost functions — of external (import and export) prices. With fixed technical coefficients, these functions are linear and homogeneous. The exogenous variables are import and export prices and the endogenous variables are domestic producer prices for the corresponding groups.

Monthly data for Germany from January 1970 to January 1983 were used, covering 37 commodity categories, to be aggregated into six groups as for a given official grouping. Since the natural way to group them is by forming weighted averages with given weights, it was most convenient to work with the price indices multiplied by their weights. Unpublished import and export price-index data of this type, called 'Wertziffern,' have been furnished by the Statistisches Bundesamt, Wiesbaden, for the Federal Republic of Germany. In the case of the internal producer-price index, *Wertziffern* were not available, and the series used were the published price indices multiplied by their weights.

Aggregation then means just summation and the nonzero elements of the grouping matrices are all ones. We considered the series of 37 commodity groups to be aggregated into six groups. There exists an official method of grouping these 37 industries into six groups which makes it possible to compare our results with results based on the official grouping. The 37 industries are two-and three-digit categories (and because of lack of data in some mining categories, combinations of some three-digit categories) called *Güterzweige*, and the six groups of industries are called *Gütergruppen*.

Starting with a random grouping, we obtained after 50,000 iterations (in about 5000 seconds for one of 10 trials on an IBM RS6000/340 workstation) a grouping with aggregation bias of 17.75, as compared with the official grouping with an aggregation bias of 72.95. This result may be suboptimal, but the general experience with Threshold Accepting and several attempts made for this problem allow for a strong feeling that it is a good solution both numerically and economically.

The official grouping may be described as a 'horizontal' one in which commodities are grouped according to stage of production, whereas the optimal grouping may be charcterized as a 'vertical' one in which groups containing certain final products also include the relevant products in the previous stages of production.*

Of course, the resulting optimized groups might be suboptimal from an economic point of view. However, not only do the statistical results advise the presented grouping, but to be fair, one should be aware of the fact that the grouping of 37 commodities in only 6 groups cannot give a completely obvious solution. Nevertheless, the method of optimal aggregation by Threshold Accepting gives a grouping superior to the official one from the econometric point of view and at least as convincing in its economic interpretation.

5. CONCLUSION

It was not the purpose of the paper to judge the method of aggregation in large-scale econometric models. But considering the computational results for the example presented one might state that results of aggregative models depend strongly on the chosen mode of aggregation. Hence, in using the common practice of aggregation, one should be aware of the possible consequences of some official modes of grouping. For linear regression models the methods presented in this paper open a way to better groupings of given variables.

The results for the application to price indices allow for the tentative conclusion that for the aggregation of these data a vertical grouping is more appropriate than the official horizontal one.

---

*For tables of the resulting groupings and a more detailed discussion of the economic meaning of the groups the reader is refered to (Chipman and Winker, 1994).

REFERENCES

Althöfer, I. and Koschnick, K.-U. (1991). On the convergence of 'Threshold Accepting'. *Appl. Math. Optim.* **24**, 183–195.

Chipman, J. S. (1975). Optimal aggregation in large-scale econometric models. *Sankhyā, Series C* **37**, 121–159.

Chipman, J. S. (1976). Estimation and aggregation in econometrics: An application of the theory of generalized inverses. In: *Generalized Inverses and Applications*, pp. 553–773, M. Z. Nashed (Ed.), Academic Press, New York.

Chipman, J. and Winker, P. (1992). Optimal aggregation by threshold accepting. Discussionpaper No. 180, SFB 178, University of Konstanz.

Chipman, J. and Winker, P. (1994). Optimal industrial classification. Discussionpaper, No. 236, SFB 178, University of Konstanz.

Dueck, G. and Scheuer, T. (1991). Threshold Accepting: A general purpose algorithm appearing superior to simulated annealing. *J. Comput. Phys.* **90**, 161–175.

Dueck, G. and Winker, P. (1992). New concepts and algorithms for portfolio choice. *Appl. Stochastic Models Data Anal.* **8**, 159–178.

Hamming, R. W. (1950). Error detecting and error correcting codes. *Bell System Technical J.* **29**, 147–160.

Winker, P. (1992). Some notes on the computational complexity of optimal aggregation. Discussionpaper No. 184, SFB 178, University of Konstanz.

Winker, P. (1994a). Identification of multivariate AR-Models by Threshold Accepting. *Comput. Statist. Data Anal.*, to appear.

Winker, P. (1994b). The tuning of the Threshold Accepting heuristic for travelling salesman problems. Mimeo, University of Konstanz.

Part Five

DISCRIMINATION AND CLASSIFICATION

Multivar. Statist., pp. 237–245
E.–M. Tiit *et al.* (Eds)

SMALL SAMPLE PROPERTIES OF RIDGE-ESTIMATE OF THE COVARIANCE MATRIX IN STATISTICAL AND NEURAL NET CLASSIFICATION

ŠARŪNAS RAUDYS and MARINA SKURIKHINA
Department of Data Analysis, Institute of Mathematics and Informatics, Akademijos 4, Vilnius 2600, Lithuania

ABSTRACT

We derive an analytical formula for the generalization error of linear adaptive classifiers trained with weight decay. Analytical and experimental results are then presented to analyse the optimal value of regularization parameters as a function of the training set size.

1. INTRODUCTION

In sociological and ecological investigations data sets have a large number of features, typically 100–200 and more. A statistical dependence between single components of the data is very often non-linear. So it can be essential to use additional polynomial features in order to evaluate non-linear dependencies mentioned. Thus dimensionality of the feature space becomes particularly large. Standard classical statistical methods such as the linear Fisher's discriminant function (DF)

$$g(\mathbf{X}) = \left[\mathbf{X} - \frac{1}{2}\left(\overline{\mathbf{X}}^{(1)} + \overline{\mathbf{X}}^{(2)}\right)\right]' \mathbf{S}^{-1}\left(\overline{\mathbf{X}}^{(1)} - \overline{\mathbf{X}}^{(2)}\right), \tag{1}$$

require the inverse of the covariance matrix; where

$\mathbf{S}$ is standard maximum likelihood estimation of the $p \times p$ covariance matrix Σ,

$\mathbf{X}$ is a p-variate vector to be classified,

$\overline{\mathbf{X}}^{(i)}$ is the sample mean vector of the i-th class.

That is principally impossible when the number of features exceeds the number of training vectors. A common technique used to overcome this difficulty is to add some constant values to diagonal elements of the estimate of the covariance matrix

$$\mathbf{S}_{\text{ridge}} = \mathbf{S} + \lambda\mathbf{I}, \tag{2}$$

where

$\mathbf{I}$ is the $p \times p$ identity matrix,

λ is a regularization parameter.

For simplicity, we analyse the case of two classes with an equal number of training vectors, N, in each class. Analysis of the general (multiclass, unequal priors) case may be carried out along the same line.

The DF (1) allocates vector $\mathbf{X}$ according to its sign to one of the 2 classes.

Using expression (2) in equation (1) we get the regularized discriminant function

$$g^{R}(\mathbf{X}) = \left[\mathbf{X} - \frac{1}{2}\left(\overline{\mathbf{X}}^{(1)} + \overline{\mathbf{X}}^{(2)}\right)\right]' (\mathbf{S} + \lambda\mathbf{I})^{-1} \left(\overline{\mathbf{X}}^{(1)} - \overline{\mathbf{X}}^{(2)}\right). \tag{3}$$

The regularized DF was analysed by Friedman (1989) and other researchers mentioned in the reference list of this paper. In order to find the best value for the regularization parameter, Friedman proposed to use a cross validation technique. For reducing the computational burden he developed updating formulas for the computation of the covariance matrix.

In extremely high dimensional case (e.g., using the second order polynomials, instead of 100 initial features we get 5150 new ones) the matrix $\mathbf{S}_{\text{ridge}}$ is ill-conditioned. In the present paper we will show that addition of the regularization term $\lambda\mathbf{W}'\mathbf{W}$ to the cost function of a linear adaptive element (ADALINE) is equivalent to computing Fisher's linear DF with a 'ridge' estimate of the covariance matrix ($\mathbf{S}_{\text{ridge}}$) instead of usual $\mathbf{S}$. The cost function of a linear neural net with a standard 'weight decay' is

$$C = \frac{1}{4N} \sum_{j=1}^{2N} \left[t_j - (\mathbf{W}'\mathbf{X}_j + w_0)\right]^2 + \frac{1}{2}\lambda\mathbf{W}'\mathbf{W}, \tag{4}$$

where

$2N$ is the number of training vectors,

t_j is a desired output (target) of the network when the j-th training vector $\mathbf{X}_j$ is presented,

w_0 is a threshold,

$\mathbf{W}$ is the weight vector of Artificial Neural Net (ANN),

λ is a regularization constant.

Thus using a standard iterative ANN training procedure we can avoid the inverse of the matrix and realize non-linear analysis of the sociological data in the very high dimensional case.

In this paper we will derive an analytical expression for the expected probability of misclassification (generalization error) as a function of training vectors and the value of the regularization parameter λ. We will show both analytically and by simulation that the optimal value of λ depends on the number of training samples and tends to zero when N increases.

2. THE STANDARD LINEAR DF AND A LINEAR PERCEPTRON

In this section we will show that minimization of (4) leads to the regularized DF (3).

Let $t_j = 1$ for the class π_1 and $t_j = -1$ for the class π_2. The gradient of the cost function (4) with respect to the threshold w_0 is

$$\frac{\partial C}{\partial w_0} = \frac{1}{2}\mathbf{W}'\left(\overline{\mathbf{X}}^{(1)} + \overline{\mathbf{X}}^{(2)}\right) + w_0. \tag{5}$$

Then the optimal value of w_0 is

$$w_0^\star = -\frac{1}{2}\mathbf{W}'\left(\overline{\mathbf{X}}^{(1)} + \overline{\mathbf{X}}^{(2)}\right). \tag{6}$$

Inserting (6) into (4) and differentiating the resulting expression with respect to $\mathbf{W}$ leads to

$$\frac{\partial C}{\partial \mathbf{W}} = -\frac{1}{2}\Delta\overline{\mathbf{X}} + \left[\mathbf{S} + \frac{1}{4}\Delta\overline{\mathbf{X}}\,\Delta\overline{\mathbf{X}}' + \lambda\mathbf{I}\right]\mathbf{W}, \tag{7}$$

where $\Delta\overline{\mathbf{X}} = \overline{\mathbf{X}}^{(1)} - \overline{\mathbf{X}}^{(2)}$.

By equating (7) to zero and solving the matrix equation we get

$$\mathbf{W} = \frac{1}{2}\left[(\mathbf{S} + \lambda\mathbf{I}) + \frac{1}{4}\Delta\overline{\mathbf{X}}\,\Delta\overline{\mathbf{X}}'\right]^{-1}\Delta\overline{\mathbf{X}}. \tag{8}$$

Below we will use Bartlett's formula (Bartlett, 1952)

$$(\mathbf{S} + \mathbf{V}\mathbf{U}')^{-1} = \mathbf{S}^{-1} - \frac{\mathbf{S}^{-1}\mathbf{V}\mathbf{U}'\mathbf{S}^{-1}}{1 + \mathbf{V}'\mathbf{S}^{-1}\mathbf{U}} \tag{9}$$

which gives the expression of the inverse of quadratic matrix. In (9) $\mathbf{S}$ is a $p \times p$ matrix, $\mathbf{V}$ and $\mathbf{U}$ are p-variate vectors.

Using (9) in (8) gives

$$\begin{aligned}\mathbf{W} &= \frac{1}{2}(\mathbf{S} + \lambda\mathbf{I})^{-1}\Delta\overline{\mathbf{X}}\left(1 - \frac{\frac{1}{4}\Delta\overline{\mathbf{X}}'(\mathbf{S} + \lambda\mathbf{I})^{-1}\Delta\overline{\mathbf{X}}}{1 + \frac{1}{4}\Delta\overline{\mathbf{X}}'(\mathbf{S} + \lambda\mathbf{I})^{-1}\Delta\overline{\mathbf{X}}}\right) \\ &= (\mathbf{S} + \lambda\mathbf{I})^{-1}\left(\overline{\mathbf{X}}^{(1)} - \overline{\mathbf{X}}^{(2)}\right)K,\end{aligned} \tag{10}$$

where

$$K = \frac{1}{2}\left(1 - \frac{\frac{1}{4}\Delta\overline{\mathbf{X}}'(\mathbf{S} + \lambda\mathbf{I})^{-1}\Delta\overline{\mathbf{X}}}{1 + \frac{1}{4}\Delta\overline{\mathbf{X}}'(\mathbf{S} + \lambda\mathbf{I})^{-1}\Delta\overline{\mathbf{X}}}\right)$$

is a scalar.

Thus the resulting DF is

$$g(\mathbf{X}) = \mathbf{W}'\mathbf{X} + w_0$$
$$= \left[\mathbf{X} - \frac{1}{2}\left(\overline{\mathbf{X}}^{(1)} + \overline{\mathbf{X}}^{(2)}\right)\right]' (\mathbf{S} + \lambda\mathbf{I})^{-1}\left(\overline{\mathbf{X}}^{(1)} - \overline{\mathbf{X}}^{(2)}\right)K. \tag{11}$$

We see that the DF (11) differs from the regularized DF (3) only by the scalar K.

Hence we conclude that the regularized DF is equivalent to using the standard weight decay term in the regularization of a linear perceptron. This is a generalization of a former result by Koford and Groner (1966) and Gallinari *et al.* (1991) — for classification. In the case of regression a similar result has been obtained by Sjöberg and Ljung (1991).

3. THE EXPECTED PROBABILITY OF MISCLASSIFICATION (PMC)

The expected PMC (generalization error) EP_N of the classification rule is the misclassification probability of an arbitrary vector $\mathbf{X}$, when the classifier has been trained using a randomly chosen training set of size $2N$.

Let $\mathbf{X}$ be a multivariate Gaussian vector and let us suppose that the two classes do have equal covariance matrices. Then for a large number of features the distribution of DF (1) will be close to Gaussian one and we can calculate (Raudys, 1972; Raudys and Pikelis, 1980)

$$EP_N = \Phi\left\{\frac{\delta}{2}\left[\frac{(2N-3)(2N-p-3)}{(2N-p-2)(2N-p-5)}\left[1+\frac{1}{N}\left(1+\frac{2p}{\delta^2}\right)+\frac{1}{N^2}\frac{p}{\delta^2}\right] + \frac{\delta^2}{2(2N-p-5)}\right]^{-1/2}\right\}, \tag{12}$$

where

δ^2 is a squared Mahalanobis distance: $\delta^2 = (\mu_1 - \mu_2)'\Sigma^{-1}(\mu_1 - \mu_2)$,

μ_1 and μ_2 are mean vectors,

$\Phi(U)$ is a standard $N(0,1)$ Gaussian cumulative distribution function.

To make expression (12) more simple for a further analysis we will consider asymptotic results when $N \to \infty$ and $p \to \infty$ with a similar rate and δ^2 being kept limited, not very small constant. Then (12) becomes

$$EP_N = P_N \to \Phi\left\{-\frac{\delta}{2}\left(\frac{2N}{2N-p}\left(1+\frac{2p}{N\delta^2}\right)\right)^{-1/2}\right\}. \tag{13}$$

It was shown by Wyman *et al.* (1990), that (13) is surprisingly accurate. Using a Taylor expansion

$$(\mathbf{S} + \lambda\mathbf{I})^{-1} = \mathbf{S}^{-1} - \lambda\mathbf{S}^{-2} + \cdots \tag{14}$$

after some simple but tedious algebra, in a way similar to (Raudys, 1972), for the regularized DF we obtain

$$EP_N \longrightarrow \Phi\left\{-\frac{\delta}{2}\left(\frac{2N}{2N-p}\left(1+\frac{2p}{N\delta^2}\right)\right)^{-1/2}\frac{\sqrt{1+\lambda B}}{1+\lambda C}\right\}, \tag{15}$$

where

$$B = \frac{2}{1-y}\beta_2 + \frac{\operatorname{tr}\mathbf{D}^{-1}}{2N}, \qquad C = \frac{1}{1-y}\beta_1, \qquad y = \frac{p}{2N},$$

$$\beta_i = \frac{\mathbf{M}'\mathbf{D}^{-1}\mathbf{M}}{\delta^2}\alpha_i + \frac{\operatorname{tr}\mathbf{D}^{-1}}{2N-p}, \qquad i = 1, 2,$$

$$\alpha_1 = 1, \qquad \alpha_2 = \left(1+\frac{2\operatorname{tr}\mathbf{D}^{-1}}{N\mathbf{M}'\mathbf{D}^{-1}\mathbf{M}}\right)\left(1+\frac{2p}{N\delta^2}\right)^{-1},$$

here $\mathbf{D}$ is the diagonal matrix of Σ eigenvalues, such that $\mathbf{T}\Sigma\mathbf{T}' = \mathbf{D}$ ($\mathbf{T}$ is a $p \times p$ orthogonal matrix), $\mathbf{M} = \mathbf{T}'(\mu_1 - \mu_2)$.

When $\lambda \to 0$ both equations (13) and (15) become identical.

4. THE OPTIMAL VALUE OF THE REGULARIZATION PARAMETER

Differentiating the integral (15) with respect to λ and equating the derivative to zero we obtain the optimal value of the regularization parameter

$$\lambda_{\text{opt}} = \frac{B - 2C}{BC}. \tag{16}$$

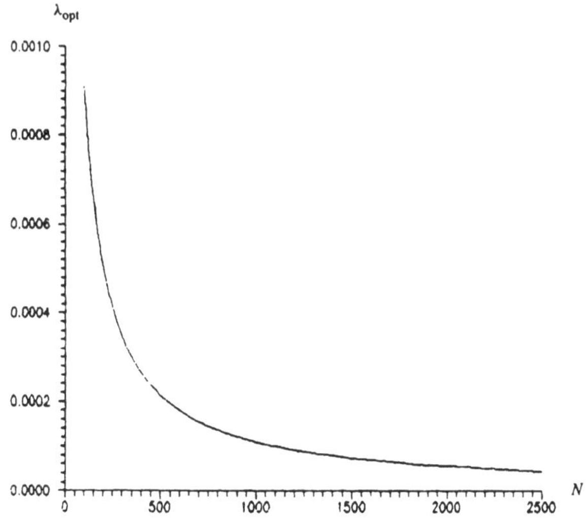

Figure 1. Dependence of λ_{opt} as a function of N.

Here λ_{opt} depends on the number N of the training vectors as well as on the parameters of the true distributions of the classes: δ^2, p, $\mathbf{M}'\mathbf{D}^{-1}\mathbf{M}$, and $\mathrm{tr}\,\mathbf{D}^{-1}$.

The graph of the dependence of λ_{opt} as a function of N, computed for $p = 8$, $\delta^2 = 24.9$, $\mathbf{M}'\mathbf{D}^{-1}\mathbf{M} = 579.6$, and $\mathrm{tr}\,\mathbf{D}^{-1} = 228.1$, is shown in Figure 1. These parameters of artificial data were also used in simulation studies presented in the next section.

We see that the optimal value of λ decreases when the number of training vectors N increases.

5. SIMULATION STUDY WITH A LINEAR ADAPTIVE CLASSIFIER

To derive the equation (15) we used the approximation (14) and made the following assumptions

(i) the number of dimensions is large,

(ii) the distribution of (3) is close to Gaussian one,

(iii) the regularization parameter λ is small.

In order to analyse the validity of the above results obtained from asymptotic expression (15) for the case of small p and large λ, simulation studies with artificial Gaussian data were performed. We generated eight-variate Gaussian data with strong dependencies between the variables and used the standard

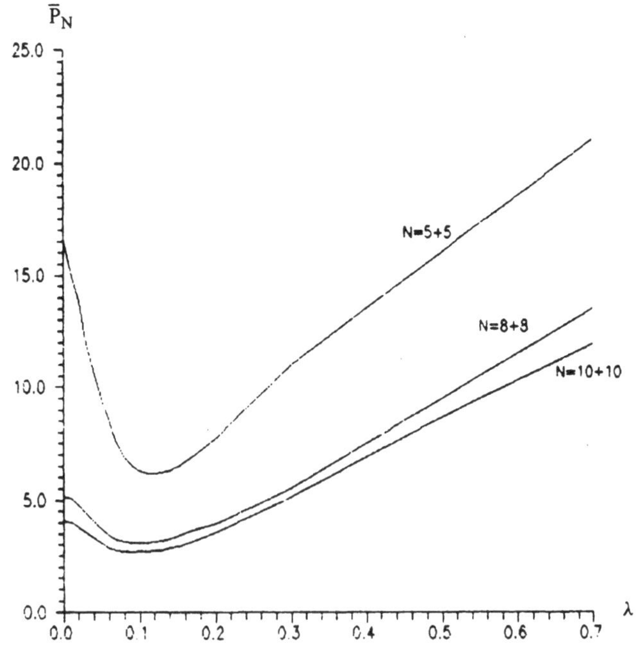

Figure 2. Generalization error EP_N versus the regularization parameter λ. Regularized discriminant analysis.

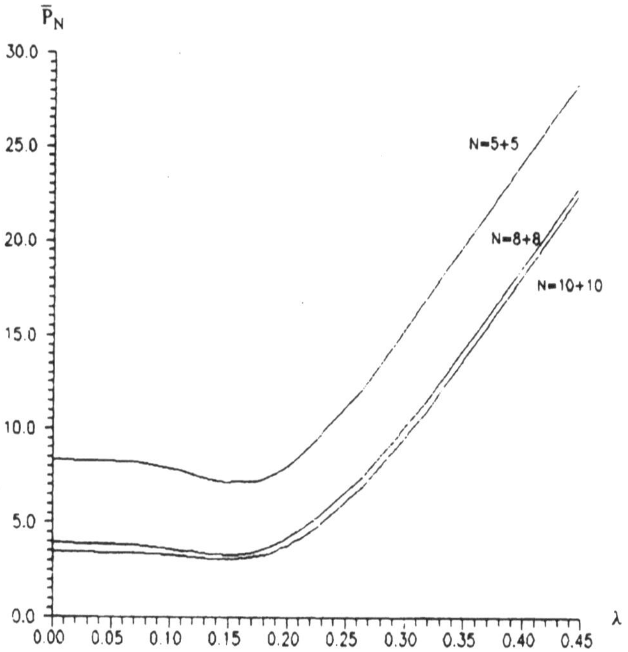

Figure 3. Generalization error EP_N versus the regularization parameter λ. Linear perceptron with weight decay term.

linear DF (1), the regularized DF (3), a linear perceptron with the regularization term (4) to classify the data.

These two series of experiments with different sizes $N = 5$, $N = 8$ and $N = 10$ of training sets were repeated 20 times. In Figures 2 and 3 we present mean values $\overline{P}_N$ obtained from 20 repetitions of the experiments.

We see that graphs obtained using these two regularization techniques are similar. They confirm that

- the two regularization techniques discussed are equivalent in the main point,
- the function, describing the dependence of the generalization error on the value of the regularization parameter, has a minimum,
- the optimal value of λ is inverse proportional to the number N of training vectors.

6. SIMULATION STUDY WITH NONLINEAR PERCEPTRONS

In order to verify that theoretical conclusions about the existence of an optimal value for the regularization parameter do apply to multilayer perceptrons as well, we have used the perceptron with 20 nodes in the hidden layer and

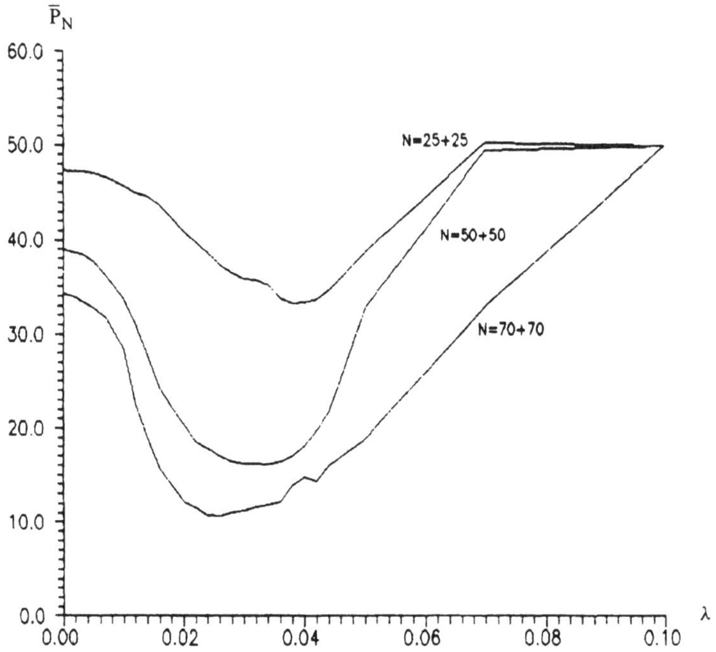

Figure 4. Generalization error EP_N versus the regularization parameter λ. Multilayer perceptron with weight decay term.

eight-variate artificial data. First class vectors were composed of eight-variate Gaussian $N(0, \mathbf{I})$ vectors $\mathbf{X} = (x_1, x_2, \ldots, x_8)$ lying inside a circle $x_1^2+x_2^2 \leqslant 1.1$, and the second class of similar vectors lying outside a circle $x_1^2+x_2^2 \geqslant 1.45$. The number of training vectors in the experiment was chosen as $N = 25$, $N = 50$ and $N = 70$. Results presented in Figure 4 were averaged over 7 repetitions on different randomly chosen training sets. We see again that the generalization error depends heavily on the value of the regularization parameter λ. It can be observed that the optimal value of λ decreases with an increase in the number of training samples.

It is worth mentioning that the experimental graphs presented in the figures show only the mean behaviour of the generalization error. Performances obtained for separate training sets differ significantly with their minimum having a large variance.

7. CONCLUSIONS

It was shown that the addition of the standard weights decay term to the mean squared error used in the training linear adaptive element is equivalent to regularized discriminant analysis. For the latter, the asymptotic equation for the generalization error was obtained, that allows to find the optimal value for the regularization parameter λ. Simulation experiments performed with artificial data with the help of linear and nonlinear perceptrons confirmed that such minima do exist for these systems, and that the optimal value of the regularization parameter decreases when the number of training samples increases. In spite of the fact that the formula obtained does not allow to calculate λ_{opt} analytically it has been shown to be useful in reducing the amount of simulations required for finding an optimal regularization.

REFERENCES

Fukunaga, K. (1972). *An Introduction to Statistical Pattern Recognition.* Academic Press, New York.

Friedman, J. M. (1989). Regularized discriminant analysis. *JASA* **84**, 165–175.

Bartlett, M. S. (1952). An inverse matrix adjustment arising in discriminant analysis. *Ann. Math. Statist.* **22**, 107.

Koford, J. S. and Groner, G. F. (1966). The use of an adaptive threshold element to design a linear optimal pattern classifier. *IEEE Trans. on Information Theory* **IT–12**, 42–50.

Gallinari, P., Thiria, S., Badran, F. and Fogelman–Soulié, F. (1991). On the relations between discriminant analysis and multi–layer Perceptrons. *Neural Networks* **4**, 349–360.

Sjöberg, J. and Ljung, L. (1991). Overtraining, regularization and searching for a minimum in neural networks. Technical Report. Department of Electrical Engineering, Linköping University.

Raudys, Š. (1972). On the amount of a priori information in designing the classification algorithm. *Tekhn. Kibernet.* **4**, 168–174.

Raudys, Š. and Pikelis, V. (1980). On dimensionality, sample size, classification error and complexity of classification algorithm in pattern recognition. *IEEE Trans. on Pattern Analysis and Machine Intelligence* **PAMI–2**, 242–252.

Wyman, F., Young, D. M. and Turner, D. W. (1990). A comparison of asymptotic error rate expansion for the sample linear discriminant function. *Pattern Recognition* **23**, 775–783.

Multivar. Statist., pp. 247–265
E.–M. Tiit *et al.* (Eds)

TRIMMED k-MEANS AND THE CAUCHY MEAN-VALUE PROPERTY*

J. A. CUESTA-ALBERTOS[1], A. GORDALIZA[2] and C. MATRÁN[2]
[1]Departamento de Matemáticas, Estadística y Computación,
Universidad de Cantabria, Spain

[2]Departamento de Estadística, Universidad de Valladolid, Spain

ABSTRACT

We analyze how a procedure of impartial trimming (determined by the data) distributes its mass on the boundary of the trimming region. This analysis leads us to study under what circumstances the multivariate Φ-means (or M-estimators) satisfies the Cauchy-mean value property.

We conclude that the mean is in some sense the only natural multivariate Φ-mean.

1. INTRODUCTION AND DEFINITIONS

Many authors have suggested the use of k-means as a tool for finding clusters in a given set of data (Hartigan, 1975; Kaufman and Rousseeuw, 1990) or as a quantization procedure in information theory (see, for instance, the special issue (IEEE Transactions on Information Theory, 1982)).

The definition of k-means is as follows. Let P be a probability measure defined on the Borel σ-algebra in $\Re^d$, $d \geqslant 1$, and let $\|\cdot\|$ be the Euclidean norm on this space. A set with k elements (k-set in the sequel) $M^P = \{m_1, \dots, m_k\} \subset \Re^d$ is said to be a *k-mean* for P if M^P minimizes the expression

$$V(M) := \int_{\Re^d} \inf_{m \in M} \|x - m\|^2 P(\mathrm{d}x)$$

over all k-sets $M \subset \Re^d$.

Properties of k-means appear, for instance, in (Arcones and Giné, 1992; Flury, 1990; Hartigan, 1978; Pollard, 1981; Pollard, 1982; Seriko and Babu, 1992). Extensions to abstract spaces have been put forward in the literature but here we are interested in the natural generalization consisting in minimizing the integral of a function of the distance (see, for instance (Cambanis and Gerr, 1983; Cuesta-Albertos and Matrán, 1988; Pärna, 1988; Pärna, 1990;

*Research partially supported by DGICYT, grants PB91-0306-C02-00, 01 and 02.

Sverdrup-Thygeson, 1981)). In this way if $\Phi\colon \Re^+ \to \Re^+$ is an increasing and convex function we say that $M_\Phi^P = \{m_1, \ldots, m_k\}$ is a *k-Φ-mean* for P if M_Φ^P minimizes the expression

$$V_\Phi(M) := \int_{\Re^d} \inf_{m \in M} \Phi\left(\|x - m\|\right) P(\mathrm{d}x) \tag{1}$$

over all k-sets $M \subset \Re^d$.

Clearly k-Φ-means have advantages because they allow great versatility in the clustering and quantization processes. Furthermore k-Φ-means are interesting from the point of view of robustness if they are regarded as centralization measures.

Let us comment on the last issue (see (Cuesta-Albertos *et al.*, 1993)). The 1-mean is the usual mean. Therefore k-means share the problems of lack of robustness that the mean has. If we follow the developments made for the mean in the one-dimensional case, a first attempt to robustify k-means could be to generalize them to 'k-M-estimators' or k-Φ-means as defined in (1).

The notion of k-Φ-means seems to be a good solution to the robustness problem because it includes the median (if $d = k = 1$ and Φ is the identity function), but unfortunately robustness is only achieved in the case $k = 1$. For instance, let us assume that $k = 2$, $\lim_{x\to\infty} \Phi(x) = \infty$, and that we have a finite sample $\{x_1, \ldots, x_n\}$ in the real line. If we add, with a positive but arbitrarily small weight, a new point x_0 to the sample in a far remote position and compute the 2-Φ-mean, M_Φ, of the new sample $\{x_0, x_1, \ldots, x_n\}$, then necessarily $x_0 \in M_\Phi$. Thus, if $k > 1$, even by choosing the generalization of the paradigm of the robustness (through $\Phi(x) = x$ the k-median) the breakdown point of the procedure is 0.

A solution to this problem, suggested in (Cuesta-Albertos *et al.*, 1993), resorts to a trimming procedure. It consists in first deleting the 'most isolated' points in the sample till a fixed proportion α of the total mass, and then computing the k-Φ-mean of the remaining points. This procedure is characterized by two facts.

1. The points to be removed from the sample are chosen by examination of the data. There is no need for fixing in advance any 'privileged' direction or zone where to look for isolated points.
2. Trimming makes it possible to exclude not only extreme points but also inner isolated points.

For isolated points, the impartial trimming procedure introduced in (Gordaliza, 1991a, 1991b) was used. This procedure, when applied to our problem works as follows (see (Cuesta-Albertos *et al.*, 1993)). Let $\alpha \in (0, 1)$ and define

$$\mathcal{T}_\alpha := \left\{\tau\colon \Re^d \to [0,1], \text{ measurable and } \int \tau \, \mathrm{d}P \geqslant 1 - \alpha\right\}.$$

The family $\mathcal{T}_\alpha$ contains all possible trimmings at level α for P because $\mathcal{T}_\alpha$ includes all possible subsets with probability $\geqslant 1 - \alpha$ along with some randomized trimmings to guarantee the existence of an optimal trimming even in those situations in which P is not continuous.

For $\tau \in \mathcal{T}_\alpha$, let $M_\Phi^\tau \subset \Re^d$ be the k-Φ-mean of P given τ, i.e., $M_\Phi^\tau \subset \Re^d$ is the k-set which minimizes

$$V_\tau(M) := \int \tau(x) \inf_{m \in M} \Phi\left(\|x - m\|\right) P(\mathrm{d}x) \tag{2}$$

on the familly of k-sets $M \subset \Re^d$. We have to find $\tau_\alpha \in \mathcal{T}_\alpha$ such that

$$V_{\tau_\alpha}\left(M_\Phi^{\tau_\alpha}\right) = \inf_{\tau \in \mathcal{T}_\alpha} V_\tau\left(M_\Phi^\tau\right).$$

We say that τ_α is an *α-optimal trimming function* for P and that $M_\Phi^{\tau_\alpha}$, or simply M_Φ^α, is an *α-trimmed k-Φ-mean.*

The preceeding considerations can be extended to define α-L_∞-optimal trimming functions and α-k-L_∞-means, by replacing the integrals in (2) by the corresponding essential supremum with respect to the probability P.

In the sequel, if there is no risk of confusion, we will often say 'trimmed k-mean' instead of α-trimmed k-Φ-mean and 'α-optimal function' instead of α-optimal trimming function.

It is proved in (Cuesta-Albertos *et al.*, 1993) that, under suitable conditions on Φ, an α-optimal function and a trimmed k-mean exist but they are in general not unique. Also known is the fact that every α-optimal function lies between the topological interior and the closure of a union of k balls whose centers form a trimmed k-mean. Moreover all of these balls have the same radius. The result is the following (see (Cuesta-Albertos *et al.*, 1993)).

PROPOSITION 1.1. *Let P be a probability measure on $\Re^d$. Let $\alpha \in (0,1)$, $k \in \mathcal{N}$ and let Φ: $\Re^+ \to \Re^+$ be a continuous, strictly increasing function such that $\Phi(0) = 0$. Then:*

1. *There exists an α-optimal function τ_α.*
2. *If τ_α is an α-optimal function, then there exist an α-trimmed k-Φ-mean, M_Φ^α, and $r > 0$ such that if we set $B := \bigcup_{m \in M_\Phi^\alpha} b(m,r)$ and $\overline{B} := \bigcup_{m \in M_\Phi^\alpha} \bar{b}(m,r)$ then*

$$I_B \leqslant \tau_\alpha \leqslant I_{\overline{B}},$$

where I_A denotes the indicator function of the set A and $b(m,r)$ (respectively, $\bar{b}(m,r)$) is the open (resp., closed) ball with centre at m and radius r.

For results on the consistency of these centralization measures see (Cuesta-Albertos *et al.*, 1993).

Here we are concerned with how P distributes its mass on the topological boundary of B and on the set of points in B which are at the same distance of two different points in the associated trimmed k-mean, i.e., on the 'internal' boundaries. We devote Section 2 to this question. The main result in Section 2 (Theorem 2.4) states that internal boundaries have probability zero and that the α-trimming set coincides a.s. with an open set or with a closed set or, else, all P-probability of its topological boundary is concentrated at a single point. The proof relies on a careful analysis of the concept of Φ-mean. Recall that given a probability P we say that $m_\Phi \in \Re^d$ is a *Φ-mean of P* if

$$\int_{\Re^d} \Phi(\|x - m_\Phi\|)P(\mathrm{d}x) = \inf_{m\in\Re^d} \int_{\Re^d} \Phi(\|x - m\|)P(\mathrm{d}x).$$

(Φ-means are the non-trimmed 1-Φ-means). The concepts of Φ-means given a trimming function or a measurable set are defined in an obvious way.

In general, Φ-means of P are not necessarily unique (consider for instance the median as the particular case where $d = 1$ and Φ is the identity). However we show (Proposition 2.1) that the Φ-mean is unique if Φ is strictly convex. The idea behind the proof of Theorem 2.4 is the following heuristic principle (where the sense of closeness has to be made precise).

Given two measurable sets A, B with positive P-probability, if we respectively denote by m_A and $m_{A\cup B}$ the Φ-means of P given A and $A \cup B$, then, unless that Φ is a crazy function, $m_{A\cup B}$ is closer to the points in B than m_A is.

We can say that in what follows we try to characterize 'non-crazy functions'. Surprisingly, the conclusion seems to be that the only convex, increasing non-crazy function is $\Phi(t) = t^2$. A possibles formalization of this principle appears in (Cuesta-Albertos *et al.*, 1994). It relies on the fact that if we assume that the Φ-mean of a set is a good indicator of the location of this set, then the relative positions of $m_{A\cup B}$ and m_B with respect to m_A should be the same. A precise formulation of this property is the following.

PROPERTY 1.2. *Given Φ: $\Re^+ \to \Re^+$ and two Borel measurable sets A, B with positive P-probability, if we respectively denote by m_A and m_B the Φ-mean of P given A and B, and $b \in \Re^d$ satisfies that $\langle b, m_B - m_A\rangle \geqslant 0$ then the Φ-mean of P given $A \cup B$, $m_{A\cup B}$, satisfies that $\langle b, m_{A\cup B} - m_A\rangle \geqslant 0$. (We denote by $\langle x, y\rangle$ the usual scalar product between $x, y \in \Re^d$).*

The first result in Section 3 (Theorem 3.1) shows that Property 1.2 is equivalent to the Cauchy mean-value property (CMVP in the sequel). It seems that this property was introduced for probabilities on $\Re$ with a finite support in (Gini, 1958) and was extended to general one-dimensional (not trimmed) Φ-means in (Brons *et al.*, 1969). It can be stated as follows.

DEFINITION 1.3. Given a convex, increasing function Φ we say that the Φ-mean possesses the *Cauchy mean-value property* if given two Borel measurable sets A, B with positive P-probability measure, if we respectively denote by m_A and m_B the Φ-mean of P given A and B, then the Φ-mean of P given $A \cup B$ is a convex linear combination of m_A and m_B.

In (Brons *et al.*, 1969, Theorem 2.3) it is shown that, in the one-dimensional case, Φ-means possess the CMVP.

In Section 3 (which includes several results taken from (Cuesta-Albertos *et al.*, 1994)) we analyze under what circumstances the Φ-means possess the CMVP if $d > 1$. The conclusions are mostly negative if we exclude the usual mean. We begin with Example 3.2 which shows that the strict CMVP, defined in an obvious way from the additional requeriment that both coefficients in the linear combination are non-zero, does not necessarily hold even in the real line when the function Φ is strictly convex. As we do not know of any reference giving conditions leading to the strict CMVP we give in Theorem 3.3 a relevant result. The interest of strict CMVP was made apparent in (Robertson and Wright, 1980).

The main result in Section 3 is Theorem 3.4 which shows that the CMVP of a multivariate Φ-mean characterizes the mean. So, from the stated point of view, it seems surprising that the only reasonable Φ-mean is the usual mean. A possible explanation for this fact has been sugested in (Cuesta-Albertos *et al.*, 1994) and is based on the idea that given a hyperplane and the associated halfspaces it may be senseless to try to decide in which halfspace lies a set when it intersects both. In this way we give a new stronger definition (Definition 3.5) just for sets A and B which belong to different halfspaces.

Investigation as to whether this definition includes Φ-means different from the usual mean is still under progress. We have obtained partial results (Theorem 3.7 and Example 3.9) which suggest that the answer is, once again, negative.

2. TRIMMED k-MEANS

In this section we analyze the way in which P distributes its mass on the boundaries of the trimming set. We first need to study the relation between Φ-means and trimmed k-means.

It is evident that, if τ_α is an α-optimal function and $M_\Phi^\alpha = \{m_1, \ldots, m_k\}$ and r are the associated trimmed k-mean of P and the real number obtained by applying Proposition 1.1, then M_Φ^α induces a partition of $\bigcup_{m \in M_\Phi^\alpha} \bar{b}(m, r)$ into k clusters in the following way: the cluster A_i consists of all points in $\bigcup_{m \in M_\Phi^\alpha} \bar{b}(m, r)$ which are closer to m_i than to the other $k-1$ points in M_Φ^α. The points at the same distance from two elements in M_Φ^α are arbitrarily assigned.

Moreover M_Φ^α also induces a splitting of the quantity

$$D_\alpha := \frac{1}{1-\alpha} \int \tau_\alpha(x) \inf_{m \in M_\Phi^P} \Phi(\|x - m\|) P(\mathrm{d}x)$$

into the contributions corresponding to each cluster

$$D_\alpha = \frac{1}{1-\alpha} \sum_{i=1}^{k} \int_{A_i} \tau_\alpha(x) \Phi(\|x - m_i\|) P(\mathrm{d}x).$$

Now m_i, $i = 1, \ldots, k$, has to be a Φ-mean of the corresponding cluster A_i, or more precisely a Φ-mean of P given τ_α and A_i, because, on the contrary, we could diminish the variation in some clusters by replacing m_i, $i = 1, \ldots, k$, by a Φ-mean of the corresponding cluster and then M_Φ^α would not be a trimmed k-mean of P.

As a consequence we conclude that the uniqueness of the trimmed k-mean depends not only on the uniqueness of the trimming function but also on the unicity of the Φ-mean.

It is not difficult to find examples where the unicity of the trimming function does not hold. For instance, if we consider $\alpha \in (0,1)$, $k = 1$, and assume that P is a uniform distribution on a ball, it is evident that we have infinite possibilities for trimming.

The unicity of the trimming function is not important to solve the main question in this section. On the other hand the unicity of the Φ-mean will be essential. We give a relevant result in the next proposition.

PROPOSITION 2.1. *If Φ is strictly convex, then the Φ-mean of P is unique.*

Proof. Let us assume that $m, m^* \in \Re^d$ are two different Φ-means of P and let $\lambda \in (0,1)$, $m_\lambda := \lambda m + (1-\lambda)m^*$. Taking into account that Φ is increasing and convex, we have the following inequalities

$$\begin{aligned} \int \Phi\big[\|x - m_\lambda\|\big] P(\mathrm{d}x) &\leqslant \int \Phi\big[\lambda\|x - m\| + (1-\lambda)\|x - m^*\|\big] P(\mathrm{d}x) \\ &\leqslant \lambda \int \Phi\big[\|x - m\|\big] P(\mathrm{d}x) + (1-\lambda) \int \Phi\big[\|x - m^*\|\big] P(\mathrm{d}x) \\ &= \int \Phi\big[\|x - m\|\big] P(\mathrm{d}x). \end{aligned} \tag{3}$$

Since Φ is strictly convex, the last inequality is strict unless $\|x - m\| = \|x - m^*\|$, P-a.s. This only happens if

$$P\left\{x + \frac{m + m^*}{2} \colon \langle x, m - m^* \rangle = 0\right\} = 1, \tag{4}$$

where $\langle \cdot, \cdot \rangle$ denotes the usual scalar product in $\Re^d$.

On the other hand, if x verifies that $\langle x, m - m^* \rangle = 0$, then

$$\left\|\left(x + \frac{m + m^*}{2}\right) - m_\lambda\right\| < \left\|\left(x + \frac{m + m^*}{2}\right) - m\right\|.$$

The strict convexity of Φ implies that Φ is strictly increasing. Therefore, if (4) holds, then the first inequality in (3) is strict, which is not possible because m is a Φ-mean of P.

Thus we not only have that the trimmed k-mean induces a partition of $\bigcup_{m\in M_\Phi^\alpha}\bar{b}(m,r)$ into k clusters but also that, if Φ is strictly convex and we exclude degenerate cases in which a cluster has P-probability zero, then the partition determines the trimmed k-mean. We summarize these considerations in the following proposition.

PROPOSITION 2.2. *With notation as above, m_i is a Φ-mean of P given τ_α and A_i, $i=1,\dots,k$. Moreover, if Φ is strictly convex and $P[A_i]>0$ then m_i is the only Φ-mean of P given τ_α and A_i, $i=1,\dots,k$.*

Another property of the Φ-means appears in the following lemma which is a simple consequence of the separating hyperplane theorem.

LEMMA 2.3. *Let us assume that Φ is strictly convex and that A is a convex set. Then the Φ-mean of P given A belongs to the closure of A.*

Note that in this lemma we can not assure that the Φ-mean of P belongs to A (see Example 2.4 in (Cuesta-Albertos *et al.*, 1994)).

By using these ideas it is possible to prove the following result which restricts the way in which P can assign probability to the boundary of the clusters. The technique that we employ in the final step in the proof is related to that used in Theorem 2.2 in (Cuesta-Albertos *et al.*, 1994).

THEOREM 2.4. *We use the same notation as in Propositions 1.1 and 2.2. If $M_\Phi^\alpha=\{m_1,\dots,m_k\}$ and we assume that Φ is strictly convex and that its first derivative exists and is continuous, then we have that*

1. $P\{x\in B\colon \|x-m_i\|=\|x-m_j\|\}=0$, *if $i\neq j$.*
2. *Let D be the topological boundary of the set $\bigcup_{m\in M_\Phi^\alpha} b(m,r)$. Then the optimal trimming function τ_α satisfies one of the following relations:*
   (a) $\int_D \tau_\alpha(x)P(\mathrm{d}x)=0$ *(i.e., the topological boundary of $\bigcup_{m\in M_\Phi^\alpha} b(m,r)$ does not lie at all in the optimal trimming set).*
   (b) $\int_D \tau_\alpha(x)P(\mathrm{d}x)=P(D)$ *(i.e., the topological boundary of $\bigcup_{m\in M_\Phi^\alpha} b(m,r)$ is contained in the optimal trimming set).*
   (c) *There exists one point x_0 in D such that all the probability mass of D is concentrated at x_0, i.e., $P[D]=P\{x_0\}$.*

Proof of Statement 1. Let $i,j\in\{1,\dots,k\}$, $i\neq j$. With the definition of trimmed k-mean in (2) we avoid the degenerate case in which $m_i=m_j$. We can assume that it is possible to construct A_i and A_j such that $\int_{A_i}\tau_\alpha(x)P(\mathrm{d}x)$ and $\int_{A_j}\tau_\alpha(x)P(\mathrm{d}x)$ are strictly positive; in other case the result would be trivial.

From Proposition 2.2 we know that m_i and m_j are the Φ-means of P given τ_α and the clusters A_i and A_j, respectively. Let us denote $B_{i,j}:=\{x\in$

B: $\|x - m_i\| = \|x - m_j\|\}$. Note that

$$\int_{A_i \cup A_j} \tau_\alpha(x) \inf_{l=1,\ldots,k} \Phi(\|x - m_l\|) P(\mathrm{d}x)$$

$$= \int_{A_i \cup A_j} \tau_\alpha(x) \inf \left[\Phi(\|x - m_i\|), \Phi(\|x - m_j\|)\right] P(\mathrm{d}x)$$

$$= \int_{A_i \cup B_{i,j}} \tau_\alpha(x) \Phi(\|x - m_i\|) P(\mathrm{d}x) + \int_{A_j - B_{i,j}} \tau_\alpha(x) \Phi(\|x - m_j\|) P(\mathrm{d}x).$$

Proposition 2.1 implies that the Φ-mean of P given τ_α and $A_i \cup B_{i,j}$, the Φ-mean of P given τ_α and $(A_i - B_{i,j})$ and m_i coincide, because, in other case, $\{m_1, \ldots, m_k\}$ would not be a trimmed k-mean for P.

Next we prove that, if $P[B_{i,j}] > 0$, then the Φ-means of P given τ_α and the sets $A_i - B_{i,j}$ and $A_i \cup B_{i,j}$ are not the same. The result follows from this.

Let m and m^1 be the Φ-means of P given τ_α and the sets $A_i - B_{i,j}$ and $B_{i,j}$, respectively. We have that

$$\int_{A_i - B_{i,j}} \tau_\alpha(x) P(\mathrm{d}x) > 0$$

because, otherwise, by Lemma 2.3, m_i would belong to the closure $\overline{B}_{i,j}$ of $B_{i,j}$, and $B_{i,j}$ would not be the boundary between A_i and A_j. As τ_α is strictly positive on $B \supset B_{i,j}$, we have that m and m^1 are well defined and $m \neq m^1$, because, otherwise $m_i = m^1$ and then m_i would belong to $\overline{B}_{i,j}$.

Consider an orthogonal basis, $\{e_1, \ldots, e_d\}$ in $\Re^d$, with $e_1 = m^1 - m$ and assume that, we fix the origin in m, so that $m^1 = (1, 0, \ldots, 0)$. Let us set $m^t = (t, 0, \ldots, 0)$, and define

$$\Phi'\left[\|x - m^{t_0}\|\right] := \frac{\mathrm{d}}{\mathrm{d}t} \Phi\left[\|x - m^t\|\right] \Big|_{t=t_0}.$$

Since $A_i \cup B_{i,j}$ is a bounded set, the continuity of the first derivative of Φ allows us to conclude that the functions

$$H_1(t) := \int_{A_i - B_{i,j}} \tau_\alpha(x) \Phi(\|x - m^t\|) P(\mathrm{d}x)$$

and

$$H_2(t) := \int_{B_{i,j}} \tau_\alpha(x) \Phi(\|x - m^t\|) P(\mathrm{d}x)$$

are differentiable with continuous derivatives given by

$$H_1'(t) = \int_{A_i - B_{i,j}} \tau_\alpha(x)\,\Phi'\big(\|x - m^t\|\big)\,P(\mathrm{d}x)$$

and

$$H_2'(t) = \int_{B_{i,j}} \tau_\alpha(x)\,\Phi'\big(\|x - m^t\|\big)\,P(\mathrm{d}x),$$

respectively.

By definition of m, $H_1'(0) = 0$. Since $B_{i,j}$ is contained in B, Proposition 1.1 implies that $\tau_\alpha(x) = 1$ for every $x \in B_{i,j}$ and a similar proof to that of Proposition 2.1 shows that H_2 is strictly convex in $[0, 1]$. Taking into account that the minimum of H_2 is reached at $t = 1$, we obtain that $H_2'(0) < 0$.

Let us finally consider the function

$$H(t) := \int_{A_i \cup B_{i,j}} \tau_\alpha(x)\,\Phi\big(\|x - m^t\|\big)\,P(\mathrm{d}x) = H_1(t) + H_2(t).$$

The preceding considerations show that H is decreasing on an interval $[0, t_0]$ with $t_0 > 0$. Thus m would not be the Φ-mean of P given $A_i \cup B_{i,j}$, so that $P[B_{i,j}] = 0$.

Proof of Statement 2. To avoid trivialities, we assume that $P[D] > 0$. Let S be the support of the conditional distribution of P given D, and suppose that

$$P\{x \in S\colon\ \tau_\alpha(x) \in (0, 1)\} > 0.$$

In this case, there exists $\epsilon > 0$ such that $P(S_\epsilon) > 0$ where

$$S_\epsilon := \{x \in S\colon\ 0 < \tau_\alpha(x) < 1 - \epsilon\}.$$

Assume that S_ϵ is an infinite set. So that, there exists i such that $S_i := S_\epsilon \cap A_i$ is also an infinite set with $P(S_i) > 0$.

Consider an orthogonal basis in $\Re^p$. At least a projection of S_i is infinite. Thus there exists a hyperplane H such that if we denote by H_+ and H_- the two half-spaces in which $\Re^p$ is divided by H, then $p := P(S_i \cap H_+) > 0$ and $q := P(S_i \cap H_-) > 0$. Let us assume that $p \geqslant q$. Let β be a real number such that $1 < \beta < \inf(1/(1-\epsilon), (p+q)/q)$. A simple computation gives us that there exists $\gamma \in (0, 1)$ such that if we define

$$\tau_1(x) = \begin{cases} \tau_\alpha(x) & \text{if } \ x \notin S_i, \\ \beta\tau_\alpha(x) & \text{if } \ x \in S_i \cap H_-, \\ \gamma\tau_\alpha(x) & \text{if } \ x \in S_i \cap H_+; \end{cases} \tag{5}$$

then $\int \tau_1(x)P(dx) = \int \tau_\alpha(x)P(dx)$ and τ_1 is a trimming function at level α.

The Φ-mean of P given a set depends on the trimming function we are considering, but in any case, similarly as in Statement 1, the Φ-mean of P given τ_α and $b(m_i, r)$ coincides with the Φ-mean of P given τ_1 and $b(m_i, r)$; but this is not possible by the same argument as in the final part of the proof for Statement 1.

Assume that S_ϵ is finite and contains at least two points. In this case, there exists a hyperplane H such that if we denote H_+ and H_- to the associated half-spaces, then there exists $x_+ \in S_\epsilon \cap H_+$ and $x_- \in S_\epsilon \cap H_-$ such that $P(x_+) > 0$ and $P(x_-) > 0$ and we can construct a trimming function at level α by a similar construction to that in (5) to get a contradiction.

Therefore, we have proved that, if the support of P given the boundary of $\bigcup_{m \in M_\Phi^\alpha} b(m, r)$ contains more than a point, the trimming function τ_α can only take the values 0 and 1 except in a point at most.

If we are in case (a) or (b) the proof is finished. In other case there exists an hyperplane H verifying that either

$$P\big[H_+ \cap \{x \in D\colon\ \tau_\alpha(x) = 1\}\big] > 0 \quad \text{and} \quad P\big[H_- \cap \{x \in D\colon\ \tau_\alpha(x) < 1\}\big] > 0$$

or

$$P\big[H_+ \cap \{x \in D\colon\ \tau_\alpha(x) = 0\}\big] > 0 \quad \text{and} \quad P\big[H_- \cap \{x \in D\colon\ \tau_\alpha(x) > 0\}\big] > 0$$

and we can make a construction similar to that in the proof of the first step to get again a contradiction.

3. CAUCHY MEAN-VALUE PROPERTY

Let us analyze the proof of Theorem 2.4. To simplify matters, we assume that $\tau_\alpha = 1$. The proof is based on the following idea. Let A, B be two sets such that $A - B$ and B are convex with positive probability and let m and m^1 denote the Φ-means of P given $A - B$ and B, respectively. We show there that if we assume that $m \neq m^1$, then there exists m^* in the segment joining m with m^1 such that

$$\int_{A \cup B} \Phi\big[\|x - m^*\|\big] P(dx) < \int_{A \cup B} \Phi\big[\|x - m\|\big] P(dx), \tag{6}$$

which contradicts Proposition 2.1.

However (6) looks reasonable according to the principle stated in the introduction and it is very close to Property 1.2. Our aim in this section is to characterize those Φ-means possessing this property.

Property 1.2 is equivalent to the CMVP. This result is taken from (Cuesta-Albertos *et al.*, 1994) and we include its proof because it exhibits the relation between both two properties.

THEOREM 3.1. *CMVP and Property* 1.2 *are equivalent.*

Proof. It is evident that CMVP implies Property 1.2.

Let Φ: $\Re^+ \rightarrow \Re^+$ a function such that the associated Φ-mean holds Property 1.2. Let A, B be two Borel measurable sets and let m_A, m_B and $m_{A\cup B}$ be the Φ-means of A, B and $A\cup B$, respectively. The result is trivial if $m_A = m_B$. So let us assume that $m_A \neq m_B$. Let $T_B := \{b \in \Re^d\colon \langle b, m_B - m_A\rangle \geqslant 0\}$ and consider the set

$$U_B := \{x \in \Re^d\colon \langle b, x - m_A\rangle \geqslant 0, \text{ for every } b \in T_B\}.$$

Note that U_B coincides with the one-dimensional cone $\{m_A + \delta(m_B - m_A)\colon \delta > 0\}$ and that the Φ-mean of P given $A \cup B$ belongs to U_B because of Property 1.2.

We analogously show that $m_{A\cup B}$ belongs to the one-dimensional cone $\{m_B + \delta(m_A - m_B)\colon \delta > 0\}$ and then we have that $m_{A\cup B}$ lies in the segment joining m_A with m_B.

After this theorem, our task in this section can be redefined by saying that we want to analyze under which conditions a Φ-mean possesses CMVP. This is not a trivial question. For instance, it may be suspected that the hypothesis on the differentiability of Φ in Theorem 2.4 is only of a technical nature. That is, the only essential points in order to obtain this result are that the Φ-mean is unique and that the functions H_1 and H_2 in the proof are strictly convex. However this is not the case, as the next example shows.

Example 3.2. Let $h > 1$ be a constant to be determined later and define

$$\Phi(x) = \begin{cases} x^2 & \text{if } \; 0 \leqslant x \leqslant 1, \\ hx^2 - (h-1) & \text{if } \; x \geqslant 1. \end{cases}$$

It is evident that Φ is strictly convex and that its first derivative exists at every point except $x = 1$.

Now let $p = (1 + 2^{1/2})^{-1}$ and let P be a probability on $\Re^2$ such that

$$\begin{aligned} &P\left[(2^{-1},0)\right] = (1-p)/2, \\ &P\left[(-2^{-1/2},2^{-1/2})\right] = P\left[(-2^{-1/2},-2^{-1/2})\right] = p/4, \\ &P\left[(2^{-1/2},2^{-1/2})\right] = P\left[(2^{-1/2},-2^{-1/2})\right] = 1/4. \end{aligned}$$

If we set

$$\begin{aligned} A &= \left\{(1/2,0), (-2^{-1/2},2^{-1/2}), (-2^{-1/2},-2^{-1/2})\right\}, \\ B &= \left\{(2^{-1/2},2^{-1/2}), (2^{-1/2},2^{-1/2})\right\}, \end{aligned}$$

then the Φ-mean of P given A is $(0,0)$ which coincides with the Φ-mean of P given $A\cup B$ if we choose h big enough, even when the functions corresponding to H_1 and H_2 are strictly convex. Therefore if we choose $A_i - B_{i,j} = A$ and $B_{i,j} = B$, (6) does not hold for this Φ.

The proof of Theorem 2.4 is based on the idea that the inclusion of a new point on a given data set makes the Φ-mean to move unless the point coincides with the Φ-mean. On the other hand, Example 3.2 works against this idea by showing that the Φ-mean can be excessively shy to changes for not too regular functions Φ.

Since this example can be immediately reconstructed in the real line, it is clear that it is necessary to impose some additional condition on Φ to assure, even in the line, that the 'Φ-mean moves'. However this additional condition may be deduced from the argument in Theorem 2.4: We must assume continuos differentiability on Φ. In this case, trivially, the CMVP of the Φ-means in the real line as stated in Theorem 2.3 in (Brons *et al.*, 1969) plus the fact that the Φ-mean can not remain unchanged allows us to conclude that, under the given conditions, strict CMVP holds. We state this in the following theorem.

THEOREM 3.3. *The Φ-mean of a probability measure on the line possesses the strict Cauchy mean-value property if Φ is a strictly convex and continuously differentiable function.*

An additional remark on the proof of Theorem 2.4. We have shown there that, under appropriate hypotheses, by adding mass in a direction, the Φ-mean moves, but nothing is said about the direction of the resulting motion. In fact, the CMVP as stated in Definition 1.3 would give complete information on this point. The accuracy given by such a definition is so large that it has a drawback: it is only true for the mean. We show this in the following theorem taken from (Cuesta-Albertos *et al.*, 1994). We include the first steps of its proof because they give a feeling for the kind of difficulty we have to handle.

THEOREM 3.4. *Let Φ: $\Re^+ \to \Re^+$ be continuous, increasing and strictly convex and let us assume that the associated Φ-mean possesses the CMVP. Let P be a probability measure such that $\int \Phi(\|x - m\|)P(\mathrm{d}x) < \infty$ for some $m \in \Re^d$ and that its support is not contained in a one-dimensional subspace. Then there exists the expectation of P and coincides with its Φ-mean.*

Proof. Let assume that the support S_P of P is finite and that the distribution of P is uniform on it. Note that, if $S_P = \{x_1, x_2\}$, then the strict convexity of Φ implies that the Φ-mean of P is $(x_1 + x_2)/2$ and result is proved in this case.

We now apply induction. Let us assume that $S_P = \{x_1, \ldots, x_n\}$ and that the result holds if we have a uniform probability supported by at most $n - 1$ points.

By hypothesis, there exist at least three extreme non aligned points y_1, y_2, y_3 in the convex hull of S_P. Let $i \in \{1, 2, 3\}$ and consider the set $S^i := S_P - \{y_i\}$. The Φ-mean of P given $\{y_i\}$ is y_i and, by the induction hypothesis, m^i, the Φ-mean of P given S^i, is the arithmetical mean of the points in S^i

The fact that y_i is an extreme point in the convex hull of S_P implies that $y_i \neq m^i$. Thus, by the CMVP, the point m_Φ lies in the segment joining the point y_i with the arithmetic mean of the points in $S_P - \{y_i\}$.

The last statement holds for $i = 1, 2, 3$. The points y_1, y_2, y_3 are not aligned, so that the segments joining the points y_i with m_i intersect in at most a point; but the Φ-mean holds the CMVP, therefore they, in fact, intersect and the intersection point is the Φ-mean. On the other hand, the mean is characterized by this property, and we have that the mean and the Φ-mean of P coincide.

The argument for general P consists in applying the consistency result in (Cuesta-Albertos, 1984) to the Φ-means of the sequence of empirical probability measures of a simple random sample of P that, by the inverse part of the SLLN, gives as a by-product the existence of the expectation of P.

In (Cuesta-Albertos *et al.*, 1994) we made an attempt to extend CMVP to the finite dimensional setting by restricting our attention to the case in which the sets A and B are separated by a hyperplane. It seems to us that no further reasonable weakening of this property is possible.

DEFINITION 3.5. We say that the Φ-mean possesses the *CMVP** if, given A, B two Borel measurable sets with strictly positive P-probability such that there exists $b \in \Re^d$ such that $\langle b, x\rangle > h$ if $x \in A$ and $\langle b, x\rangle < h$ if $x \in B$, then the Φ-means m_A and $m_{A\cup B}$ of P given A and $A \cup B$, respectively, satisfy that $\langle b, m_{A\cup B} - m_A\rangle \leqslant 0$.

We want to remark that if we replace the last inequality by a strict one, we would have a *strict* CMVP* which we know does not hold in general (see Example 3.2). However, it is also evident that the mean possesses that property.

As stated in the introduction, the investigation of Φ-means possessing CMVP* is yet unfinished. Till now we have only derived some partial results relative to p-means, $1 \leqslant p \leqslant \infty$. These results, which are included below, suggest again that the mean is the only Φ-mean possessing CMVP*.

The *p-means* when $1 \leqslant p < \infty$ are the particular cases of Φ-means obtained by taking $\Phi(t) = t^p$. An ∞*-mean* is any element in $\Re^d$ minimizing the function

$$V_\infty(m) := \text{ess inf}\, \|X - m\|,$$

where X is a random vector with distribution P.

We first analyze the case $p = 1$. This is a generalization of the median, *the spatial median,* which is widely employed and has been studied by several authors. A good survey of this and other possible generalizations of the median is (Small, 1990).

It is well known that, if $d > 1$, the spatial median is uniquely determined if the support of P is not contained in a line and that if we suitably move a single point from the initial location, then the spatial median varies.

However the spatial median does not vary if we move any data point just along the line joining its initial location with the spatial median. A nice mathematical explanation for this fact can be obtained by analyzing the gradient of the function to be minimized. The analogy of this gradient with the one obtained in the one-dimensional case leads to extend the definition of sign to the multidimensional case ((Hettmansperger *et al.*, 1987) is a good survey on the available possibilities when defining a sign in the multidimensional case).

To be precise, given $x \in \Re^p$, we define $\text{sign}(x) := x/\|x\|$ and now, if the sample is not contained in a line, the spatial median is

1. a point in the sample,
2. the only point, m_0, in which the weighted sum of the signs of the differences between points in the sample and m_0 vanishes. The weights in the sum coincide with the probabilities of the points in the sample.

This idea was used in (Kemperman, 1987) to give a physical interpretation, which we are going to quote, of the stability of the spatial median. While the interpretation works also in more general situations, let us assume that $d = 2$ and that thc support of P contains three points x_1, x_2, x_3 not in the same line with probabilities $p_i := P\{x_i\} > 0$, $i = 1, 2, 3$. Take three ropes tied together with a single knot by one of their extremes and let the free ends hang through pulleis placed at each point x_i, $i = 1, 2, 3$. If we put a weight of mass p_i, $i = 1, 2, 3$, at every free end, then there exists just a position for the knot in which it is in equilibrium. This position coincides with the spatial median m of P. It is evident that the point of equilibrium does not vary if we move any pulley along the corresponding straight line $\{m+\lambda(x_i-m)\colon \lambda > 0\}$, $i = 1, 2, 3$.

Let us consider a geometric interpretation of the spatial median which relates it to the so-called point of Steiner in a triangle. Let $x_1, x_2, \ldots, x_k$ be points in the plane. Any point y in the plane determines a curve passing through y corresponding to the set of points z such that

$$\sum_{i=1}^{k} \Phi(\|z - x_i\|) = \sum_{i=1}^{k} \Phi(\|y - x_i\|).$$

This curve is the Φ-contour curve of y with respect to $\{x_1, \ldots, x_k\}$. A point y at which the Φ-contour curves with respect to $\{x_1, x_2\}$ and $\{x_3\}$ cross cannot be a Φ-mean of the uniform distribution over $\{x_1, x_2, x_3\}$, because every point in the region inner to both curves would be better than y. The argument is general, so every good candidate y to be a Φ-mean of a finite set $\{x_1, \ldots, x_k\}$ has to satisfy that if we consider a hyperplane dividing the set $\{x_1, \ldots, x_k\}$ into two subsets $\{x_{i_1}, \ldots, x_{i_j}\}$ and $\{x_{i_{j+1}}, \ldots, x_{i_k}\}$, then y must be a tangency point of the Φ-contour curves of y with respect to the sets $\{x_{i_1}, \ldots, x_{i_j}\}$ and $\{x_{i_{j+1}}, \ldots, x_{i_k}\}$.

As an illustrative example let us compute the spatial median of a uniform probability on a set of four points $\{x_1, x_2, x_3, x_4\}$ contained in $\Re^2$ where we assume that x_1 is adjacent to x_2 and x_3 in the square $x_1x_2x_3x_4$. The Φ-contour curve of the point y with respect to $\{x_1, x_2\}$ is an ellipse passing through y with x_1 and x_2 as foci. Therefore every candidate y to be a spatial median must be a tangency point of two ellipses with foci in x_1, x_2 and x_3, x_4 respectively.

Now the mirror property of the ellipses implies that the common tangent to both curves has to be orthogonal to the bisectrices of the angles $\widehat{x_1yx_2}$ and $\widehat{x_3yx_4}$ and then these bisectrices coincide.

The same argument, when applied to the points x_1, x_3 and x_2, x_4, shows that the angles $\widehat{x_1yx_3}$ and $\widehat{x_2yx_4}$ also share their bisectrix, which in turn is orthogonal to that of the angles $\widehat{x_1yx_2}$ and $\widehat{x_3yx_4}$. Therefore the angles $\widehat{x_1yx_2}$ and $\widehat{x_3yx_4}$ are of equal magnitude. Or, in other words, y is the only common point to the diagonals of these four points.

A slight modification of the argument characterizes the median of a set of three points: y must be a tangency point of an ellipse with foci at x_1 and x_2 and a circumference with center at x_3. The mirror property now implies that x_3 has to be in the bisectrix of the angle $\widehat{x_1yx_2}$. By repeating the argument for the three possible pairs of curves it is inmediate to characterize the median as the one point, y_0, which makes the amplitude of the angles $\widehat{x_iy_0x_j}$, $i, j \in \{1, 2, 3\}$, $i \neq j$, to be 120°, if possible. Else one of the angles of the triangle defined by the three points is greater than 120° and the corresponding vertex is the median.

In particular, if we consider a fixed point m and three half-lines r_1, r_2 and r_3 starting from m, with angles of 120° between them, we choose a point on each one, x_1, x_2, and x_3 and define $P\{x_i\} = 1/3$ the spatial median of these points is m, independently of their exact location.

The previous fact allows us to construct a family of examples to show that, even in the simple case that the support of P contains three points, the spatial median does not hold the CMVP*.

Example 3.6. Let $m \in \Re^2$ and e_1, e_2, e_3 three vectors in $\Re^2$ such that $\|e_i\| = 1$, $i = 1, 2, 3$, and that the angle between any two of them is 120°. Let $r_i = \{m + \delta e_i\colon\ \delta \geqslant 0\}$, $i = 1, 2, 3$. Let $0 < \delta_1 < \delta_2$, consider the points $x_i = m + \delta_i e_i$, $i = 1, 2$, and define $P\{x_i\} = 1/3$, $i = 1, 2, 3$, where exact location of x_3 on r_3 is fixed later. We are going to construct a counterexample to the CMVP* by taking Φ as the identity function, $A = \{x_1, x_2\}$ and $B = \{x_3\}$. The median of P given $\{x_1, x_2\}$ is not uniquely defined, but we are going to choose the point $m^* = (x_1 + x_2)/2$. This selection is justified because we are interested in considering this case as the limit of the p-means when $p \to 1+$ and if $p > 1$, then $\Phi(t) = t^p$ is strictly convex and m^* is the p-mean of P given $\{x_1, x_2\}$. Let us now consider the vector $v = m - m^*$. Since $\delta_1 < \delta_2$, v and e_3 are not parallel. Moreover the angle between these vectors is less than 120° and then there exist $\alpha, \delta > 0$ such that $m + \delta e_3 = x_2 + \alpha v$. Let $\delta_3 > \delta^* > \delta$, $x_3 = m + \delta_3 e_3$ and consider v^* a vector orthogonal to $(m + \delta^* e_3) - x_2$ and such that $\langle x_3 - x_2, v^* \rangle > 0$. By construction we have that $\langle x_i - x_2, v^* \rangle \leqslant 0$, $i = 1, 2$, while $\langle m - m^*, v^* \rangle < 0$. Therefore the spatial median does not possess the CMVP*.

Moreover, as the spatial median, when it is unique, coincides with the limit as $p \to 1+$ of the p-means (see (Cuesta-Albertos and Matrán, 1989)), this example also proves that there exist some p-means with $p < 2$ which do not either possess the CMVP*.

We have included Example 3.6 because it contains a constructive counterexample for the limit case $p = 1$. Note that it is the asymmetry between the points x_1 and x_2 that permit us to construct the counterexample.

We next include a result which shows that every p-mean does not possess CMVP* if $1 \leqslant p < 2$ even in the case in which the three points are placed symmetrically. We only present the two-dimensional case; the extension to higher dimensions is trivial. For a proof of this result see Proposition 2.5 and Corollary 2.6 in (Cuesta-Albertos *et al.*, 1994).

THEOREM 3.7. *Let us assume that $d = 2$ and that P is the uniform distribution on the set $\{(-1,0),(1,0),(0,t)\}$ for any $t > 0$. Let $p \in [1,2)$. Then there exists $a > 0$ ($a = a(p)$) such that, if $t < a$, then the p-mean of P does not possess the CMVP*.*

We next analyze the situation in the case $p > 2$. Here, we are not aware of a general result like Theorem 3.7, but we show the state of the art at our best knowledge. First we include a result taken from (Cuesta-Albertos *et al.*, 1994) which shows that it is not possible to construct a counterexample with just three points, if P is the uniform distribution on them.

PROPOSITION 3.8. *Let $p > 2$ and let us assume that P is the uniform distribution on the set $S_p := \{x_1, x_2, x_3\}$. Given $A \subset S_p$ we denote by Π_A the p-mean of P given A and by Π_p the p-mean of P. Let $v \in \Re^d$ and $h \in \Re$ such that $\langle v, x\rangle \geqslant h$ if $x \in A$ and $\langle v, x\rangle < h$ if $x \in S_p - A$. Then $\langle v, \Pi_p\rangle < \langle v, \Pi_A\rangle$.*

The preceeding proposition and the technique of proving Theorem 3.4 could suggest that it is impossible to find a counterexample for CMVP* for the p-mean when $p > 2$, but this is not true as shown with next appealing example. In this example the ∞-mean of a given set A coincides with the origin, then we add a point with positive mass on the upper halfspace and the ∞-mean moves to the lower halfspace. Moreover, it moves not just a bit but a half of the coordinate of the added point in the upper halfspace.

The main interest of this example lies in that the fact that the limit properties of the p-means guarantee that the CMVP* does not hold at least for p-means with big enough values of p.

Example 3.9. Let P be the uniform distribution on

$$x_1 = (1,0,0), \qquad x_2 = (-1,0,0),$$
$$x_3 = \left(0, -3^{1/2}/2, -1/2\right), \quad x_4 = (0,1,0.3).$$

The points x_1, x_2, x_3 belong to the unit sphere with center at the origin and x_1 and x_2 are symmetric with respect to the origin. Thus the (unique) ∞-mean of P given $A := \{x_1, x_2, x_3\}$ is $(0,0,0)$. However $\|x_4\| = 1.044$. So that x_4 does not belong to the unit ball and the ∞-mean of P can not be any more $(0,0,0)$. In this case the ∞-mean of P is determined by the pair of points whose relative distance is greatest because these points are x_3 and x_4 and $m := (x_3 + x_4)/2 = (0, 0.068, -0.156)$ and a simple computation gives us that

$$\|x_3 - m\|^2 = \|x_4 - m\|^2 = 1.0305$$

and

$$\|x_1 - m\|^2 = \|x_2 - m\|^2 = 1.0144.$$

Thus we conclude that m is the ∞-mean of P, what gives the desired counterexample because the hyperplane $\{(x, y, z)\colon\ z = 0\}$ separates the points in A from x_4.

As an illustration, we have carried out several computations with MATHEMATICA which give, that if you handle another separating hyperplanes between A and x_4 then this P provides counterexamples for $p \geqslant 3.5$. P does not give counterexamples for $p \leqslant 3$.

The preceeding results refer only to the Euclidean norm. It is also interesting to study how the CMVP behaves if we employ other norms which can be suggested by the nature of the problem to compute the Φ-mean or other centralization measures.

For instance, given a convex function Φ instead of using the Φ-mean as defined in (1) we can employ the associated Orlitz-norm and compute the centralization measure with respect to this norm. In particular, in the case of the spatial median, this would lead us to minimize the expression

$$V_1(m) := \int_{\Re^d} \|x - m\|_1 P(\mathrm{d}x)$$

where $\|\cdot\|_1$ denotes the L_1-norm in $\Re^d$. This question is solved in (Cuesta-Albertos *et al.*, 1994, Corollary 2.4) where the following proposition is proved.

PROPOSITION 3.10. *Let $\mathcal{P}$ be a family of probability measures defined on the Borel σ-algebra on $\Re^d$. Let us assume that $\mathcal{P}$ contains every probability measure with a finite support.*

Let $T\colon \mathcal{P} \to \Re^d$ a centralization measure satisfying that

1. *$T\left[\frac{1}{2}(\delta_{x_1} + \delta_{x_2})\right] = \frac{1}{2}(x_1 + x_2)$, for every $x_1, x_2 \in \Re^d$, where δ_x is the Dirac measure on the point x.*
2. *Given $P \in \mathcal{P}$, if $\{P_n\}$ is a sequence of empirical measures obtained from a simple random sample of P, then the sequence $\{T(P_n)\}_n$ converges to $T(P)$ almost surely.*
3. *If $\{X_n\}$ and X are random vectors such that $P_{X_n},\ P_X \in \mathcal{P}$, $n = 1, 2, \ldots$, and that the sequence $\{\sup \|X_n - X\|\}_n$ converges to zero a.s., then $\lim_n T(P_{X_n}) = T(P_X)$.*
4. *T holds CMVP.*

Then, if $P \in \mathcal{P}$ and its support is not contained in a single line there exists the expectation of P and coincides with $T(P)$.

However we would like to remark that it is possible that a centralization measure possesses the CMVP in a particular direction and not in other. To explain this let us consider the usual p-norm on $\Re^d$, $\|\cdot\|_p$, $1 \leqslant p < \infty$,

$$\|x\|_p := \left[\sum_{1 \leqslant i \leqslant d} |x_i|^p\right]^{1/p}$$

and let us define the *p^*-mean* of the probability P as the element $m_p^* \in \Re^p$ such that

$$\int \|x - m_p^*\|_p^p P(\mathrm{d}x) \leqslant \int \|x - m\|_p^p P(\mathrm{d}x), \qquad \text{for all} \quad m \in \Re^d.$$

Note that the p^*-mean and the p-mean do not necessarily coincide and that m_p^* is the d-dimensional vector whose components are the one-dimensional p-means of the marginal distributions of P.

Therefore if we consider two sets A, B such that the p^*-means m_A and m_B of P given A and B respectively, satisfy that their difference is parallel to a vector in the basis we are considering, then the p^*-mean of P given $A \cup B$ lies in the segment joining m_A with m_B because the one-dimensional p-mean possesses the CMVP. This property is related to the fact (mentioned, for instance, in (Hettmansperger *et al.*, 1987)) that p^*-means are not rotational invariant if $p \neq 2$.

Many questions remain open. We mention two of them:

1. We do not know whether it is possible to extend the results related to CMVP* to every $p > 2$ or to general Φ-means. We conjecture that the answer is positive

2. We do not know when it is possible to find a counterexample to the CMVP* for $p > 2$ in the two dimensional space (involving, of course, more than three points).

REFERENCES

Arcones, M. A. and Giné, E. (1992). On the bootstrap of M-estimators and other statistical functionals. In: *Exploring the Limits of Bootstrap*, pp. 13–47, R. Lepage and L. Billard (Eds), Wiley, New York.

Brons, H. K., Brunk, H. D., Franck, W. E. and Hanson, D. L. (1969). Generalized means and associated families of distributions. *Ann. Math. Stat.* **40**, 339–355.

Cambanis, S. and Gerr, N. L. (1983). A simple class of asymptotically optimal quantizers. *IEEE Trans. Inform. Theory* **IT-29**, 664–676.

Cuesta-Albertos, J. A. (1984). Medidas de centralización multidimensionales (Ley Fuerte de los Grandes Números). *Trab. Est. I.O.* **35**, 3–16.

Cuesta-Albertos, J. A., Gordaliza, A. and Matrán, C. (1993). Trimmed k-means: An attempt to robustify quantizers. Preprint.

Cuesta-Albertos, J. A., Gordaliza, A. and Matrán, C. (1994). On the behavior of the multidimensional M-estimators. Preprint.

Cuesta-Albertos, J. A. and Matrán, C. (1988). The strong law of large numbers for k-means and best possible nets of Banach valued random variables. *Probab. Th. Related Fields* **78**, 523–534.

Cuesta-Albertos, J. A. and Matrán, C. (1989). Uniform consistency of r-means. *Stat. Probab. Lett.* **6**, 65–71.

Flury, B. A. (1990). Principal points. *Biometrika* **77**, 33–41.

Gini, C. (1958). *Le Medie.* Unione Tipografico-Editrice Torinese.

Gordaliza, A. (1991a). Best approximations to random variables based on trimming procedures. *J. Approx. Theory* **64**, 162–180.

Gordaliza, A. (1991b). On the breakdwon point of multivariate location estimators based on trimming procedures. *Stat. Probab. Lett.* **11**, 387–394.

Hartigan, J. (1975). *Clustering Algorithms.* Wiley, New York.

Hartigan, J. (1978). Asymptotic distributions for clustering criteria. *Ann. Stat.* **6**, 117–131.

Hettmansperger, T. P., Nyblom, J. and Oja, H. (1987). On multivariate notions of sign and rank. In: *Statistical Data Analysis Based on the L_1-norm and Related Methods.*, pp. 217–230, Y. Dodge (Ed.), North-Holland, Amsterdam.

IEEE Transactions on Information Theory (1982). **IT-28**, No. 2.

Kaufman, L. and Rousseeuw, P. J. (1990). *Finding Groups in Data. An Introduction to Cluster Analysis.* Wiley, New York.

Kemperman, J. H. B. (1987). The median of a finite measure on a Banach space. In: *Statistical Data Analysis Based on the L_1-norm and Related Methods*, pp. 217–230, Y. Dodge (Ed.), North-Holland, Amsterdam.

Pärna, K. (1988). On the stability of k-means clustering in metric spaces. *Tartu Riikl. Ülik. Toimetised* **798**, 19–36.

Pärna, K. (1990). On the existence and weak convergence of k-centres in Banach spaces. *Tartu Riikl. Ülik. Toimetised* **893**, 17–28.

Pollard, D. (1981). Strong Consistency of k-means clustering. *Ann. Stat.* **9**, 135–140.

Pollard, D. (1982). A central limit theorem for k-mean clustering. *Ann. Probab.* **10**, 919–926.

Robertson, T. and Wright, F. T. (1980). Algorithms in order restricted inference and the Cauchy mean value property. *Ann. Stat.* **8** (3), 645–651.

Serinko, R. J. and Babu, G. J. (1992). Weak limit theorems for univariate k-mean clustering under a nonregular condition. *J. Multivariate Anal.* **41**, 273–296.

Small, C. G. (1990). A survey on multidimensional medians. *Int. Stat. Rev.* **58**, 263–277.

Sverdrup-Thygeson, H. (1981). Strong law of large numbers for measures of central tendency and dispersion of random variables in compact metrics spaces. *Ann. Stat.* **9**, 141–145.

Multivar. Statist., pp. 267–278
E.–M. Tiit *et al.* (Eds)

ON SOME PROPERTIES OF k-VARIANCE

K. PÄRNA and J. LEMBER

Department of Mathematical Statistics, University of Tartu,
Liivi 2, EE2400 Tartu, Estonia

ABSTRACT

The k-centre of a distribution P is defined as a set of k points which lies closest to the random variable $X \sim P$ in the sense of quadratic loss function (the case $k = 1$ reduces to the mean value of X). Corresponding minimum value of the loss function, $D_k(X)$, is called the *k-variance* of X (or P). The k-variance shares several basic properties with classical variance DX, but it is not additive with respect to independent summands X and Y, except some special cases. Instead, the relationship $D_k(X+Y) \geqslant D_k(X) + D_k(Y)$ is conjectured, and several closely related inequalities are obtained.

1. PRELIMINARIES

Let X be a m-variate random vector defined on a probability space $(\Omega, \mathcal{F}, \mathbf{P})$. Throughout this paper, P and F will denote the law and the distribution function (c.d.f.) of X, respectively. Whenever needed, it will be implicitly assumed that the second moment of X is finite. The Euclidean norm of $x \in \Re^m$ will be denoted by $\|x\|$. We attempt to approximate P by a finite set.

DEFINITION 1.1. Let $A = \{a_1, \dots, a_k\} \subset \Re^m$ denote a set of k points. The function

$$W_k(A, F) = \int_{\Re^m} \min_{a_i \in A} \|x - a_i\|^2 \, \mathrm{d}F$$

is called the loss function of the distribution of F by the set A.

If we define the distance from $x \in \Re^m$ to A by

$$\|x - A\| = \min_{a_i \in A} \|x - a_i\|,$$

then we have

$$W_k(A, F) = E\,\|X - A\|^2 .$$

From the Definition 1.1 follows that

$$W_k(A, F) = \sum_{i=1}^{k} \int_{S_i(A)} \|x - a_i\|^2 \, \mathrm{d}F, \tag{1.1}$$

where

$$S_i(A) = \{x\colon\ \|x - a_i\| < \|x - a_j\|,\ \forall j < i,\ \|x - a_i\| \leqslant \|x - a_j\|,\ \forall j \geqslant i\}.$$

Hence the set $S_i(A) \subset \Re^m$ consists of all $x \in \Re^m$ having a_i as their nearest point in A. The partition $S(A) = \{S_1(A), \ldots, S_k(A)\}$ is called the *Voronoi partition generated by the set A*. Let $a(S_i) = E(X \mid S_i)$ denote the conditional mean of X given the set S_i. Now it is easy to transform (1.1) into

$$W_k(A, F) = E\,\|X\|^2 + \sum_{i=1}^{k} \left(\|a(S_i) - a_i\|^2 - \|a(S_i)\|^2\right) P(S_i), \tag{1.2}$$

where $S_i = S_i(A)$.

Similarly the loss function by a partition can be defined. In order to do that let $S = \{S_1, \ldots, S_k\}$ be a Borel-measurable partition of $\Re^m$, i.e.,

$$\bigcup_{i=1}^{k} S_i = \Re^m, \qquad S_i \bigcap S_j = \varnothing, \quad \forall i \neq j.$$

We will call two partitions $S = \{S_1, \ldots, S_k\}$ and $S' = \{S'_1, \ldots, S'_k\}$ *P-equivalent* if for suitable relabelling $i_1, \ldots, i_k$, $P(S_j \triangle S'_{i_j}) = 0$, $j = 1, \ldots, k$. The set of all equivalence-classes of k-partitions of $\Re^m$ is denoted by S. For a partition $S = \{S_1, \ldots, S_n\}$ the set $A(S) = \{a(S_1), \ldots, a(S_k)\}$ is called *the conditional means set of X given S.*

DEFINITION 1.2. The function

$$W_k(S, F) = \sum_{i=1}^{k} \int_{S_i} \|x - a(S_i)\|^2\, \mathrm{d}F$$

is called the loss function of the distribution F by the partition S.

Using the definition of conditional means it is not difficult to check the equality

$$W_k(S, F) = E\|X\|^2 - \sum_{i=1}^{k} \|a(S_i)\|^2 P(S_i), \qquad S \in \mathcal{S}, \tag{1.3}$$

(see, e.g. (Pärna, 1988)).

DEFINITION 1.3. A set of k points $A^* = \{a_1^*, \ldots, a_k^*\}$ is called the k-center of the distribution F if it minimizes $W_k(A, F)$ over all k-sets A of $\Re^m$.

DEFINITION 1.4. A partition $S^* = \{S_1^*, \ldots, S_k^*\}$ is called optimal for the distribution F if it minimizes $W_k(S, F)$ over all k-partitions of $\Re^m$.

The following duality properties are well known: *1) the Voronoi partition by a k-center is optimal and, 2) the conditional means set of an optimal partition is a k-center* (e.g. (Bock, 1989; Pärna, 1988)).

Consequently, the minimization of the function $W_k(S, F)$ is equivalent to the minimization of the function $W_k(A, F)$ and $\inf_A W_k(A, F) = \inf_S W_k(S, F)$.

2. k-VARIANCE AND k-APPROXIMATION

In this section we define k-variance and k-approximation for a random variable X. Then we will give a list of their properties which in several cases are already mentioned in the literature (Cuesta and Matràn, 1988; Cuesta-Albertos *et al.*, 1995; Flury, 1990; Pärna, 1988). Therefore, we prove only the less-known or unknown results.

DEFINITION 2.1. The joint minimum value of the loss functions $W_k(A, F)$ and $W_k(S, F)$ is called the k-variance of the distribution F and will be denoted by $D_k(X)$.

By the duality properties mentioned above

$$D_k(X) = W_k(S^*, F) = W_k(A^*, F) = E\|X - A^*\|^2,$$

where A^* and S^* are a k-center and an optimal partition respectively.

PROPERTY 2.1. $D_1(X) = \operatorname{tr}\Sigma$, *where* Σ *is the variance-covariance matrix of* X.

In the univariate case $D_1(X) = D(X) \equiv \operatorname{Var}(X)$ (see also (Flury, 1990)).

PROPERTY 2.2. $D_k(X) = 0$ *iff with probability one* X *takes at most* k *different values.*

PROPERTY 2.3. $D_k(X + C) = D_k(X)$, $C \in \Re^m$.

PROPERTY 2.4. $D_k(cX) = c^2 D_k(X)$, $c \in \Re^1$.

PROPERTY 2.5. $P(\|X - A^*\| \geqslant \varepsilon) \leqslant D_k(X)/\varepsilon^2$ *for each* $\varepsilon > 0$ (*the generalized Chebyshev's inequality*).

PROPERTY 2.6. *For any* k-*center* A^*, *and for any optimal partition* S^*, *the following equalities hold*

$$D_k(X) = E\|X\|^2 - \sum_{i=1}^{k} \|a_i^*\|^2 P(S_i(A^*)),$$

$$D_k(X) = E\|X\|^2 - \sum_{i=1}^{k} \|a(S_i^*)\|^2 P(S_i^*).$$

Proof. The first inequality comes from (1.2) since $a(S_i(A^*)) = a_i^*$, the second inequality follows from (1.3).

PROPERTY 2.7. *The sequence of* k-*variances of* F *goes monotonically to zero,*

$$D_k(X) \searrow 0 \quad \text{if} \quad k \to \infty.$$

Proof. First, show that $D_k(X) > D_{k+1}(X)$ if $D_k(X) > 0$. Let S^* be an optimal partition, then

$$D_k(X) = W_k(S^*, F) = \sum_{i=1}^{k} \int_{S_i^*} \|x - a(S_i^*)\|^2 \, \mathrm{d}F.$$

Since $D_k(X) > 0$, there must exist an optimal region $S_i^* \in S^*$ which verifies

$$\int_{S_i^*} \|x - a(S_i^*)\|^2 \, \mathrm{d}F > 0. \tag{2.1}$$

Choose $b \in S_i^*$ so that $P\big(B(b)\big) > 0$, where the ball $B(b)$ is defined as

$$B(b) = \big\{x \in S_i^*\colon \|x - b\| < \|x - a(S_i^*)\|\big\}.$$

If there were not such $a, b \in S_i^*$, the probability mass of the region S_i^* would be concentrated on the single point $a(S_i^*)$ — a contradiction with (2.1). Hence

$$\begin{aligned} \int_{S_i^*} \|x - a(S_i^*)\|^2 \, \mathrm{d}F &= \int_{S_i^* \setminus B(b)} \|x - a(S_i^*)\|^2 \, \mathrm{d}F + \int_{B(b)} \|x - a(S_i^*)\|^2 \, \mathrm{d}F \\ &> \int_{S_i^* \setminus B(b)} \|x - a(S_i^*)\|^2 \, \mathrm{d}F + \int_{B(b)} \|x - b\|^2 \, \mathrm{d}F, \end{aligned}$$

implying

$$D_k(X) = W_k(A(S), F) > W_{k+1}\big(\{A(S), b\}, F\big) \geqslant D_{k+1}(X).$$

Thus the sequence $\{D_k(X)\}$ is decreasing.

Let $B_n \subset \Re^m$, $n = 1, 2, \ldots$, be a sequence of concentric spheres with radius 2^n. Let A_n be a 2^{-n} lattice in the sphere B_n, consisting of $\nu(n)$ elements. Obviously, $\nu(n) < \infty$. From Definition 2.1 we have

$$\begin{aligned} D_{\nu(n)}(X) \leqslant W_{\nu(n)}(A_n, F) = &\int_{B_n} \min_{a_i \in A_n} \|a - a_i\|^2 \, \mathrm{d}F \\ &+ \int_{\Re^m \setminus B_n} \min_{a_i \in A_n} \|a - a_i\|^2 \, \mathrm{d}F. \end{aligned}$$

Since $\|x - a_i\|^2 \leqslant 2^{-2n}$,

$$\int_{B_n} \min_{a_i \in A_n} \|a - a_i\|^2 \, \mathrm{d}F \leqslant 2^{-2n} \longrightarrow 0, \qquad \text{if} \quad n \to \infty.$$

The finiteness of the second moment of X implies

$$\int_{\Re^m \setminus B_n} \|x - a_i\|^2 \, \mathrm{d}F \longrightarrow 0, \qquad \text{if} \quad n \to \infty.$$

Hence, the sequence $\{D_k(X)\}$ contains a subsequence $\{D_{\nu(n)}(X)\}$ which converges to zero. Now Property 2.7 follows.

DEFINITION 2.2. Let $A^* = \{a_1^*, \ldots, a_k^*\}$ be a k-center of F, $X \sim F$. The random variable

$$X_k = \sum_{i=1}^{k} a_i^* I_{\{\omega:\ X(\omega) \in S_i(A^*)\}}$$

is called a k-approximation of X (generated by A^*).

The domain of the k-approximation of X is the set A^*, the probability of each value a_i^* is the probability of the corresponding optimal region $S_i(A^*)$. Since generally the k-center of a distribution F is not unique, the k-approximation is not unique neither.

In the following some properties of k-approximations are presented. Except the property 2.14, the proofs are omitted. The skipped proofs are easy to restore — Properties 2.10 and 2.11 follow from corresponding properties of k-variance; Properties 2.8, 2.9, 2.12 and 2.13 follow from the Definition 2.2 and (1.3).

PROPERTY 2.8. *Any k-approximation X_k can be presented as a version of the conditional expectation*

$$X_k = E(X \mid \mathcal{G}_{S^*}) \qquad \text{a.s.},$$

where $\mathcal{G}_{S^*}$ is the sub-σ-algebra of F generated by the sets $\{\omega\colon X(\omega) \in S_i(A^*)\}$, $i = 1, 2, \ldots, k$.

In the following the σ-algebra $\mathcal{G}_{S^*}$ will be called *the σ-algebra generated by the partition $S(A^*)$.*

PROPERTY 2.9. $EX_k = EX$.

PROPERTY 2.10. *Let $C \in \Re^m$. For any k-approximation $(X + C)_k$ of the random vector $X+C$ there exists a k-approximation X_k of X such that $(X + C)_k = X_k + C$.*

PROPERTY 2.11. *Let $c \in \Re^1$. For any k-approximation $(cX)_k$ of the random vector cX there exists a k-approximation X_k of X such that $(cX)_k = cX_k$.*

PROPERTY 2.12. $(X_k)_k = X_k$.

PROPERTY 2.13. *If the random vectors X and Y are independent then any two k-approximations X_k and Y_k are also independent.*

PROPERTY 2.14. *For any sequence $X_1, X_2, \ldots$ of k-approximations the convergence $X_k \to X$ a.s. holds.*

Proof. Let $X(\omega)$ be an element of support of P (i.e., minimal closed subset of $\Re^m$ having probability 1). Then for each $\varepsilon > 0$, the closed sphere $B(X(\omega), \varepsilon)$ has a positive P-measure, $P(B(X(\omega), \varepsilon)) > 0$. We have to show that any such sphere contains at least one value of X_k for all k large enough.

Indeed, if (contraversially) for infinitely many values of k X_k has all its values outside $B(X(\omega), \varepsilon)$, then for infinitely many k

$$D_k(X) = \int_{\Re^m} \min_{a^*_{ik} \in A^*_k} \|x - a^*_{ik}\|^2 \, \mathrm{d}F \geqslant \int_{B(X(\omega),\varepsilon)} \min_{a^*_{ik} \in A^*_k} \|x - a^*_{ik}\|^2 \, \mathrm{d}F$$
$$\geqslant \varepsilon^2 P\big(B(X(\omega), \varepsilon)\big) > 0, \tag{2.2}$$

where $A^*_k = \{a^*_{1k}, \dots, a^*_{kk}\}$ is the domain of X_k. Since $\varepsilon^2 P\big(B(X(\omega), \varepsilon)\big)$ does not depend on k, (2.2) contradicts to Property 2.7. The proof is completed.

From Property 2.7 it immediately follows that $X_k \to X$ also in quadratic mean.

Note that the formula (1.3) is equivalent to

$$D_k(X) = E\|X\|^2 - E\|X_k\|^2,$$

which in the univariate case reduces to

$$D_k(X) = EX^2 - EX_k^2 = DX - DX_k,$$

implying a decomposition formula for the variance:

$$DX = D_k(X) + DX_k. \tag{2.3}$$

In the definition of X_k an optimal partition $S^* = S(A^*)$ was used. If $S = \{S_1, \dots, S_k\}$ is an arbitrary partition, then, again, we define

$$X_k^S = E(X \mid \mathcal{G}_S) \qquad \text{a.s.}$$

where $\mathcal{G}_S$ is the σ-algebra generated by S. Now the formula (1.3) can be rewritten as

$$W_k(S, F) = E\|X\|^2 - E\|X_k^S\|^2. \tag{2.4}$$

In the univariate case we have

$$DX = DX_k^S + W_k(S, F) \tag{2.5}$$

which, due to $W_k(S, F) \geqslant D_k(X)$, implies $DX_k^S \leqslant DX_k$.

We now generalize the concept of the partition.

DEFINITION 2.3. A distribution P is called k-stratified if P is presented as a sum of k unnormed non-negative measures P^i

$$P = \sum_{i=1}^{k} P^i, \qquad \int_{\Re^m} P^i(\mathrm{d}x) = p^i, \quad \sum_{i=1}^{k} p^i = 1.$$

In the following the measure P^i (on $\Re^m$) will be called a *stratum* of P, and the k-stratification will be denoted by $K = \{P^1, \dots, P^k\}$. Let $F^i(t) = P^i(-\infty, t)$, then $F = \sum_{i=1}^k F^i$.

In the special case, when there exists a partition $\{S_1, \dots, S_k\}$ of $\Re^m$ such that $P^i(S^i) = p^i$ for each $i = 1, 2, \dots, k$, the stratification $K = \{P^1, \dots, P^k\}$ is called a *k-partition of P*.

DEFINITION 2.4. Let K be a k-stratification of P. The function

$$W_k(K,F) = \sum_{i=1}^{k} \int_{\Re^m} \|x - a(P^i)\|^2 \, dF^i \tag{2.6}$$

is called the loss function of F by the stratification K. Here $a(P^i)$ is the conditional mean of P^i, i.e.,

$$a(P^i) = \frac{1}{p^i} \int_{\Re^m} x \, P^i(dx).$$

(For a similar concept, see also (Degens, 1978; Yang and Yu, 1990). It is important to observe that a necessary condition for a k-stratification K to be a minimizer of (2.6) is that K must be a k-partition of P. Furthermore, since there exists one-to-one correspondence between the k-partitions of P and the elements of S (the set of P-equivalence-classes of k-partitions of $\Re^m$), the previous sentence may be reformulated as follows:

LEMMA 2.1. *Any optimal partition S^* minimizes the loss function by the stratification.*

Sometimes it is useful to regard the k-stratification as a joint distribution of an $(m+1)$-dimensional random vector (X, X_{m+1}), where X_{m+1} — the *stratifier* — is a simple random variable with values $1,\dots,k$, the probability of i being p^i. The random variable X_{m+1}, having the distribution described above, can be defined in many different ways, and not necessarily on the initial probability space $(\Omega, \mathcal{F}, \mathbf{P})$. But the latter can be modified in an obvious way to get the random vector (X, X_{m+1}) defined on that modified space $(\Omega', \mathcal{F}', \mathbf{P}')$. For simplicity, we still write $(\Omega, \mathcal{F}, \mathbf{P})$ instead of $(\Omega', \mathcal{F}', \mathbf{P}')$. Now the notion of k-approximation can be generalized for all k-stratifications: for a k-stratification K and for a stratifier X_{m+1} the k-approximation X_k^K is defined as follows

$$X_k^K = \sum_{i=1}^{k} a(P^i) I_{\{X_{m+1}=i\}} \quad \text{or} \quad X_k^K = E(X \mid X_{m+1}) \quad \text{a.s.} \tag{2.7}$$

Now the formula (2.4) can be generalized for the case of stratifications K:

$$W_k(K,F) = E\|X\|^2 - E\|X_k^K\|^2,$$

which in the univariate case reduces to

$$DX = DX_k^K + W_k(K,F). \tag{2.8}$$

3. k-VARIANCE OF SUM

In this section we are mainly interested in k-variance of the sum of two independent random variables. For simplicity the univariate case ($m = 1$) is considered only.

Let X, Y be two independent random variables defined on a probability space $(\Omega, \mathcal{F}, \mathbf{P})$. Let T be the distribution of $X+Y$, R the distribution of the vector $Z = (X, Y)$, and F_Z the c.d.f. of Z.

Generally speaking, the well-known property of the variance $D(X+Y) = D(X)+D(Y)$ does not hold for k-variances. To see that, it is enough to take X and Y as two independent Bernoulli variables. Then both separate 2-variances, $D_2(X)$ and $D_2(Y)$, are zero, but since the distribution of $X+Y$ is concentrated at three points, the 2-variance $D_2(X+Y)$ is positive. Therefore, we conjecture the inequality

$$D_k(X+Y) \geqslant D_k(X) + D_k(Y). \tag{3.1}$$

An equivalent form for (3.1) can be obtained by means of the decomposition formula (2.3):

$$\begin{aligned} D(X+Y) &= DX + DY = D_k(X) + D_k(Y) + D(X_k + Y_k), \\ D(X+Y) &= D_k(X+Y) + D(X+Y)_k. \end{aligned} \tag{3.2}$$

So the hypothesis (3.1) is equivalent to:

$$D(X+Y)_k \leqslant D(X_k + Y_k).$$

Actually, the strict equality in (3.1) holds in some cases. The issue is related to the stability of the distributions of X and Y. Suppose, for instance, that X and Y are i.i.d. random variables with a stable distribution F. Then the distribution of $X+Y$ is of the same type as F, namely F streched $\sqrt{2}$ times. According to Property 2.4, the k-variance of $X+Y$ is then twice that of X or Y, i.e.,

$$D_k(X+Y) = D_k(X) + D_k(Y). \tag{$*$}$$

Since the normal distribution is stable, the same reasoning gives:

PROPOSITION. *Let X and Y be independent, normally distributed random variables. Then, for each $k = 1, 2, \ldots$, the equality $(*)$ holds.*

Moreover, we conjecture that the additivity $(*)$ is a characteristic property of the normal distribution.

HYPOTHESIS. *Let X and Y be i.i.d. random variables with a distribution F. Then the equality $(*)$ holds for each $k = 1, 2, \ldots$ if and only if the distribution F is normal.*

We now prove some lemmas, in lieu of (3.1) and $(*)$, related to the k-variance of sum. All of them will be formulated in terms of stratifications.

LEMMA 3.1. *Let X_p and Y_q be arbitrary p- and q-approximations of the independent random variables X and Y, respectively. Let l be the number of different values of $X_p + Y_q$ $(p+q-1 \leqslant l \leqslant pq)$. Then there exist an l-stratification K of the distribution T, and its l-approximation $(X+Y)_l^K$ such that*

$$(X+Y)_l^K = X_p + Y_q \qquad \text{a.s.}$$

Proof. Let $s_1, \ldots, s_l$ be the values of $X_p + Y_q$, and let $\mathcal{B}$ denote the Borel σ-algebra on $\mathcal{R}^1$. We show that the stratification K defined as

$$T^i(B) = \mathbf{P}\{X + Y \in B \wedge X_p + Y_q = s_i\}, \qquad i = 1, \ldots, l, \quad B \in \mathcal{B},$$

and its l-approximation defined as

$$(X+Y)_l^K = E(X + Y \mid X_p + Y_q)$$

are those we need. Indeed, for the relationship between σ-algebras $\sigma(X_p+Y_q) \subseteq \sigma(X_p, Y_q)$, and for the independence of X_p and Y_q, the well-known properties of the conditional expectation give us

$$\begin{aligned}(X+Y)_l^K &= E(X+Y \mid X_p+Y_q) = E\big(E(X+Y \mid X_p, Y_q) \mid X_p + Y_q\big)\\ &= E\big(E(X \mid X_p, Y_q) + E(Y \mid X_p, Y_q) \mid X_p + Y_q\big)\\ &= E\big(E(X \mid X_p) + E(Y \mid Y_q) \mid X_p + Y_q\big) = E(X_p + Y_q \mid X_p + Y_q)\\ &= X_p + Y_q \qquad \text{a.s.}\end{aligned}$$

The proof is completed.

Using Lemma 3.1, relationships (3.2), (2.8) and Lemma 2.1, the following corollary can be easily verified.

COROLLARY 3.1. *Let l be the number of different values of some approximations X_p and Y_q. Then*

$$D_pX + D_qY \geqslant D_l(X+Y) \geqslant D_{pq}(X+Y)$$

or

$$DX_p + DY_q \leqslant D(X+Y)_l \leqslant D(X+Y)_{pq}.$$

In the case when $p = k,\ q = k$, the Corollary 3.1 reduces to

$$D_kX + D_kY \geqslant D_l(X+Y) \geqslant D_{k^2}(X+Y)$$

or

$$DX_k + DY_k \leqslant D(X+Y)_l \leqslant D(X+Y)_{k^2}.$$

If $p = k,\ q = 1$, we get

$$DY + D_kX \geqslant D_k(X+Y) \quad \text{or} \quad DX_k \leqslant D(X+Y)_k.$$

LEMMA 3.2. *Let X_p and Y_q be some p- and q-approximations of the independent X and Y, respectively. Then there exists a k-stratification K of T, and its k-approximation $(X+Y)_k^K$ such that*

$$(X+Y)_k^K = (X_p + Y_q)_k \qquad \text{a.s.}$$

Proof. The proof is similar to that of Lemma 3.1. Let $v_1, \ldots, v_k$ be the values of $(X_p + Y_q)_k$. We show that the stratification K defined as

$$T^i(B) = \mathbf{P}\{X + Y \in B \wedge (X_p + Y_q)_k = v_i\}, \qquad i = 1, \ldots, k, \quad B \in \mathcal{B},$$

and its k-approximation defined as

$$(X+Y)_k^K = E\big(X + Y \mid (X_p + Y_q)_k\big)$$

are those we need.

Indeed, by $\sigma((X_p + Y_q)_k) \subseteq \sigma(X_p + Y_q)$ and Lemma 3.1, we have

$$\begin{aligned} E\big(X + Y \mid (X_p + Y_q)_k\big) &= E\big(E(X + Y \mid X_p + Y_q) \mid (X_p + Y_q)_k\big) \\ &= E\big(X_p + Y_q \mid (X_p + Y_q)_k\big) = (X_p + Y_q)_k \qquad \text{a.s.} \end{aligned}$$

Hence the Lemma 3.2 follows.

Using Lemma 3.2, the relations (3.2), (2.8), and the Lemma 2.1, the following corollary can be easily proved.

COROLLARY 3.2. *If X and Y are independent, then for any integers k, p, q*

$$\begin{aligned} &D_k(X+Y) \leqslant D_p X + D_q Y + D_k(X_p + Y_q), \\ &D(X+Y)_k \geqslant D(X_p + Y_q)_k. \end{aligned}$$

In the case when $p = k$ and $q = k$, the Corollary 3.2 reduces to

$$D_k(X+Y) \leqslant D_k Y + D_k X + D_k(X_k + Y_k)$$

or

$$D(X+Y)_k \geqslant D(X_k + Y_k)_k.$$

LEMMA 3.3. *Let X and Y be random variables, $X \sim P$, $Y \sim Q$. Then there exist k-stratifications of P and Q, say K^P and K^Q, and their k-approximations $\tilde{X}_k$ and $\tilde{Y}_k$ such that*

$$\tilde{X}_k + \tilde{Y}_k = (X+Y)_k \qquad \text{a.s.}$$

Note that the independence of X and Y is not assumed in this lemma.

Proof. Let us denote the values of $(X+Y)_k$ by $s_1,\dots,s_k$. We now specify the required stratifications $K^P=(P^1,\dots,P^k)$, $K^Q=(Q^1,\dots,Q^k)$ and their k-approximations.

For each $i=1,\dots,k$ and for any Borel set $B\in\mathcal{B}$ define the stratifications

$$P^i(B)=\mathbf{P}\{X\in B\wedge(X+Y)_k=s_i\},$$
$$Q^i(B)=\mathbf{P}\{Y\in B\wedge(X+Y)_k=s_i\},$$

and their k-approximations

$$\tilde{X}_k=E(X\mid(X+Y)_k)\qquad\text{a. s.},$$
$$\tilde{Y}_k=E(Y\mid(X+Y)_k)\qquad\text{a. s.}$$

Show now that they verify the equality stated in the lemma:

$$\begin{aligned}\tilde{X}_k+\tilde{Y}_k&=E\big(X\mid(X+Y)_k\big)+E\big(Y\mid(X+Y)_k\big)\\&=E\big(X+Y\mid(X+Y)_k\big)=(X+Y)_k\qquad\text{a.s.}\end{aligned}$$

The lemma is proved.

COROLLARY 3.3. *For any X,Y and k*

$$D_k(X+Y)+2\mathrm{cov}\left(\tilde{X}_k,\tilde{Y}_k\right)\geqslant D_k(X)+D_k(Y)+2\mathrm{cov}\,(X,Y)\tag{3.3}$$
$$D(X+Y)_k\leqslant DX_k+DY_k+2\mathrm{cov}\left(\tilde{X}_k,\tilde{Y}_k\right),\tag{3.4}$$

where $\tilde{X}_k$ and $\tilde{Y}_k$ are defined as in Lemma 3.3.

Proof. Show first that

$$D\tilde{X}_k\leqslant DX_k,\tag{3.5}$$
$$D\tilde{Y}_k\leqslant DY_k.\tag{3.6}$$

Indeed, from the relationships $DX=D_k(X)+DX_k$, $DX=W_k(K^P,F)+D\tilde{X}_k$, and $W_k(K^P,F)\geqslant D_k(X)$ the inequality (3.5) follows and, by analogy, we get (3.6). By (2.3), (2.5) and (2.8), the variance of $X+Y$ can be decomposed in different ways:

$$D(X+Y)=D_kX+D_kY+DX_k+DY_k+2\mathrm{cov}\,(X,Y),$$
$$D(X+Y)=W_k(K^P,F)+W_k(K^Q,G)+D\tilde{X}_k+D\tilde{Y}_k+2\mathrm{cov}\,(X,Y),$$
$$D(X+Y)=D_k(X+Y)+D(X+Y)_k.$$

Adding the term $2\mathrm{cov}\,(\tilde{X}_k,\tilde{Y}_k)$ to the equalities above and using

$$D\tilde{X}_k+D\tilde{Y}_k+2\mathrm{cov}\left(\tilde{X}_k,\tilde{Y}_k\right)=D(X+Y)_k\qquad\text{(Lemma 3.3)}$$

together with the relations (3.5) and (3.6), the inequalities (3.3) and (3.4) are obtained. The proof of Corollary 3.3 is completed.

In the case when X and Y are independent, the inequality (3.4) implies

$$D_k(X+Y) \geqslant D_k(X) + D_k(Y) - 2\mathrm{cov}\left(\tilde{X}_k, \tilde{Y}_k\right).$$

Hence, in the case of independent X and Y, Corollaries 3.2 and 3.3 provide us with following boundaries for $D_k(X+Y)$:

$$\begin{aligned} D_k(X) + D_k(Y) - 2\mathrm{cov}\left(\tilde{X}_k, \tilde{Y}_k\right) &\leqslant D_k(X+Y) \\ &\leqslant D_pX + D_qY + D_k(X_p + Y_q), \\ D(X_k + Y_k) + 2\mathrm{cov}\left(\tilde{X}_k, \tilde{Y}_k\right) &\geqslant D(X+Y)_k \geqslant D(X_p + Y_q)_k. \end{aligned}$$

REFERENCES

Bock, H.-H. (1989). Probabilistic aspects in cluster analysis. In: *Conceptual and Numerical Analysis of Data. Proceedings of the 13th Conference of the Gesellschaft für Klassifikation e.V.*, pp. 12–44, O. Opitz (Ed.), Springer, Berlin, Heidelberg, New York.

Cuesta, J. and Matrán C. (1988). The strong law of large numbers for k-means and best possible nets of Banach valued random variables. *Probab. Th. Related Fields* **78**, 523–534.

Cuesta-Albertos, J. A., Gordaliza, A. and Matrán, C. (1995). Trimmed k-means and the Cauchy mean-value property. In: *Proceedings of Fifth Tartu Conference on Multivariate Analysis*, pp. 247–265, E.-M. Tiit *et al.* (Eds), VSP, Utrecht/ TEV, Vilnius.

Degens, P. O. (1978). *Clusteranalyse auf Topologisch–Maßtheoretischer Grundlage*. Universität München. Dissertation.

Flury, B. A. (1990). Principal points. *Biometrika* **77**, 33–41.

Pärna, K. (1988). On the stability of k-means clustering in metric spaces. *Tartu Riikl. Ülik. Toimetised* **798**, 19–36.

Yang, M.-S. and Yu, K. F. (1990). On stochastic convergence theorems for the fuzzy c-means clustering procedure. *Int. J. General Systems* **16**, 397–411.

Multivar. Statist., pp. 279–286
E.-M. Tiit *et al.* (Eds)

RANDOM MULTIGRAPHS, CLASSIFICATION AND CLUSTERING

B. HARRIS[1], E. GODEHARDT[2] and A. HORSCH[2]
[1]University of Wisconsin, Madison

[2]Heinrich Heine Universität, Düsseldorf

1. INTRODUCTION AND SUMMARY

A data set of n vectors with k coordinates is given. The purpose of classification and clustering techniques is to partition the n vectors into meaningful classes. In present usage, these classes are referred to as *clusters*. Contemporary clustering methods permit the various coordinates to assume values in different domains, for example, some of the coordinates may be real valued and taking values in $(-\infty, \infty)$, some may be restricted to integer values, and some may be binary, assuming only the values zero and one. This latter situation arises if the presence or absence of some attribute is being indicated.

The present report is regarded as being primarily expository. Following a brief historical summary of classification and clustering, the notion of random multigraphs is introduced to show how techniques based on the theory of random multigraphs can be utilized to determine if the clusters obtained are *meaningful* or due to *chance variation*.

2. HISTORICAL SUMMARY

In two celebrated papers Fisher (1936, 1940) proposed the basic techniques for *classification*. Two multivariate normal populations with known mean vectors and with known equal covariance matrices are given. n vectors are given and presumed to be observations from one of the specified populations. The experimenter wishes to specify which population each observation comes from in such a way that the probability of misclassification is small. These techniques have been extended in many directions, including extensions to k populations and to decision theoretic methods of classification. If the population means are unknown, the problem is solved by taking random samples from each of the populations and using these to estimate the unknown mean vectors. Extensive references and discussions can be found in standard textbooks and reference books on Multivariate Analysis, such as (Anderson, 1984).

During the late 1960's and the 1970's, the related techniques of cluster analysis were developed. In general, these methods require more extensive computation and it is therefore reasonable to believe that the use of high speed computers played a significant role in their development. Many clustering methods can be described as follows. A measure of similarity (or similarly, a distance measure) between vectors is introduced. In principle, vectors which are sufficiently similar are placed in the same cluster. A substantial fraction of the extensive literature on clustering methodology is concerned with computer algorithms for generating the clusters. General discussion of the methodology of clustering can be found in (Bock, 1974; van Ryzin, 1977; Hartigan, 1975), as well as others.

3. CLUSTERING TECHNIQUES

The vectors $x_1, x_2, \ldots, x_n$ are given. As previously noted, the components may be continuous, discrete, binary, and so forth. The purpose of the investigation is to divide the n vectors into groups, each group containing those data points which 'belong together'. The number of such groups need not be specified in advance.

One very common technique may be described as follows. A distance measure $\rho(x, y)$ is selected, where $x, y \in \overline{X} = \{x_1, x_2, \ldots, x_n\}$. This distance measure need not necessarily satisfy the usual postulates for a metric. A threshold d is chosen. If $\rho(x, y) \leqslant d$, then x and y are placed in the same cluster. We say that there is a *path* from x_i to x_j, if there is a sequence of vectors $x_i, x_{i+1}, \ldots, x_j$ such that $\rho(x_{i+r}, x_{i+r+1}) \leqslant d$, $r = 0, 1, \ldots, j-i-1$. If there is a path from x_i to x_j, then x_i and x_j are in the same cluster. Clusters formed by this method are called *single linkage clusters.* If d is chosen too small, then one may obtain very many small clusters. If d is chosen too large, then too few clusters will be obtained. Also, *bridging* may occur, that is, one may have two disjoint sets of vectors, A and B, each of which has a 'small' diameter and $\rho(A, B)$ is 'large'. However, there exists an $x \in A$ and a $y \in B$, with a path from x to y. Then, using the method described above, A and B are subsets of the same cluster.

Single linkage clustering can be extended to form k-linkage clusters, $k > 1$. In a k-linkage cluster, every element in a cluster is 'related' to k other elements in the same cluster. Clearly, the shape of the clusters depends on k. In general, if k is small, the clusters are likely to have a stretched and twisted appearance. If k is large, the clusters will have a tendency to be compact and ellipsoidal in shape.

Many techniques for clustering, including the linkage clustering techniques summarized above can be formulated as graph theoretical clustering techniques. Accordingly, in the next section, we will describe some basic ideas from the Theory of Graphs.

4. GRAPH THEORETICAL METHODS

Let V be a set of elements with $|V| = n$, (for any set A, $|A|$ denotes the cardinality of A). Let E be a subset of the $\binom{n}{2}$ pairs of the elements of V. Then (V, E) is a *graph*. The set V is called the set of vertices (the terms nodes and points are also used) of the graph. The set E is called the set of edges (or arcs) of the graph.

It is customary to represent a graph by listing the vertices and edges. For example, let G be the pair (V, E) where

$$V = \{1, 2, 3, 4, 5, 6, 7, 8, 9\}$$

and

$$E = \{(1,2), (1,4), (1,6), (2,4), (2,7), (3,4), (3,6), (7,8)\}.$$

This is usually represented pictorially by plotting the vertices V as points in the plane and the edges E as lines connecting the pairs of point in E.

A graph G is a method of representing the binary relation R on the elements of V. E is a listing of the pairs of points of V which satisfy the indicated relation. Thus $(i, j) \in E$ if and only if iRj, where $i, j \in V$.

H is a *subgraph* of G if $H = (V_H, E_H)$, with $V_H \subset V$ and $E_H = \{E^* \subset E,\ E^* = (i,j),\ i \in V_H,\ j \in V_H\}$.

A *path* P is an alternating sequence of vertices and edges, beginning and ending in a vertex, such as $v_1, (v_1, v_2), v_2, \ldots, (v_{m-1}, v_m), v_m$ where $v_i \in V$, $(v_{i_1}, v_{i+1}) \in E$. Some writers require that the elements of E in P are distinct. A *cycle* is a path with $v_m = v_1$. Thus, $\{(1,4), 4, (4,3), 3, (3,6), 6, (6,1)\}$ is a cycle.

A graph is a *complete graph* if $|E| = \binom{n}{2}$.

M is a *multigraph* if $M = \{V, E, T\}$, where $V = \{1, 2, \ldots, n\}$ is the set of vertices, $E = \{(i,j)^{(k)},\ i < j,\ k \in T\}$ is the collection of unordered pairs of elements of V, and $T = \{0, 1, \ldots, t\}$ is the set of possible multiplicities for each edge. That is, the multigraph may have as many as t edges between any pair of vertices. For example, let $n = 4$. Then,

$$M = \left\{(1,2)^{(2)}, (1,3)^{(0)}, (1,4)^{(1)}, (2,3)^{(4)}, (2,4)^{(0)}, (3,4)^{(1)}\right\}$$

is a multigraph. The multiplicities of the edges are indicated as exponents assigned to each possible edge. If each such exponent is 0 or 1, then M is a graph.

The *S-projection* of a multigraph is the graph $G_S^{(M)}$ with

$$V(M) = V\big(G_S^{(M)}\big) \quad \text{and} \quad E\big(G_S^{(M)}\big) = \{(i,j),\ t(i,j) \geqslant S\}.$$

The objective of cluster analysis is to partition the data vectors into classes, that is, to write $C = (C_1, C_2, \ldots)$. (Sometimes the number of classes is specified in advance.) Elements in the same class should be 'homogeneous' and elements in different classes should be 'substantially different'. In order to establish whether two vectors are sufficiently similar or sufficiently different, measures of 'distance' or 'similarity' are introduced. These measures include L_ρ distances, the Mahalanobis distance, and the Hamming distance, among others. In addition many scaling methods and normalizations are utilized. There are a large number of computer algorithms in use. For example, the books by Godehardt (1990) and Bock (1974) provide extensive discussions of these methods. To use graph theoretical techniques, we proceed as follows:

Define a distance (or equivalently, a similarity measure) on the vectors of the data set. If $\rho(x_i, x_j) \leqslant d$, then place (i, j) in E (for similarity measures, if $\tau(x_i, x_j) \geqslant d$, place (i, j) in E).

Then choose a characteristic of a graph. For example, if we choose a path, then if there is a path from vertex i to vertex j, place the vectors corresponding to vertices i and j in the same cluster. Another possibility would be to define maximal complete subgraphs as clusters. (A complete subgraph is maximal if it is not a proper subgraph of a complete subgraph.) There are many other possibilities as well.

To utilize the theory of multigraphs, we proceed as follows. We denote the data by $X_i = (X_{i1}, X_{i2}, \ldots, X_{im})$, where $i = 1, 2, \ldots, n$. As above, we compare X_{il} and X_{jl} with respect to distance or similarity, $l = 1, 2, \ldots, m$. The number of possible multigraphs that can be obtained in this manner is $2^{m\binom{n}{2}}$, provided the m layers are kept distinct.

5. RANDOM GRAPHS AND RANDOM MULTIGRAPHS

There are many possible probabilistic models that one might use in formulating a theory of random graphs and random multigraphs. For random graphs, two are very widely utilized. They are:

1. *The Binomial Model.* Choose p, $0 < p < 1$, and conduct $\binom{n}{2}$ Bernoulli trials, one for each possible member of E (the ordered pairs of vertices). If a 'success' is obtained (with probability p), then an edge is drawn for that pair of vertices. Proceeding in this manner, a random number of edges are obtained. The number of edges $|E|$ will have the binomial distribution with parameters $(\binom{n}{2}, p)$.
2. *The Hypergeometric Model.* Select M, $0 \leqslant M \leqslant \binom{n}{2}$. Choose M of the $\binom{n}{2}$ pairs of vertices at random (sampling without replacement). These are the elements of E.

For random multigraphs, the following models may be proposed.

Let m be the maximum number of possible edges for each pair of vertices (in general, this will be the dimension of the vectors). The above models can now be extended as follows:

1. *The Binomial Model.* Let p_i, $0 \leqslant p_i \leqslant 1$, be given. We require that at least one p_i not be zero or one. Then the binomial random graph model

is employed for each of the m layers, resulting in $m\binom{n}{2}$ Bernoulli trials, where the probability of forming an edge between two specified vertices is constant within each layer, but may be different for different layers.

2. *The Hypergeometric Model.* Choose $M_1, M_2, \ldots, M_m$ vertices for each of the layers by sampling without replacement from the n vertices available.
3. *The Symmetric Hypergeometric Model.* Choose N edges without replacement from the $m\binom{n}{2}$ possible edges.

Note that the first two models postulate stochastic independence of the layers and both can be extended when independence is not assumed or the data has not been transformed to obtain independence.

The purpose of using models from random graphs and random multigraphs is to provide a mechanism for testing whether the clusters obtained are 'real' or merely 'chance occurrences'. Also, the theory of random graphs has an extensive literature. Hence, much of the methodology which is appropriate for this purpose is presently available. While there is a 'prehistory', the rapid growth and development of the theory of random graphs can be traced in large measure to two celebrated papers of Erdös and Rényi (1959, 1960).

In this brief exposition, it will not be possible to exhibit more than a few of the potential test statistics which might be used for this purpose.

A graph (or a subgraph) is said to be *connected* if for every pair of vertices, (v_i, v_j), there is a path from v_i to v_j. A *tree* is a connected graph (or subgraph) without cycles. The results which are stated below for trees apply to both the hypergeometric and binomial models, since these are asymptotically equivalent when the number of edges is 'small' and the number of vertices is 'large'.

Then if

$$p = \frac{1}{n}\left(\frac{1}{n}\log n + \frac{n-1}{n}\log\log n + c + \mathrm{o}(1)\right),$$

the number of trees with m vertices has an asymptotic Poisson distribution with mean

$$\lambda = \mathrm{e}^{-mc}/mm!.$$

Similarly, if

$$pn^{m/(m-1)} \longrightarrow c > 0, \qquad \text{as} \quad n \to \infty,$$

then the number of trees has an asymptotic Poisson distribution with mean

$$\lambda = c^{m-1}m^{m-2}/m!.$$

The *degree* of a vertex is the number of edges incident to it. Then, for both the binomial and the hypergeometric models, the distribution of the number of vertices of degree $m-1$ is asymptotically Poisson with mean

$$\lambda = n\binom{n}{m-1}p^{m-1}(1-p)^{n-m}$$

provided

$$N(n) = \left[\frac{1}{2}n\left(\frac{1}{m}\log n + \frac{m-1}{n}\log\log n + c + \mathrm{o}(1)\right)\right],$$

where N is the number of edges. For the binomial model use

$$N = \binom{n}{2}p.$$

The distribution of the degrees of the vertices is asymptotically Poisson with mean c provided

$$p = \frac{N}{\binom{n}{2}}, \qquad N(n) = \frac{n}{2}(c + \mathrm{o}(1)).$$

Under identical conditions, the distribution of the number of isolated cycles, the number of maximal connected subgraphs (components), the number of cycles (not necessarily isolated) of cardinality m are all asymptotically Poisson distributed with means

$$\lambda = \frac{\mathrm{e}^m}{2m}\,\mathrm{e}^{-cm},$$
$$\mu = \lambda\left(1 + m + \frac{m^2}{2!} + \cdots + \frac{m^{m-2}}{(m-3)!}\right),$$
$$v = \frac{c^m}{2m},$$

respectively.

To insure that each vertex belongs to no more than one cluster, we utilize s-projections. Then, for the symmetric hypergeometric model with $N = N(n), 1 \leqslant s \leqslant t$, the number of s-isolated vertices has an asymptotically Poisson distribution with mean

$$\lambda = \mathrm{e}^{-c},$$

where

$$N(n) = \frac{t}{2\binom{t}{s}^{1/s}}\, n^{2-1/s}\big(\log n + c + \mathrm{o}(1)\big)^{1/s}.$$

Under the same conditions, the s-projection will have one 'giant component' and a modest number of s-isolated vertices almost surely.

The above is a small subset of the many results of this type which are known at present, see (Godehardt, 1990). The purpose of the present manuscript is not to list the various possibilities, but to suggest the following application for the above subset and for many others not cited here. That is, if the clusters are formed using graph theoretic (or multigraph theoretic) methods, then comparing the graph obtained with one or more of the above probability distributions provides a statistical test for the hypothesis that the graph could have been obtained by chance and therefore that the clusters observed are chance occurrences. To construct hypothesis tests, assume the graph is random. For the binomial model, p is determined by the similarity measure. Then if, for example, the degrees of the vertices appear to be a random sample from the indicated asymptotic Poisson distribution, there is little reason to believe that the clusters are meaningful.

In particular, the authors have been investigating the use of *interval graphs*. These may be described as follows. Let $I_1, I_2, \ldots, I_n$ be intervals on the real line. Let $V = \{1, 2, \ldots, n\}$ be the vertices of a graph. For $1 \leqslant i, j \leqslant n$, (i, j) is in E whenever $I_i \cap I_j \neq \phi$.

Example. Let $I_1 = (0, 1]$, $I_2 = (1, 3]$, $I_3 = (1/2, 3/2]$, $I_4 = (-1, 6]$ and $I_5 = (2, 4]$. Then the set of edges E is given by (1, 3), (1, 4), (2, 3), (2; 4), (2, 5), (3, 4), (4, 5).

The mean and variance of the number of edges in a random interval graph is calculated below.

Let $X_1, X_2, \ldots, X_n$ be independent, identically distributed random variables with cumulative distribution function $F_X(x)$; where $F_X(x)$ is absolutely continuous with respect to Lebesgue measure.

Let X_i, X_j be two of these random variables and let $d > 0$ be given. If $|X_i - X_j| < d$, then the vertices $(i, j) \in E$.

$$\begin{aligned} P\big[|X_i - X_j| < d\big] &= \int_{-\infty}^{\infty} \int_{x_j - d}^{x_j + d} f(x_i) f(x_j) \, \mathrm{d}x_i \, \mathrm{d}x_j \\ &= \int_{-\infty}^{\infty} \big[F_X(x_j + d) - F_X(x_j - d)\big] f(x_j) \, \mathrm{d}x_j \\ &= E_X\big[F_X(x + d) - F_X(x - d)\big] = E_x[H(X)]. \end{aligned}$$

Specifically, for the uniform distribution on $[0, 1]$,

$$\begin{aligned} E_X[H(X)] = {} & E_X\big[F_X(X + d) \mid (X < d)\big] P\{X < d\} \\ & + E_X[F(X + d) - F(X - d) \mid d < X < 1 - d] \\ & \times P\{d < X < 1 - d\} \\ & + E_X[1 - F(X - d) \mid X > 1 - d] P\{X > 1 - d\} \\ = {} & d\left(d + \frac{d}{2}\right) + (1 - 2d)(2d) + \left(1 - \left(1 - \frac{d}{2} - d\right)\right) d = 2d - d^2. \end{aligned}$$

Therefore the expected number of edges is $\binom{n}{2}(2d - d^2)$.

The second moment of the number of edges in E is

$$\binom{n}{2}\binom{n-2}{2}\left(E\left(F(X+d)-F(X-d)\right)\right)^2$$
$$+2\binom{n}{2}\binom{n-2}{1}E\left\{\left[F(X+d)-F(X-d)\right]^2\right\}$$
$$+\binom{n}{2}E\{F(X+d)-F(X-d)\}$$
$$-\binom{n}{2}^2\left[E\{F(X+d)-F(X-d)\}\right]^2,$$

which for 'small' d is approximately $6n^3d^2$.

The above moments enable the experimenter to obtain preliminary evidence as to whether the number of edges is compatible with the assumed model.

The methodology using interval graphs is closely related to the theory of spacings, see, for example (Barbour *et al.*, 1992; Pyke, 1965).

6. CONCLUDING REMARKS

The purpose of this report is to suggest some techniques which may be utilized in studying clustering problems. The objective is not the construction of clusters, but rather the examination of clusters which have been obtained in order to determine if they are 'meaningful' or possibly due to the inherent randomness in the data set. Such methods appear to be quite useful in many applications which arise in medicine, the biological and the social sciences. This report describes a continuing research project and it is expected that definitive results (some of which are now at hand) will be reported in future communications.

REFERENCES

Anderson, T. W. (1984). *An Introduction to Multivariate Statistical Analysis.* 2nd edn. Wiley, New York.

Barbour, A. D., Holst, L. and Janson, S. (1992). *Poisson Approximation.* Oxford University Press, New York.

Bock, H. H. (1974). *Automatische Klassifikation.* Vandenhoeck and Ruprecht, Göttingen.

Erdös, P. and Rényi, A. (1959). On random graphs, I. *Publicationes Mathematicae* **9**, 290–297.

Erdös, P. and Rényi, A. (1960). On the evolution of random graphs. *Publ. Math. Inst. Hungarian Academy of Sci.* **5**, 17–60.

Fisher, R. A. (1936). The use of multiple measurements in taxonomic problems. *Ann. Eugen.* **7**, 179-188.

Fisher, R. A. (1940). The precision of discriminant functions. *Ann. Eugen.* **10**, 422-429.

Godehardt, E. (1990). *Graphs as Structural Models.* 2nd edn. Vieweg, Braunschweig.

Hartigan, J. A. (1975). *Clustering Algorithms.* Wiley, New York.

Pyke, R. (1965). Spacings (with discussion). *J. Roy. Statist. Soc., Series B* **27**, 395-449.

Van Ryzin, J. (Ed.). (1977). *Clustering and Classification.* Academic Press, New York.

Part Six

MATRICES IN STATISTICS

Multivar. Statist., pp. 289–311
E.–M. Tiit *et al.* (Eds)

SHORTED MATRICES AND THEIR APPLICATIONS IN LINEAR STATISTICAL MODELS: A REVIEW

SUJIT KUMAR MITRA[1,*], SIMO PUNTANEN[2,**] and GEORGE P. H. STYAN[3,***]

[1]Indian Statistical Institute, 7, S. J. S. Sansanwal Marg,
New Delhi — 110016, India

[2]Department of Mathematical Sciences, University of Tampere,
P. O. Box 607, FIN-33101 Tampere, Finland

[3]Department of Mathematics and Statistics, McGill University,
805 ouest rue Sherbrooke, Montréal, Québec, Canada H3A 2K6

ABSTRACT

We collect together, in a historical perspective, the various properties of shorted matrices and shorted operators that have appeared recently in the literature. Most proofs are omitted. Special attention is given to the role played by the shorted matrix in linear statistical models. An extensive bibliography on shorted matrices is also included.

1. INTRODUCTION

The concept of shorted operator was first introduced, at least implicitly and in Hilbert space, by Krein (1947), and later rediscovered and explicitly studied in finite-dimensional vector spaces by Anderson (1971), who introduced the term 'shorted operator'. Anderson (1971) used an explicit matrix construction as a definition of the shorted operator and showed (Theorem 1) that the shorted operator was a solution to a certain maximization problem. Anderson and Trapp (1975) generalized these results to an arbitrary Hilbert space using this maximization property as definition. The concept of parallel sum of two operators, introduced by Anderson and Duffin (1969), was utilized frequently by Anderson (1971) and by Anderson and Trapp (1975).

---

*Research supported in part by the Council of Scientific and Industrial Research, Government of India, under its Emeritus Scientist Scheme.

**Research conducted while a Senior Researcher of the Academy of Finland.

***Research supported in part by the Natural Sciences and Engineering Research Council of Canada.

Mitra and Puri (1979) obtained explicit expressions for the shorted operator of a non-negative definite symmetric operator (in a finite-dimensional linear space) in terms of a minimum-seminorm generalized inverse and a semi-least-squares inverse of a complex matrix. This paper by Mitra and Puri (1979) may be considered, therefore, as the starting point for statistical applications of the shorted matrix and the shorted operator. As we will see in Section 3 the shorted matrix plays a key role in linear statistical models.

Anderson (1971) was motivated by the theory of resistive n-port electrical networks which, as pointed out by Anderson and Trapp (1975) and Mitra (1990a), can be viewed as a black box with n pairs of terminals or ports. Voltage and current measurements are taken at each port thus resulting in a current vector $\boldsymbol{i}$ and a voltage vector $\boldsymbol{v}$, each with n components. These two vectors are related by the equation $\boldsymbol{v} = \boldsymbol{Z}\boldsymbol{i}$, where $\boldsymbol{Z}$ is the impedance matrix; the inner product $\boldsymbol{v}'\boldsymbol{i}$ is the power dissipated by the network. When the network consists only of resistors, the impedance matrix $\boldsymbol{Z}$ is real, symmetric and non-negative definite; when some of the ports are shorted (so that no voltage exists between these ports), the resulting impedance matrix of the network is our 'shorted matrix'.

In this paper boldface italic capital letters denote matrices, and for a matrix $\boldsymbol{A}$, the symbols $\boldsymbol{A}'$ and $\mathcal{M}(\boldsymbol{A})$ denote, respectively, the transpose and the column space (or range or manifold) of $\boldsymbol{A}$. Moreover, for a given $m \times n$ matrix $\boldsymbol{A}$ and a given $m \times p$ matrix $\boldsymbol{B}$, the symbol $(\boldsymbol{A} : \boldsymbol{B})$ denotes the $m \times (n+p)$ partitioned matrix with $\boldsymbol{A}$ placed next to $\boldsymbol{B}$. Capital italic letters are used to denote certain matrix-valued functions, e.g., $S(\boldsymbol{V} \mid \boldsymbol{X})$ denotes the shorted matrix of $\boldsymbol{V}$ with respect to $\boldsymbol{X}$. The vector space of real n-tuples is denoted by $\mathbb{R}_n$ and the set of $m \times n$ matrices with real elements by $\mathbb{R}_{m \times n}$. Boldface italic lower-case letters denote column vectors and all row vectors are primed. Having statistical applications in mind, we consider only real matrices and vectors in this paper; our results, however, may be easily extended to complex matrices, and most of our results also hold for matrices with elements in an arbitrary field.

We denote the cone of real symmetric non-negative definite $n \times n$ matrices by $\mathfrak{L}_n$ and in Definition 1 we consider the Löwner partial ordering of matrices in $\mathfrak{L}_n$. If $\boldsymbol{A}, \boldsymbol{B} \in \mathbb{R}_{n \times n}$, we write $\boldsymbol{A} \leqslant_{\mathrm{L}} \boldsymbol{B}$ whenever $\boldsymbol{B} - \boldsymbol{A} \in \mathfrak{L}_n$, i.e., whenever $\boldsymbol{B} - \boldsymbol{A} = \boldsymbol{F}\boldsymbol{F}'$ for some conformable matrix $\boldsymbol{F}$; we may also write $\boldsymbol{B} - \boldsymbol{A} \geqslant_{\mathrm{L}} \mathbf{0}$ or $\mathbf{0} \leqslant_{\mathrm{L}} \boldsymbol{B} - \boldsymbol{A}$. For most of our applications, however, we also have $\boldsymbol{A}, \boldsymbol{B} \in \mathfrak{L}_n$ so that both $\boldsymbol{A} \geqslant_{\mathrm{L}} \mathbf{0}$ and $\boldsymbol{B} \geqslant_{\mathrm{L}} \mathbf{0}$.

DEFINITION 1. Let $\boldsymbol{V} \in \mathfrak{L}_n$, $\boldsymbol{X} \in \mathbb{R}_{n \times p}$, and let

$$\mathfrak{C}_1 = \{\boldsymbol{C}\colon\ \mathbf{0} \leqslant_{\mathrm{L}} \boldsymbol{C} \leqslant_{\mathrm{L}} \boldsymbol{V},\ \mathcal{M}(\boldsymbol{C}) \subset \mathcal{M}(\boldsymbol{X})\}. \tag{1.1}$$

The shorted matrix $S(\boldsymbol{V} \mid \boldsymbol{X})$ of $\boldsymbol{V}$ with respect to $\boldsymbol{X}$ is the maximal element $\boldsymbol{C} \in \mathfrak{C}_1$ in the Löwner partial ordering.

As shown by Anderson (1971) and Anderson and Trapp (1975), the set $\mathfrak{C}_1$ in (1.1) has a maximal element and it, the shorted matrix, is unique. The shorted matrix of $\boldsymbol{V}$ with respect to $\boldsymbol{X}$ is, therefore, the unique nonnegative definite matrix which is 'as close to $\boldsymbol{V}$ as possible' in the Löwner partial ordering, but whose column space is in that of $\boldsymbol{X}$. We note that, in general, the concept of closeness between matrices is not uniquely defined, but closeness in the Löwner sense is quite natural, especially from the point of view of statistical applications.

Let us now consider a general linear model $\{\boldsymbol{y}, \boldsymbol{X\beta}, \boldsymbol{V}\}$, where $\boldsymbol{y}$ is an $n \times 1$ observable random vector with $n \times 1$ expectation vector $\boldsymbol{X\beta}$ and $n \times n$ dispersion (or covariance) matrix $\boldsymbol{V}$. Here the $n \times p$ design (or model) matrix $\boldsymbol{X}$ has $\operatorname{rank}(\boldsymbol{X}) = r > 0$, and is known, while the $p \times 1$ vector $\boldsymbol{\beta}$ is unknown and the $n \times n$ dispersion matrix $\boldsymbol{V}$ is symmetric non-negative definite with $\operatorname{rank}(\boldsymbol{V}) > r$ and known. We assume that the model satisfies the consistency condition

$$\boldsymbol{y} \in \mathcal{M}(\boldsymbol{X}\colon \boldsymbol{V}) \qquad \text{with probability 1.} \tag{1.2}$$

Then $\boldsymbol{Ay}$ is said to be the best linear unbiased estimator (BLUE) of $\boldsymbol{X\beta}$ whenever the dispersion matrix of $\boldsymbol{Ay}$ is smallest (in the Löwner sense) among all linear unbiased estimators of $\boldsymbol{X\beta}$. One characterization (Drygas, 1970, p. 55; Rao, 1973, p. 282) of the BLUE($\boldsymbol{X\beta}$) is:

$$\boldsymbol{Ay} = \mathrm{BLUE}(\boldsymbol{X\beta}) \quad \text{under} \quad \{\boldsymbol{y}, \boldsymbol{X\beta}, \boldsymbol{V}\} \Longleftrightarrow \boldsymbol{A}(\boldsymbol{X}\colon \boldsymbol{VM}) = (\boldsymbol{X}\colon \boldsymbol{0}); \tag{1.3}$$

here $\boldsymbol{M} = \boldsymbol{I} - \boldsymbol{H}$, with $\boldsymbol{H}$ the 'hat matrix' or orthogonal projector onto the column space $\mathcal{M}(\boldsymbol{X})$.

As a simple example, let us consider a linear model $\{\boldsymbol{y}, \boldsymbol{1\beta}, \boldsymbol{V}\}$, where the model matrix $\boldsymbol{X} = \boldsymbol{1} = (1, 1, ..., 1)'$ and $\boldsymbol{V}$ is positive definite. What is now the shorted matrix $S(\boldsymbol{V} \mid \boldsymbol{1}) = \boldsymbol{S}$, say? Since $\boldsymbol{S}$ is non-negative definite we must have $\boldsymbol{S} = \boldsymbol{UU}'$ for some $\boldsymbol{U}$ of full column rank. Further, the condition $\mathcal{M}(\boldsymbol{C}) \subset \mathcal{M}(\boldsymbol{1})$ implies that $\boldsymbol{U} = a\boldsymbol{1}$ for some scalar a and hence $\boldsymbol{S} = a^2\boldsymbol{11}'$. Our objective is to find a so that the difference $\boldsymbol{V} - a^2\boldsymbol{11}'$ is minimized in the Löwner sense; this minimum is to be found under the condition

$$a^2\boldsymbol{11}' \leqslant_{\mathrm{L}} \boldsymbol{V}. \tag{1.4}$$

By Baksalary and Kala (1983, Theorem 1) (see also Lemma 2.2 below), (1.4) holds if and only if $a^2 \leqslant (\boldsymbol{1}'\boldsymbol{V}^{-1}1(\boldsymbol{V}^{-1}\boldsymbol{1}))^{-1}$ and therefore the shorted matrix $S(\boldsymbol{V} \mid \boldsymbol{1})$ is

$$\boldsymbol{S} = \left(\boldsymbol{1}'\boldsymbol{V}^{-1}\boldsymbol{1}\right)^{-1}\boldsymbol{11}'. \tag{1.5}$$

We now observe that $\boldsymbol{S}$ in (1.5) is actually the dispersion matrix of the BLUE at $\boldsymbol{1\beta}$ under $\{\boldsymbol{y}, \boldsymbol{1\beta}, \boldsymbol{V}\}$; the BLUE($\boldsymbol{1\beta}$) itself is

$$\mathrm{BLUE}(\boldsymbol{1\beta}) = \boldsymbol{1}\left(\boldsymbol{1}'\boldsymbol{V}^{-1}\boldsymbol{1}\right)^{-1}\boldsymbol{1}'\boldsymbol{V}^{-1}\boldsymbol{y} \tag{1.6}$$

(cf. (3.10) below). It is interesting from a statistical point of view, that also in the general case, the dispersion matrix of the BLUE($\boldsymbol{X\beta}$) under $\{\boldsymbol{y}, \boldsymbol{X\beta}, \boldsymbol{V}\}$ is, as we will show in Section 3, equal to the shorted matrix of $\boldsymbol{V}$ with respect to $\boldsymbol{X}$.

Definition 1 applies only to symmetric nonnegative definite matrices. The generalization to rectangular matrices was made by Mitra and Puri (1982); see also (Mitra, 1982a, 1982b). They gave the following definition based on a minimal rank property.

DEFINITION 2. Let $\boldsymbol{A}$ be a given $m \times n$ matrix, and let $\mathcal{S}$ and $\mathcal{T}$, respectively, be given subspaces of $\mathbb{R}_m$ and $\mathbb{R}_n$. Consider the set

$$\mathfrak{C} = \{\boldsymbol{C} \in \mathbb{R}_{m \times n}\colon\ \mathcal{M}(\boldsymbol{C}) \subset \mathcal{S} \text{ and } \mathcal{M}(\boldsymbol{C}') \subset \mathcal{T}\}. \tag{1.7}$$

If there exists a unique matrix $\boldsymbol{C} = \boldsymbol{C}_s \in \mathfrak{C}$ such that

$$\operatorname{rank}(\boldsymbol{A} - \boldsymbol{C}_s) \leqslant \operatorname{rank}(\boldsymbol{A} - \boldsymbol{C}) \qquad \text{for all } \boldsymbol{C} \in \mathfrak{C}_2, \tag{1.8}$$

then $\mathfrak{C}_s$ is the shorted matrix $S(\boldsymbol{A} \mid \mathcal{S}, \mathcal{T})$ of $\boldsymbol{A}$ with respect to $\mathcal{S}$ and $\mathcal{T}$.

The 'generalized' shorted matrix $S(\boldsymbol{A} \mid \mathcal{S}, \mathcal{T})$ in Definition 2 does not, unfortunately, always exist as opposed to the unique 'classic' shorted matrix $S(\boldsymbol{V} \mid \boldsymbol{X})$ in Definition 1. Necessary and sufficient conditions, however, for the existence and uniqueness of $S(\boldsymbol{V} \mid \mathcal{S}, \mathcal{T})$ are established by Mitra and Puri (1982). See also Definitions 3 and 4 in Section 4 below.

Let us consider again our simple example, i.e., the linear model $\{\boldsymbol{y}, \boldsymbol{1}\beta, \boldsymbol{V}\}$, and let us find the shorted matrix $S(\boldsymbol{V} \mid \mathcal{S}, \mathcal{T}) = \boldsymbol{T}$, say, according to Definition 2 when $\mathcal{S} = \mathcal{T} = \mathcal{M}(\boldsymbol{1})$. We first note that (1.4) implies that $\boldsymbol{T} = b^2\boldsymbol{1}\boldsymbol{1}'$ for some scalar b and hence we have to find b so that the rank of the difference $\boldsymbol{V} - b^2\boldsymbol{1}\boldsymbol{1}'$ attains its minimum value. It is easy to show that $\operatorname{rank}(\boldsymbol{V} - b^2\boldsymbol{1}\boldsymbol{1}') = \operatorname{rank}(\boldsymbol{V}) - 1 + \operatorname{rank}(1 - b^2\boldsymbol{1}'\boldsymbol{V}^{-1}\boldsymbol{1})$, and hence the minimum of $\operatorname{rank}(\boldsymbol{V} - b^2\boldsymbol{1}\boldsymbol{1}')$ occurs when $b^2 = (\boldsymbol{1}'\boldsymbol{V}^{-1}\boldsymbol{1})^{-1}$. Therefore the shorted matrix $S(\boldsymbol{V} \mid \mathcal{S}, \mathcal{T}) = \boldsymbol{T}$ under Definition 2 coincides with the shorted matrix under Definition 1, cf. (1.5).

Our objective in this paper is to present together, in a historical perspective, various properties of shorted matrices and shorted operators that have appeared recently in the literature. Most proofs are omitted. Special attention is given to the role played by the shorted matrix in linear statistical models. In Sections 2 and 3 we concentrate on the shorted matrix in the sense of Definition 1, while in Section 4 we will briefly consider the generalized shorted matrix of Definition 2.

2. PRELIMINARY RESULTS

There will be several situations where we are interested in the invariance of a product of the type $AB^{-}C$ with respect to the choice of generalized inverse B^{-}, i.e., any solution B^{-1} to the equation $BB^{-}B = B$. The following important result is proved by Rao and Mitra (1971, Lemma 2.2.4 and Complement 14, p. 43); see also (Rao, Mitra and Bhimasankaram, 1972, Lemma 1; Hartwig, 1975, p. 273; Mitra and Odell, 1986, Lemma 1.1; Baksalary and Markiewicz, 1995; Puntanen *et al.*, 1995).

LEMMA 2.1. *Let $A \neq 0$ and $C \neq 0$. Then $AB^{-}C$ is invariant with respect to the choice of generalized inverse B^{-} if and only if the matrices A, B, and C satisfy the two set inclusions:*

$$\mathcal{M}(A') \subset \mathcal{M}(B') \quad \textit{and} \quad \mathcal{M}(C) \subset \mathcal{M}(B). \tag{2.1}$$

2.1. *Two matrix partial orderings*

The most popular matrix partial ordering in statistics is certainly the Löwner ordering of symmetric nonnegative definite matrices. For a proof of the following result see, e.g. (Baksalary *et al.*, 1989, p. 61).

LEMMA 2.2. *Let the $n \times n$ matrices A and B be symmetric and non-negative definite. Then*

$$A \leqslant_{\mathrm{L}} B \Longleftrightarrow \mathcal{M}(A) \subset \mathcal{M}(B) \quad \textit{and} \quad AB^{-}A \leqslant_{\mathrm{L}} A. \tag{2.2}$$

The matrix $AB^{-}A$ in (2.2) is invariant with respect to the choice of generalized inverse B^{-} since both A and B are symmetric and $\mathcal{M}(A) \subset \mathcal{M}(B)$, cf. Lemma 2.1 above.

Another important matrix partial ordering, applicable to rectangular matrices, is the rank-subtractivity (or 'minus') partial ordering introduced by Hartwig (1980) and by Nambooripad (1980); for the two $m \times n$ matrices A and B, we will write:

$$\begin{aligned} A \leqslant_{\mathrm{rs}} B \Longleftrightarrow A^{-}A &= A^{-}B, \\ AA^{-} = BA^{-} \quad &\text{for some} \quad A^{-} \in \{A^{-}\} \end{aligned} \tag{2.3}$$

or equivalently, cf. Lemma 2.3 below,

$$A \leqslant_{\mathrm{rs}} B \Longleftrightarrow \operatorname{rank}(B - A) = \operatorname{rank}(B) - \operatorname{rank}(A). \tag{2.4}$$

In (2.3), $\{A^{-}\}$ denotes the set of all generalized inverses of $A \in \mathbb{R}_{m \times n}$, i.e., $\{A^{-}\} = \{G \in \mathbb{R}_{n \times m}\colon\ AGA = A\}$.

The following lemma contains some useful characterizations of the rank-subtractivity partial ordering (Hartwig and Styan, 1986).

LEMMA 2.3. *For the two $m \times n$ matrices $\boldsymbol{A}$ and $\boldsymbol{B}$, let the rank-subtractivity partial ordering $\boldsymbol{A} \leqslant_{\mathrm{rs}} \boldsymbol{B}$ be defined by* (2.3) *and* (2.4). *Then the following six statements are equivalent*:

(a) $\boldsymbol{A} \leqslant_{\mathrm{rs}} \boldsymbol{B}$, *i.e.*, $\boldsymbol{A}^-\boldsymbol{A} = \boldsymbol{A}^-\boldsymbol{B}$ *and* $\boldsymbol{A}\boldsymbol{A}^- = \boldsymbol{B}\boldsymbol{A}^-$ *for some* $\boldsymbol{A}^- \in \{\boldsymbol{A}^-\}$;

(b) $\operatorname{rank}(\boldsymbol{B}-\boldsymbol{A}) = \operatorname{rank}(\boldsymbol{B}) - \operatorname{rank}(\boldsymbol{A})$;

(c) $\boldsymbol{B}-\boldsymbol{A} \leqslant_{\mathrm{rs}} \boldsymbol{B}$;

(d) $\{\boldsymbol{B}^-\} \subset \{\boldsymbol{A}^-\}$;

(e) $\mathcal{M}(\boldsymbol{A}) \cap \mathcal{M}(\boldsymbol{B}-\boldsymbol{A}) = \{\boldsymbol{0}\}$ *and* $\mathcal{M}(\boldsymbol{A}') \cap \mathcal{M}(\boldsymbol{B}'-\boldsymbol{A}') = \{\boldsymbol{0}\}$;

(f) $\mathcal{M}(\boldsymbol{A}) \subset \mathcal{M}(\boldsymbol{B})$, $\mathcal{M}(\boldsymbol{A}') \subset \mathcal{M}(\boldsymbol{B}')$ *and* $\boldsymbol{A}\boldsymbol{B}^-\boldsymbol{A} = \boldsymbol{A}$.

If (f) *holds for some* $\boldsymbol{B}^- \in \{\boldsymbol{B}^-\}$ *then* (f) *holds for all* $\boldsymbol{B}^- \in \{\boldsymbol{B}^-\}$.

Property (b) was proved by Hartwig (1980) and the equivalence of (b) and (e) by Marsaglia and Styan (1972) and Mitra (1972, Lemma 2.1). From (Marsaglia and Styan, 1974, p. 288; Cline and Funderlic, 1979, p. 195) it follows that (b) is equivalent to (f). The result (d) $\Leftrightarrow$ (a) is given by Mitra (1986, Theorem 2.2).

Condition (e) in Lemma 2.3 means that the row and column spaces of the matrices $\boldsymbol{A}$ and $\boldsymbol{B}-\boldsymbol{A}$ are virtually disjoint; following Mitra (1972) we then say that the two matrices $\boldsymbol{A}$ and $\boldsymbol{B}-\boldsymbol{A}$ are 'disjoint' and write

$$\boldsymbol{B} = \boldsymbol{A} \oplus (\boldsymbol{B}-\boldsymbol{A}). \tag{2.5}$$

We note that the two matrices $\boldsymbol{A}$ and $\boldsymbol{B}-\boldsymbol{A}$ are disjoint if and only if they are rank additive,

$$\operatorname{rank}(\boldsymbol{B}) = \operatorname{rank}(\boldsymbol{A}) + \operatorname{rank}(\boldsymbol{B}-\boldsymbol{A}), \tag{2.6}$$

(Marsaglia and Styan, 1972, 1974 and Mitra, 1972). When the two matrices $\boldsymbol{A}, \boldsymbol{B} \in \mathfrak{L}_n$, the condition (f) in Lemma 2.3 on the rank-subtractivity partial ordering becomes

$$\mathcal{M}(\boldsymbol{A}) \subset \mathcal{M}(\boldsymbol{B}) \quad \text{and} \quad \boldsymbol{A}\boldsymbol{B}^-\boldsymbol{A} = \boldsymbol{A} \tag{2.7}$$

while condition (b) in Lemma 2.2 on the Löwner partial ordering is

$$\mathcal{M}(\boldsymbol{A}) \subset \mathcal{M}(\boldsymbol{B}) \quad \text{and} \quad \boldsymbol{A}\boldsymbol{B}^-\boldsymbol{A} \leqslant_{\mathrm{L}} \boldsymbol{A}. \tag{2.8}$$

We see, therefore, that when the matrices $\boldsymbol{A}, \boldsymbol{B} \in \mathfrak{L}_n$, then the rank-subtractivity partial ordering implies the Löwner partial ordering (Baksalary and Hauke, 1984, p. 35), i.e.,

$$\boldsymbol{A} \leqslant_{\mathrm{rs}} \boldsymbol{B} \Longrightarrow \boldsymbol{A} \leqslant_{\mathrm{L}} \boldsymbol{B} \quad \text{whenever} \quad \boldsymbol{A}, \boldsymbol{B} \in \mathfrak{L}_n. \tag{2.9}$$

In both (2.7) and (2.8) the product $\boldsymbol{A}\boldsymbol{B}^-\boldsymbol{A}$ is invariant for all $\boldsymbol{B}^- \in \{\boldsymbol{B}^-\}$, since $\boldsymbol{A} = \boldsymbol{A}'$ and $\mathcal{M}(\boldsymbol{A}) \subset \mathcal{M}(\boldsymbol{B})$, cf. Lemma 2.1 above.

2.2. *Two special kinds of generalized inverses*

We consider the following two special kinds of generalized inverses:

(1) The Moore-Penrose inverse $\boldsymbol{A}^+$ is the unique solution $\boldsymbol{G}$ to the four equations

$$\boldsymbol{AGA} = \boldsymbol{A},\ \boldsymbol{GAG} = \boldsymbol{G},\ (\boldsymbol{AG})' = \boldsymbol{AG}, \quad \text{and} \quad (\boldsymbol{GA})' = \boldsymbol{GA}. \tag{2.10}$$

(2) Let $\boldsymbol{V} \in \mathfrak{L}_n$, i.e., $\boldsymbol{V} \geqslant_{\mathrm{L}} \boldsymbol{0}$. Then a matrix $\boldsymbol{G} \in \mathbb{R}_{n\times m}$ is a minimum-$\boldsymbol{V}$-seminorm generalized inverse of $\boldsymbol{A} \in \mathbb{R}_{m\times n}$ whenever $\boldsymbol{G}$ satisfies the pair of equations

$$\boldsymbol{AGA} = \boldsymbol{A} \quad \text{and} \quad (\boldsymbol{VGA})' = \boldsymbol{VGA}. \tag{2.11}$$

We denote such a minimum-$\boldsymbol{V}$-seminorm generalized inverse $\boldsymbol{G}$ by $\boldsymbol{A}^-_{m(\boldsymbol{V})}$ and the class of such generalized inverses by $\{\boldsymbol{A}^-_{m(\boldsymbol{V})}\}$.

In the following lemma we present some properties of a minimum-$\boldsymbol{V}$-seminorm generalized inverse $\boldsymbol{A}^-_{m(\boldsymbol{V})}$ of $\boldsymbol{A} \in \mathbb{R}_{m\times n}$ as defined in (2.11) above. Following Theorem 2.1 in (Mitra and Puri, 1979), we have

LEMMA 2.4. *Let $\boldsymbol{V} \in \mathfrak{L}_n$, so that $\boldsymbol{V} \geqslant_{\mathrm{L}} \boldsymbol{0}$, and let $\boldsymbol{G} \in \mathbb{R}_{n\times m}$ be a minimum-$\boldsymbol{V}$-seminorm generalized inverse of $\boldsymbol{A} \in \mathbb{R}_{m\times n}$, as defined in* (2.11). *Then*

(a) *$\boldsymbol{VGA}$ is unique with respect to the choice of $\boldsymbol{G} \in \{\boldsymbol{A}^-_{m(\boldsymbol{V})}\}$;*

(b) *$\boldsymbol{VGA} \geqslant_{\mathrm{L}} \boldsymbol{0}$ and $\boldsymbol{V} - \boldsymbol{VGA} \geqslant_{\mathrm{L}} \boldsymbol{0}$;*

(c) $\mathcal{M}(\boldsymbol{V} - \boldsymbol{VGA}) \cap \mathcal{M}(\boldsymbol{A}') = \{\boldsymbol{0}\}$;

(d) $\mathcal{M}(\boldsymbol{V}) = \mathcal{M}(\boldsymbol{VGA}) \oplus \mathcal{M}(\boldsymbol{V} - \boldsymbol{VGA})$;

(e) $\mathcal{M}(\boldsymbol{VGA}) = \mathcal{M}(\boldsymbol{V}) \cap \mathcal{M}(\boldsymbol{A}')$.

2.3. *The parallel sum*

A key concept in the study of shorted matrices is the 'parallel sum' of two matrices. Following (Rao and Mitra, 1971, p. 189) we will say that the two $m \times n$ matrices $\boldsymbol{A}$ and $\boldsymbol{B}$ are 'parallel summable' whenever $\boldsymbol{A}(\boldsymbol{A}+\boldsymbol{B})^-\boldsymbol{B}$ is invariant with respect to the choice of generalized inverse $(\boldsymbol{A}+\boldsymbol{B})$. If $\boldsymbol{A}$ and $\boldsymbol{B}$ are parallel summable, then we call $\boldsymbol{A}(\boldsymbol{A}+\boldsymbol{B})^-\boldsymbol{B}$ the 'parallel sum' of $\boldsymbol{A}$ and $\boldsymbol{B}$, which we write as $P(\boldsymbol{A}, \boldsymbol{B})$. The concept of parallel sum of two matrices $\boldsymbol{A}$ and $\boldsymbol{B}$ was introduced by Anderson and Duffin (1969) when both $\boldsymbol{A}$ and $\boldsymbol{B}$ are symmetric non-negative definite.

Using Lemma 2.1 we see that the non-null $m \times n$ matrices $\boldsymbol{A}$ and $\boldsymbol{B}$ are parallel summable if and only if $\boldsymbol{A}$ and $\boldsymbol{B}$ satisfy the two set inclusions

$$\mathcal{M}(\boldsymbol{A}) \subset \mathcal{M}(\boldsymbol{A}+\boldsymbol{B}) \quad \text{and} \quad \mathcal{M}(\boldsymbol{A}') \subset \mathcal{M}(\boldsymbol{A}'+\boldsymbol{B}'), \tag{2.12}$$

or equivalently (Rao and Mitra, 1971, p. 189),

$$\mathcal{M}(\boldsymbol{B}) \subset \mathcal{M}(\boldsymbol{A}+\boldsymbol{B}) \quad \text{and} \quad \mathcal{M}(\boldsymbol{B}') \subset \mathcal{M}(\boldsymbol{A}'+\boldsymbol{B}'). \tag{2.13}$$

We note that the first set inclusion in (2.12) is equivalent to the first set inclusion in (2.13), and similarly for the second set inclusions. The conditions (2.12) and (2.13) may also be expressed in terms of rank:

$$\operatorname{rank}(\boldsymbol{A}+\boldsymbol{B}) = \operatorname{rank}(\boldsymbol{A}:\boldsymbol{B}) = \operatorname{rank}(\boldsymbol{A}':\boldsymbol{B}'). \tag{2.14}$$

For useful properties of the parallel sum, see (Anderson and Duffin, 1969; Mitra and Puri, 1973; Rao and Mitra, 1971).

2.4. *Schur complements*

The concept of Schur complement is closely related to shorted operators, e.g. (Ando, 1979; Carlson, 1986, 1987). Consider the partitioned matrix

$$\boldsymbol{A} = \begin{pmatrix} \boldsymbol{B} & \boldsymbol{C} \\ \boldsymbol{D} & \boldsymbol{E} \end{pmatrix}. \tag{2.15}$$

A (generalized) Schur complement of $\boldsymbol{E}$ in $\boldsymbol{A}$ relative to a generalized inverse $\boldsymbol{E}^-$ is defined as

$$\boldsymbol{B} - \boldsymbol{C}\boldsymbol{E}^-\boldsymbol{D} = (\boldsymbol{A}/\boldsymbol{E}), \tag{2.16}$$

say, e.g. (Ouellette, 1981; Styan, 1985). By Lemma 2.1 we see that such a matrix $(\boldsymbol{A}/\boldsymbol{E})$ is unique if and only if the matrices $\boldsymbol{C}, \boldsymbol{D}$, and $\boldsymbol{E}$ satisfy the two set inclusions

$$\mathcal{M}(\boldsymbol{C}') \subset \mathcal{M}(\boldsymbol{E}') \quad \text{and} \quad \mathcal{M}(\boldsymbol{D}) \subset \mathcal{M}(\boldsymbol{E}) \tag{2.17}$$

or equivalently

$$\operatorname{rank}(\boldsymbol{C}':\boldsymbol{E}') = \operatorname{rank}(\boldsymbol{D}:\boldsymbol{E}) = \operatorname{rank}(\boldsymbol{E}). \tag{2.18}$$

Whenever the Schur complement $(\boldsymbol{A}/\boldsymbol{E})$ is unique then rank is additive in the following sense:

$$\operatorname{rank}(\boldsymbol{A}) = \operatorname{rank}(\boldsymbol{E}) + \operatorname{rank}(\boldsymbol{A}/\boldsymbol{E}), \tag{2.19}$$

e.g. (Ouellette, 1981; Styan, 1985). We note that (2.19) is always satisfied when the partitioned matrix $\boldsymbol{A}$ is symmetric and non-negative definite.

3. SHORTED MATRICES

3.1. *The shorted matrix and best linear unbiased estimation*

We start this section by showing that the dispersion matrix of the best linear unbiased estimator (BLUE) of $\boldsymbol{X\beta}$ in the general linear model $\{\boldsymbol{y}, \boldsymbol{X\beta}, \boldsymbol{V}\}$ is the shorted matrix of V with respect to X, cf. Definition 1. Our proof here follows that of Mitra and Puntanen (1991).

THEOREM 3.1. *In the general linear model* $\{y, X\beta, V\}$ *the dispersion matrix of the* BLUE *of* $X\beta$ *is the shorted matrix of* V *with respect to* X.

Proof. Let Ay be the BLUE of $X\beta$ under $\{y, X\beta, V\}$. We may consider such an A which satisfies $\mathcal{M}(A) \subset \mathcal{M}(X)$, i.e., $A = XB$ for some B; cf. e.g. (Rao, 1971, Theorem 5.1; Mitra and Moore, 1973, Theorem 2.1). The unbiasedness of XBy means that B is a generalized inverse of X, i.e., $B = X^-$, say. The dispersion matrix of XX^-y is $XX^-VX^{-\prime}X' = D$, say. Since y itself is unbiased for $X\beta$, we see that the dispersion matrix

$$\mathcal{D}\text{BLUE}(X\beta) = D = XX^-VX^{-\prime}X' \leqslant_{\mathrm{L}} V. \tag{3.1}$$

Because $\mathcal{M}(D) \subset \mathcal{M}(X)$, the matrix D belongs to the set $\mathfrak{C}_1$ defined in (1.1). Consider now an arbitrary C in $\mathfrak{C}_1$. Since $C \leqslant_{\mathrm{L}} V$ and $\mathcal{M}(C) \subset \mathcal{M}(X)$, we see that

$$C = XX^-CX^{-\prime}X' \leqslant_{\mathrm{L}} XX^-VX^{-\prime}X' = D, \tag{3.2}$$

and hence D is the unique maximal element in $\mathfrak{C}_1$. □

A direct consequence of Theorem 3.1 is that the shorted matrix of V with respect to X is the minimum (in the Löwner sense) of $XGVG'X'$ over all generalized inverses G of X, i.e.,

$$\min_{G \in \{X^-\}} XGVG'X' = \mathcal{S}(V \mid X). \tag{3.3}$$

On the other hand, since the set of matrix products XG is the set $\mathcal{P}$, say, of projectors onto $\mathcal{M}(X)$, we can write (3.3) as (Mitra and Puri, 1979, Theorem 2.2),

$$\min_{P \in \mathcal{P}} PVP' = \mathcal{S}(V \mid X) \tag{3.4}$$

or equivalently as

$$\min_{FX = X} FVF' = \mathcal{S}(V \mid X). \tag{3.5}$$

In this context, we note that Chipman (1968, p. 120, Lemma 1.2; 1976, p. 562, Lemma 1.2) obtained the following 'generalized Schwarz inequality': Let $V \in \mathfrak{L}_n$, $X \in \mathbb{R}_{n \times p}$ and let $X^\sim$ satisfy the two conditions

$$XX^\sim X = X \quad \text{and} \quad XX^\sim V = (XX^\sim V)'. \tag{3.6}$$

Then for any $A \in \mathbb{R}_{p \times n}$,

$$AXX^\sim VX^{\sim\prime}X'A' \leqslant_{\mathrm{L}} AVA' \tag{3.7}$$

with equality if and only if

$$AV = AXX^{\sim}V. \tag{3.8}$$

The inequality (3.7) coincides with (3.1) except that there may be different generalized inverses X^- and $X^\sim$ involved.

We are now approaching the concept of a minimum-V-seminorm generalized inverse. Namely, in view of (2.11), the two equations in (3.6) define precisely that

$$X^{\sim\prime} \in \{(X')^-_{m(V)}\}. \tag{3.9}$$

Actually we can prove the following theorem, which gives an interesting relationship between the Löwner and rank-subtractivity partial orderings.

THEOREM 3.2. *Let* $V \in \mathfrak{L}_n$, $X \in \mathbb{R}_{n\times p}$ *and* $G \in \{X^-\}$. *Then the following statements are equivalent*:

(a) $XGVG'X' \leqslant_{\rm L} V$;

(b) G' *is a minimum-*V*-seminorm generalized inverse of* X';

(c) XGy *is the* BLUE *of* $X\beta$;

(d) $XGVG'X' \leqslant_{\rm rs} V$;

(e) $\mathcal{S}(V \mid X) = XGVG'X'$.

Proof. The equivalence of (a) and (b) follows directly from (Baksalary *et al*., 1983), where Corollary 2 asserts that if an $n \times n$ matrix B is such that $B^2 = B$, and if C is an $n \times n$ symmetric non-negative definite matrix, then the Löwner ordering $B'CB \leqslant_{\rm L} C$ holds if and only if $B'C = CB$, with the difference matrix $C - B'CB$ otherwise being indefinite. The matrix B now corresponds to XG in (a) and since XG is idempotent we get (a) $\Leftrightarrow$ (b). On the basis of (1.3), condition (b) implies (c), and clearly (c) implies (a). Furthermore, by means of condition (d) in Lemma 2.4, condition (b) implies that $\mathcal{M}(XGVG'X') \cap \mathcal{M}(V - XGVG'X') = \{0\}$ and hence (d) holds. Due to (2.9), we see that (d) $\Rightarrow$ (a), while by Theorem 3.1, (c) $\Leftrightarrow$ (e). □

Our Theorem 3.2 is related to a result in (Baksalary and Puntanen, 1990, p. 366; Rao, 1971, Theorem 5.1). Baksalary and Puntanen (1990) characterized Ay as the BLUE of $X\beta$ if and only if the two conditions $AX = X$ and $\mathcal{M}(AV) \subset \mathcal{M}(X)$ hold along with any one of the three conditions: $AV = VA'$, $AVA' \leqslant_{\rm L} V$, and $AVA' \leqslant_{\rm rs} V$. Putting $A = XG$ with $G \in \{X^-\}$ establishes the equivalence of (a), (b), (c), and (d).

Note also that putting $G = X^+$ yields interesting characterizations of the equality between the ordinary least squares estimator (OLSE) and the BLUE; e.g. (Puntanen and Styan, 1989; Baksalary and Puntanen, 1990).

3.2. *Explicit representation of the shorted matrix*

Let us now consider a closed-form representation for the shorted matrix. When the dispersion matrix V is positive definite then it is well known that the best linear unbiased estimator of $X\beta$ is

$$\text{BLUE}(X\beta) = X(X'V^{-1}X)^{-}X'V^{-1}y. \tag{3.10}$$

When the dispersion matrix V is non-negative definite and possibly singular, we have, for example, the following representations for the BLUE$(X\beta)$:

$$\begin{aligned} \text{BLUE}(X\beta) &= [H - HVM(MVM)^{-}M]y && \text{(3.11a)} \\ &= \text{OLSE}(X\beta) - HVM(MVM)^{-}My && \text{(3.11b)} \\ &= [I - VM(MVM)^{-}M]y, && \text{(3.11c)} \end{aligned}$$

where the hat matrix $H = XX^{+}$, $M = I - H$, and $Hy = \text{OLSE}(X\beta)$, the ordinary least squares estimator of $X\beta$ (Drygas, 1970, p. 78; Albert, 1973, p. 182).

We note that the matrix A such that Ay is BLUE is unique if and only if $\mathcal{M}(X\colon\ V) = \mathbb{R}_n$; the numerical value, however, of Ay is unique with probability one. From (3.10) we find that the dispersion matrix

$$\mathcal{D}\text{BLUE}(X\beta) = S(V \mid X) = X(X'V^{-1}X)^{-}X', \tag{3.12}$$

and from (3.11b) and (3.11c) it follows at once that the dispersion matrix

$$\begin{aligned} \mathcal{D}\text{BLUE}(X\beta) &= S(V \mid X) \\ &= HVH - HVM(MVM)^{-}MVH && \text{(3.13a)} \\ &= V - VM(MVM)^{-}MV. && \text{(3.13b)} \end{aligned}$$

The matrix expression (3.13a), which is invariant for all choices of generalized inverse $(MVM)^{-}$, is the Schur complement of the residual dispersion matrix $\mathcal{D}(My) = MVM$ in the joint dispersion matrix of the OLSE$(X\beta) = Hy$ and the residual vector My

$$\mathcal{D}\begin{pmatrix} Hy \\ My \end{pmatrix} = \begin{pmatrix} HVH & HVM \\ MVH & MVM \end{pmatrix}. \tag{3.14}$$

Similarly (3.13b), which is also invariant for all choices of $(MVM)^{-}$, is the Schur complement of the residual dispersion matrix $\mathcal{D}(My) = MVM$ in the joint dispersion matrix of the response vector y — as opposed to the OLSE$(X\beta) = Hy$ before — and the residual vector My

$$\mathcal{D}\begin{pmatrix} y \\ My \end{pmatrix} = \begin{pmatrix} V & VM \\ MV & MVM \end{pmatrix}. \tag{3.15}$$

It is interesting to consider the special canonical case where, with $n > p$,

$$X = \begin{pmatrix} I_p \\ 0 \end{pmatrix} = X_0, \tag{3.16}$$

say (Anderson, 1971). Partitioning

$$\mathcal{D}y = V = \begin{pmatrix} V_{11} & V_{12} \\ V_{21} & V_{22} \end{pmatrix} \tag{3.17}$$

we obtain

$$\mathcal{D}\text{BLUE}\,(X_0\beta) = \mathcal{S}(V \mid X_0) = \begin{pmatrix} V_{11} - V_{12}V_{22}^{-}V_{21} & 0 \\ 0 & 0 \end{pmatrix} \tag{3.18}$$

with the Schur complement of V_{22} in V, i.e., $(V/V_{22}) = V_{11} - V_{12}V_{22}^{-}V_{21}$, in the top left-hand corner. Ando (1979) calls the full matrix (3.18) a 'generalized' Schur complement since the Schur complement (V/V_{22}) in the top left-hand corner is bordered by zero matrices. But the bordered matrix (3.18) is a 'regular' Schur complement as well, being the Schur complement of the residual dispersion matrix

$$\mathcal{D}M_0y = M_0VM_0 = \begin{pmatrix} 0 & 0 \\ 0 & V_{22} \end{pmatrix} \tag{3.19}$$

in the joint dispersion matrix of the response vector y and the residual vector M_0y

$$\mathcal{D}\begin{pmatrix} y \\ M_0y \end{pmatrix} = \begin{pmatrix} V_{11} & 0 & 0 & V_{12} \\ 0 & 0 & 0 & 0 \\ 0 & 0 & 0 & 0 \\ V_{21} & 0 & 0 & V_{22} \end{pmatrix}. \tag{3.20}$$

In (3.19) and (3.20) the residual matrix

$$M_0 = \begin{pmatrix} 0 & 0 \\ 0 & I_{n-p} \end{pmatrix}. \tag{3.21}$$

Another useful expression for the dispersion matrix of the BLUE is the following:

$$\mathcal{D}\text{BLUE}\,(X\beta) = X\left[X'(V + XUX')^{-}X\right]^{-}X' - XUX', \tag{3.22}$$

where U is any $p \times p$ matrix such that $\mathcal{M}(V + XUX') = \mathcal{M}(X\colon\ V)$ (Rao, 1971). For an extensive study of these and several other representations for the dispersion matrix of the BLUE$(X\beta)$, see (Baksalary *et al.*, 1990). A key component is a certain property of the canonical correlations between the random vectors $\boldsymbol{Hy}$ and $\boldsymbol{My}$. Writing cc_i for the i-th canonical correlation we note that

$$\text{all} \quad \mathrm{cc}_i(\boldsymbol{Hy}, \boldsymbol{My}) < 1 \iff \boldsymbol{HVV}^+\boldsymbol{M} = \boldsymbol{0} \tag{3.23}$$

and then the shorted matrix of $\boldsymbol{V}$ with respect to $\boldsymbol{X}$ may be expressed as

$$\mathcal{S}(\boldsymbol{V} \mid \boldsymbol{X}) = (\boldsymbol{HV}^+\boldsymbol{H})^+, \tag{3.24}$$

or as

$$\mathcal{S}(\boldsymbol{V} \mid \boldsymbol{X}) = \boldsymbol{Z}(\boldsymbol{Z}'\boldsymbol{V}^+\boldsymbol{Z})^+\boldsymbol{Z}' \tag{3.25}$$

for any matrix $\boldsymbol{Z}$ whose columns form an orthonormal basis for $\mathcal{M}(\boldsymbol{X})$.

Furthermore,

$$\mathcal{M}(\boldsymbol{XX}'\boldsymbol{V}) \subset \mathcal{M}(\boldsymbol{V}) \iff \mathcal{S}(\boldsymbol{V} \mid \boldsymbol{X}) = \boldsymbol{X}(\boldsymbol{X}'\boldsymbol{V}^+\boldsymbol{X})^+\boldsymbol{X}' \tag{3.26}$$

(Baksalary *et al.*, 1990, Theorem 5); when $\mathcal{M}(\boldsymbol{X}) \subset \mathcal{M}(\boldsymbol{V})$ the Moore–Penrose inverses in (3.26) may be replaced by arbitrary generalized inverses since $\boldsymbol{X}(\boldsymbol{X}'\boldsymbol{V}^-\boldsymbol{X})^-\boldsymbol{X}'$ is then invariant with respect to the choice of $\boldsymbol{V}^-$ and $(\boldsymbol{X}'\boldsymbol{V}^-\boldsymbol{X})^-$, cf. Lemma 2.1. Hence (Zyskind and Martin, 1969, p. 1196),

$$\mathcal{M}(\boldsymbol{X}) \subset \mathcal{M}(\boldsymbol{V}) \iff \mathcal{S}(\boldsymbol{V} \mid \boldsymbol{X}) = \boldsymbol{X}(\boldsymbol{X}'\boldsymbol{V}^-\boldsymbol{X})^-\boldsymbol{X}' \tag{3.27}$$

for all $\boldsymbol{V}^-$ and $(\boldsymbol{X}'\boldsymbol{V}^-\boldsymbol{X})^-$.

In the special case when $\mathcal{M}(\boldsymbol{VX}) \subset \mathcal{M}(\boldsymbol{X})$, the $\mathrm{BLUE}(\boldsymbol{X\beta})$ coincides with the $\mathrm{OLSE}(\boldsymbol{X\beta})$ and the common dispersion matrix simplifies to $\boldsymbol{HVH}$, and so

$$\mathcal{M}(\boldsymbol{VX}) \subset \mathcal{M}(\boldsymbol{X}) \iff \mathcal{S}(\boldsymbol{V} \mid \boldsymbol{X}) = \boldsymbol{HVH} \tag{3.28}$$

(Puntanen and Styan, 1989).

Another interesting way of characterizing the shorted matrix is through the popular Inverse Partitioned Matrix (IPM) method introduced by Rao (1971); see also (Werner, 1987). Consider the fundamental bordered matrix of linear estimation (Khatri, 1968; Rao, 1971, 1972, 1973; Hall and Meyer, 1975),

$$\boldsymbol{F} = \begin{pmatrix} \boldsymbol{V} & \boldsymbol{X} \\ \boldsymbol{X}' & \boldsymbol{0} \end{pmatrix}, \tag{3.29}$$

and let $\boldsymbol{G}$ denote a generalized inverse of $\boldsymbol{F}$, partitioned according to (3.29):

$$\boldsymbol{G} = \begin{pmatrix} \boldsymbol{C}_1 & \boldsymbol{C}_2 \\ \boldsymbol{C}_3 & -\boldsymbol{C}_4 \end{pmatrix} \in \{\boldsymbol{F}^-\}. \tag{3.30}$$

Then it follows that $\boldsymbol{C}_2$ and $\boldsymbol{C}_3'$ are minimum-$\boldsymbol{V}$-seminorm generalized inverses of $\boldsymbol{X}'$. By Theorem 3.2, we see that $\boldsymbol{VC}_2\boldsymbol{X}'$ and $\boldsymbol{XC}_3\boldsymbol{V}$ are invariant under the choice $\boldsymbol{G}$ of generalized inverse of $\boldsymbol{F}$, and that the shorted matrix

$$\mathcal{S}(\boldsymbol{V} \mid \boldsymbol{X}) = \boldsymbol{VC}_2\boldsymbol{X}' = \boldsymbol{XC}_3\boldsymbol{V} = \boldsymbol{XC}_4\boldsymbol{X}'. \tag{3.31}$$

We now collect together various properties of the shorted matrix (Theorem 3.3 below); these properties follow from the results of Section 2 and Theorems 3.1 and 3.2; see also (Goller, 1986).

THEOREM 3.3. *Let* $V \in \mathfrak{L}_n$, $X \in \mathbb{R}_{n\times p}$, $M = I - XX^+$ *and let*

$$\mathfrak{C}_1 = \{C\colon\ 0 \leqslant_{\mathrm{L}} C \leqslant_{\mathrm{L}} V,\ \mathcal{M}(C) \subset \mathcal{M}(X)\}, \tag{1.1}$$

$$\mathfrak{C}_* = \{C\colon\ \mathcal{M}(C) \subset \mathcal{M}(X),\ \mathcal{M}(C') \subset \mathcal{M}(X)\}, \tag{3.32}$$

$$\mathfrak{C}_{\mathrm{rs}} = \{C\colon\ \mathcal{M}(C) \subset \mathcal{M}(X),\ \mathcal{M}(C') \subset \mathcal{M}(X),\ C \leqslant_{\mathrm{rs}} V\}. \tag{3.33}$$

Then the following six statements are equivalent:

(a) $S = S(V \mid X)$, i.e., S *is the maximal element of* $\mathfrak{C}_1$ *in the Löwner partial ordering*;

(b) $0 \leqslant_{\mathrm{L}} S \leqslant_{\mathrm{L}} V$, $\mathcal{M}(S) \subset \mathcal{M}(X)$, $\mathcal{M}(V - S) \cap \mathcal{M}(X) = \{0\}$;

(c) $S = S'$, $\mathcal{M}(S) \subset \mathcal{M}(X)$, $\mathcal{M}(V - S) \cap \mathcal{M}(X) = \{0\}$;

(d) $S \in \mathfrak{C}_*$ *and* $\operatorname{rank}(V - S) \leqslant \operatorname{rank}(V - C)$ *for all* $C \in \mathfrak{C}_*$;

(e) S *is the maximal element of* $\mathfrak{C}_{\mathrm{rs}}$ *in the rank-subtractivity partial ordering*;

(f) $S = V - VM(MVM)^- MV$ *for some (and hence for all)* $(MVM)^- \in \{(MVM)^-\}$.

Moreover, the shorted matrix $S = S(V \mid X)$ *has the following properties*:

(g1) $\mathcal{M}(S) = \mathcal{M}(X) \cap \mathcal{M}(V)$;

(g2) $\mathcal{M}(V) = \mathcal{M}(S) \oplus \mathcal{M}(V - S)$;

(g3) $\operatorname{rank}(V) = \operatorname{rank}(S) + \operatorname{rank}(V - S)$;

(g4) $SV^+(V - S) = 0$;

(g5) $V^- \in \{S^-\}$ *for some (and hence for all)* $V^- \in \{V^-\}$, *i.e.*, $\{V^-\} \subset \{S^-\}$;

(g6) $\{S^-\} = \{V^- + U\}$, *where* V^- *is an arbitrary generalized inverse of* V *and* U *is an arbitrary solution to the equation* $XUX' = 0$;

(g7) $S(V^+ \mid X) \leqslant_{\mathrm{L}} [S(V \mid X)]^+$;

(g8) $S(\lambda V \mid X) = \lambda S(V \mid X)$ *for all* $\lambda > 0$;

(g9) $S[S(V \mid X) \mid Z] = S[S(V \mid Z) \mid X] = S(V \mid T)$ *whenever* $\mathcal{M}(T) = \mathcal{M}(X) \cap \mathcal{M}(Z)$;

(g10) $S(V \mid X) \leqslant_{\mathrm{L}} S(V \mid Z)$ *whenever* $\mathcal{M}(X) \subset \mathcal{M}(Z)$;

(g11) $S(V \mid X) \leqslant_{\mathrm{L}} S(W \mid X)$ *for all* $V \leqslant_{\mathrm{L}} W \in \mathfrak{L}_n$;

(g12) $S(V \mid X) + S(W \mid X) \leqslant_{\mathrm{L}} S(V + W \mid X)$ *for all* $W \in \mathfrak{L}_n$;

(g13) $(1 - \lambda)S(V \mid X) + \lambda S(W \mid X) \leqslant_{\mathrm{L}} S[(1 - \lambda)V + \lambda W \mid X]$ *for all* $\lambda \in (0, 1)$, $W \in \mathfrak{L}_n$.

When the shorted matrix $S = S(V \mid X) \neq 0$ *then*

(h1) *the non-zero proper generalized eigenvalues of* V *with respect to* H *are the non-zero eigenvalues of* S *and*

(h2) *if there exists a matrix* S_0 *such that* $\{S_0^-\} = \{V^- + U\}$, *where* V^- *is an arbitrary generalized inverse of* V *and* U *is an arbitrary solution to the equation* $XUX' = 0$, *then* $S_0 = S$.

3.4. *The shorted matrix and the parallel sum*

Mitra and Puri (1982, Theorem 2.3) gave a representation for the shorted matrix as a limit involving a parallel sum:

THEOREM 3.4. *The shorted matrix $S(\boldsymbol{V} \mid \boldsymbol{X})$ can be expressed as the limit*

$$S(\boldsymbol{V} \mid \boldsymbol{X}) = \lim_{\lambda \to 0^+} \frac{1}{\lambda} P(\lambda \boldsymbol{V}, \boldsymbol{X}\boldsymbol{X}') = \lim_{\lambda \to 0^+} \boldsymbol{V}(\lambda \boldsymbol{V} + \boldsymbol{X}\boldsymbol{X}')^- \boldsymbol{X}\boldsymbol{X}', \tag{3.34}$$

where $P(\boldsymbol{A}, \boldsymbol{B})$ is the parallel sum $\boldsymbol{A}(\boldsymbol{A} + \boldsymbol{B})^- \boldsymbol{B}$.

Proof. We first observe that $\boldsymbol{V}(\lambda \boldsymbol{V} + \boldsymbol{X}\boldsymbol{X}')^-(\lambda \boldsymbol{V} + \boldsymbol{X}\boldsymbol{X}') = \boldsymbol{V}$, and hence

$$\begin{aligned} \boldsymbol{V}(\lambda \boldsymbol{V} + \boldsymbol{X}\boldsymbol{X}')^- \boldsymbol{X}\boldsymbol{X}' &= \boldsymbol{V} - \lambda \boldsymbol{V}(\lambda \boldsymbol{V} + \boldsymbol{X}\boldsymbol{X}')^- \boldsymbol{V} \\ &= \boldsymbol{V} - \boldsymbol{V}\left(\boldsymbol{V} + \lambda^{-1} \boldsymbol{X}\boldsymbol{X}'\right)^- \boldsymbol{V}, \end{aligned} \tag{3.35}$$

which is invariant with respect to the choice of the generalized inverse, cf. Lemma 2.1. Following (Albert, 1972, Corollary 4.9.5), we have the following expansion:

$$\begin{aligned} \left(\boldsymbol{V} + \lambda^{-1} \boldsymbol{X}\boldsymbol{X}'\right) &= (\boldsymbol{M}\boldsymbol{V}\boldsymbol{M})^+ + \lambda\left[\boldsymbol{I} - (\boldsymbol{M}\boldsymbol{V}\boldsymbol{M})^+ \boldsymbol{M}\boldsymbol{V}\right] \\ &\quad \times (\boldsymbol{X}\boldsymbol{X}')^+ \left[\boldsymbol{I} - \boldsymbol{V}\boldsymbol{M}(\boldsymbol{M}\boldsymbol{V}\boldsymbol{M})^+\right] + \boldsymbol{O}(\lambda^2), \end{aligned} \tag{3.36}$$

for sufficiently small positive λ; the matrix $\boldsymbol{O}(\lambda^2)$ in (3.36) converges to $\boldsymbol{0}$ quadratically as $\lambda \to 0^+$. Substituting (3.36) into (3.35) and letting $\lambda \to 0^+$, we obtain

$$\begin{aligned} &\lim_{\lambda \to 0^+} \boldsymbol{V}(\lambda \boldsymbol{V} + \boldsymbol{X}\boldsymbol{X}')^- \boldsymbol{X}\boldsymbol{X}' \\ &\quad = \boldsymbol{V} - \boldsymbol{V}(\boldsymbol{M}\boldsymbol{V}\boldsymbol{M})^+ \boldsymbol{V} \\ &\quad = \boldsymbol{V} - \boldsymbol{V}\boldsymbol{M}(\boldsymbol{M}\boldsymbol{V}\boldsymbol{M})^+ \boldsymbol{M}\boldsymbol{V} = S(\boldsymbol{V} \mid \boldsymbol{X}), \end{aligned} \tag{3.37}$$

cf. (3.13b). □

The limit representation (3.34) for the shorted matrix was established by Anderson and Trapp (1975, Theorem 12) for the special case when $\boldsymbol{X}\boldsymbol{X}'$ is the orthogonal projector $\boldsymbol{X}\boldsymbol{X}^+ = \boldsymbol{H}$. Mitra and Puri (1982, p. 60) point out that this assumption is not necessary and refer to Ando (1976), who used a similar limit to present a Lebesgue-type decomposition of positive operators. Furthermore, Mitra and Puri (1982, p. 60) mentioned that a direct proof could be constructed using simultaneous diagonalization of a pair of non-negative definite matrices (see, e.g. (Mitra and Rao, 1968)).

Similarly, using (3.36), we obtain

$$\begin{aligned} \lim_{\lambda \to 0^+} \boldsymbol{X}'(\lambda \boldsymbol{V} + \boldsymbol{X}\boldsymbol{X}')^+ &= \boldsymbol{X}'(\boldsymbol{X}\boldsymbol{X}')^+[\boldsymbol{I} - \boldsymbol{V}\boldsymbol{M}(\boldsymbol{M}\boldsymbol{V}\boldsymbol{M})^+] \\ &= \boldsymbol{X}^+[\boldsymbol{I} - \boldsymbol{V}\boldsymbol{M}(\boldsymbol{M}\boldsymbol{V}\boldsymbol{M})^+] \\ &= \boldsymbol{G}, \end{aligned} \tag{3.38}$$

say. Furthermore, it is easy to see that $\boldsymbol{XGy} = \text{BLUE}(\boldsymbol{X\beta})$ and that $\boldsymbol{G}'$ is a minimum-$\boldsymbol{V}$-seminorm generalized inverse of $\boldsymbol{X}'$. Mitra (1975) considered related questions using a different approach; application of his Corollary 5.1 yields, in parallel to our (3.38), the limit

$$\lim_{\lambda \to 0+} (\lambda \boldsymbol{V} + \boldsymbol{XX}')^{+}\boldsymbol{X} \in \{(\boldsymbol{X}')^{-}_{m(\boldsymbol{V})}\}. \tag{3.39}$$

Stone and Wang (1989) obtained related results in the context of a Bayesian linear model using a coordinate-free approach. Let us assume that $\boldsymbol{\beta}$ itself is a random vector with expectation $\mathbf{0}$ and dispersion matrix $\tau^2\boldsymbol{I}$. Then the Bayes estimator of $\boldsymbol{\beta}$ is

$$\boldsymbol{\beta}_b(\tau) = \boldsymbol{X}'\left(\tau^{-2}\boldsymbol{V} + \boldsymbol{XX}'\right)^{-}\boldsymbol{y}. \tag{3.40}$$

Note that due to the consistency condition

$$\boldsymbol{y} \in \mathcal{M}(\boldsymbol{X} : \boldsymbol{V}) \qquad \text{with probability 1,} \tag{1.2}$$

the formula (3.40) is invariant with respect to the choice of generalized inverse $(\tau^{-2}\boldsymbol{V}+\boldsymbol{XX}')^{-}$; we may, therefore, use the Moore–Penrose inverse in (3.40), and then using (3.38) with $\lambda = \tau^{-2}$, yields

$$\lim_{\tau \to \infty} \boldsymbol{X\beta}_b(\tau) = \boldsymbol{XGy} = \text{BLUE}(\boldsymbol{X\beta}), \tag{3.41}$$

since $\boldsymbol{G}' \in \{(\boldsymbol{X}')^{-}_{m(\boldsymbol{V})}\}$, cf. our Theorem 3.2. However (Stone and Wang, 1989, Section 2) established (3.41) in a different way.

For further applications of the shorted matrix in a statistical context, see (Gaffke, 1985a, 1985b; Krafft, 1983; Krafft and Mitra, 1982; Mathew, 1983; Mathew and Mitra, 1984; Mitra, 1982c; Pukelsheim, 1993, p. 411). Here the shorted matrix $S(\boldsymbol{V} \mid \boldsymbol{X})$ appears to be tailor-made for the problems under investigation on account of its properties (g1) and (g5) in Theorem 3.3 and the one described in (3.3). In (Mitra, 1982c; Mathew and Mitra, 1984), the authors addressed themselves to the following problems: To what extent can one recover the matrix $\boldsymbol{V} \in \mathfrak{L}_n$ given the shorted versions $S(\boldsymbol{V} \mid \boldsymbol{X})$? They point out the use of a Balanced Incomplete Block Design (BIBD) to pick out certain $\boldsymbol{X}$ matrices that will ensure the identifiability of $\boldsymbol{V}$.

4. SOME COMMENTS ON EXTENDED CONCEPTS OF THE SHORTED MATRIX

As we have pointed out, our Definition 1 for the shorted matrix applies only to symmetric non-negative definite matrices. Mitra and Puri (1982), in generalizing this definition to rectangular matrices, used two approaches, one where the key point of the development is a theorem of Anderson and Trapp (1975) exhibiting the shorted non-negative matrix as the limit of a sequence of parallel-sum matrices, and the other through a minimum rank property as in our Definition 2.

We recall here briefly the definition of the shorted matrix through the parallel sum and some of its properties, cf. Theorem 3.4. Let $\boldsymbol{A}$, $\boldsymbol{B}$ be a pair of matrices in $\mathbb{R}_{m\times n}$, and for some non-null real number c let $\boldsymbol{A}$ and $c\boldsymbol{B}$ be parallel summable and suppose that the limit matrix

$$\lim_{\lambda\to 0+} \frac{1}{\lambda} P(\lambda\boldsymbol{A}, \boldsymbol{B}) = \lim_{\lambda\to 0+} \boldsymbol{A}(\lambda\boldsymbol{A}+\boldsymbol{B})^{-}\boldsymbol{B} = \boldsymbol{C}, \tag{4.1}$$

say, exists and is finite. Mitra and Puri (1982) called $\boldsymbol{C}$ the 'matrix $\boldsymbol{A}$ shorted by $\boldsymbol{B}$' and denoted it by $S(\boldsymbol{A} \mid \boldsymbol{B})$; cf. Theorem 3.4 with $\boldsymbol{A} = \boldsymbol{V}$ and $\boldsymbol{B} = \boldsymbol{X}\boldsymbol{X}'$.

This matrix $\boldsymbol{C}$ now has several interesting properties. In particular, let $\boldsymbol{E}$ be any matrix such that $\mathcal{M}(\boldsymbol{E}) \subset \mathcal{M}(\boldsymbol{B})$, $\mathcal{M}(\boldsymbol{E}') \subset \mathcal{M}(\boldsymbol{B}')$; then

$$\operatorname{rank}(\boldsymbol{A}-\boldsymbol{C}) \leqslant \operatorname{rank}(\boldsymbol{A}-\boldsymbol{E}) \tag{4.2}$$

with equality holding if and only if $\boldsymbol{E} = \boldsymbol{C}$. It was just this property that tempted (Mitra and Puri, 1982) to introduce the definition (cf. Definition 2 of Section 1):

DEFINITION 3. *Let $\boldsymbol{A} \in \mathbb{R}_{m\times n}$, $\boldsymbol{X} \in \mathbb{R}_{m\times p}$ and $\boldsymbol{Y} \in \mathbb{R}_{q\times n}$ be given matrices and consider the set*

$$\mathfrak{C}_3 = \{\boldsymbol{C} \in \mathbb{R}_{m\times n}\colon\ \mathcal{M}(\boldsymbol{C}) \subset \mathcal{M}(\boldsymbol{X}) \quad \text{and} \quad \mathcal{M}(\boldsymbol{C}') \subset \mathcal{M}(\boldsymbol{Y}')\}. \tag{4.3}$$

If there exists a unique matrix $\boldsymbol{A}_* \in \mathfrak{C}_3$ such that

$$\operatorname{rank}(\boldsymbol{A}-\boldsymbol{A}_*) \leqslant \operatorname{rank}(\boldsymbol{A}-\boldsymbol{C}) \qquad \text{for all } \boldsymbol{C} \in \mathfrak{C}_3, \tag{4.4}$$

then $\boldsymbol{A}_*$ is the shorted matrix $S(\boldsymbol{A} \mid \boldsymbol{X}, \boldsymbol{Y})$ of $\boldsymbol{A}$ with respect to $\boldsymbol{X}$ and $\boldsymbol{Y}$.

It is worth noting that there always exists a matrix $\boldsymbol{C} \in \mathfrak{C}_3$ for which $\operatorname{rank}(\boldsymbol{A}-\boldsymbol{C})$ is minimal but it may not be unique; an example is given by Mitra and Puri (1982, p. 69).

Here are some comments on the computation of the shorted matrix. Define the bordered matrix

$$\boldsymbol{F} = \begin{pmatrix} \boldsymbol{A} & \boldsymbol{X} \\ \boldsymbol{Y} & \boldsymbol{0} \end{pmatrix}. \tag{4.5}$$

Mitra and Puri (1982, Lemma 5.1, Theorem 4.1) and Mitra (1982a) showed that the shorted matrix $S(\boldsymbol{A} \mid \boldsymbol{X}, \boldsymbol{Y})$ in Definition 3 exists, i.e., the unique solution to the minimum rank condition (4.4) exists, if

$$\operatorname{rank}(\boldsymbol{F}) = \operatorname{rank}(\boldsymbol{A}'\colon\ \boldsymbol{Y}') + \operatorname{rank}(\boldsymbol{X}) = \operatorname{rank}(\boldsymbol{A}\colon\ \boldsymbol{X}) + \operatorname{rank}(\boldsymbol{Y}). \tag{4.6}$$

It was also shown that the condition (4.6) is also necessary unless precisely one of the matrices $\boldsymbol{X}$ and $\boldsymbol{Y}$ is the null matrix. (The situation where both $\boldsymbol{X} = \mathbf{0}$ and $\boldsymbol{Y} = \mathbf{0}$ is covered under (4.6).)

Let $\boldsymbol{G}$ be a generalized inverse of $\boldsymbol{F}$:

$$\boldsymbol{G} = \begin{pmatrix} \boldsymbol{C}_1 & \boldsymbol{C}_2 \\ \boldsymbol{C}_3 & -\boldsymbol{C}_4 \end{pmatrix} \in \{\boldsymbol{F}^-\}. \tag{4.7}$$

Then, if (4.6) holds, the matrix $\boldsymbol{X}\boldsymbol{C}_4\boldsymbol{Y}$, which can be written equivalently as

$$\boldsymbol{A}\boldsymbol{C}_2\boldsymbol{Y} = \boldsymbol{X}\boldsymbol{C}_3\boldsymbol{A} = \boldsymbol{X}\boldsymbol{C}_4\boldsymbol{Y} = \boldsymbol{A} - \boldsymbol{A}\boldsymbol{C}_1\boldsymbol{A}, \tag{4.8}$$

is the shorted matrix $S(\boldsymbol{A} \mid \boldsymbol{X}, \boldsymbol{Y})$, and the expressions in (4.8) are invariant with respect to the choice of generalized inverse $\boldsymbol{G}$.

Mitra (1986) used the rank-subtractivity partial ordering to redefine the shorted matrix as follows:

DEFINITION 4. *Let $\boldsymbol{A}, \boldsymbol{X}, \boldsymbol{Y}$ and the set $\mathfrak{C}_3$ be defined as in Definition 3. Consider a subset*

$$\mathfrak{C}_4 = \{\boldsymbol{C} \in \mathfrak{C}_3 \colon \boldsymbol{C} \leqslant_{\text{rs}} \boldsymbol{A}\} \tag{4.9}$$

of $\mathfrak{C}_3$. If $\mathfrak{C}_4$ has a unique maximal element $\boldsymbol{A}_\# \in \mathfrak{C}_4$ under the rank-subtractivity partial ordering such that

$$\boldsymbol{C} \leqslant_{\text{rs}} \boldsymbol{A}_\# \qquad \textit{for all} \quad \boldsymbol{C} \in \mathfrak{C}_4, \tag{4.10}$$

then $\boldsymbol{A}_\#$ is the shorted matrix $S(\boldsymbol{A} \mid \boldsymbol{X}, \boldsymbol{Y})$ of $\boldsymbol{A}$ with respect to $\boldsymbol{X}$ and $\boldsymbol{Y}$.

Furthermore, Mitra (1986, Theorem 3.1) showed that when the condition (4.6) holds, then $\mathfrak{C}_4$ has a unique maximal element under the rank-subtractivity partial ordering, and that the condition (4.6) is also necessary unless $\mathfrak{C}_4 = \{\mathbf{0}\}$.

We may now ask whether our Definitions 3 and 4 do indeed lead to the same shorted matrix $S(\boldsymbol{A} \mid \boldsymbol{X}, \boldsymbol{Y})$. In a series of papers, Mitra (1986, p. 26, 1993, 1994, Section 5) provides an interesting discussion related to this question. When the condition (4.6) holds then Definitions 3 and 4 do lead to the same end product, which we denote as $S(\boldsymbol{A} \mid \boldsymbol{X}, \boldsymbol{Y})$. The unique shorted matrix $S(\boldsymbol{A} \mid \boldsymbol{X}, \boldsymbol{Y})$ in either case can be computed using any one of the four expressions covered by the string of equalities (4.8). The shorted matrix computed under condition (4.6) is called a 'regular' shorted matrix. A regular shorted matrix has many attractive properties, e.g.,

$$\mathcal{M}[S(\boldsymbol{A} \mid \boldsymbol{X}, \boldsymbol{Y})] = \mathcal{M}(\boldsymbol{X}) \cap \mathcal{M}(\boldsymbol{A}), \tag{4.11}$$

$$\mathcal{M}\{[S(\boldsymbol{A} \mid \boldsymbol{X}, \boldsymbol{Y})]'\} = \mathcal{M}(\boldsymbol{Y}') \cap \mathcal{M}(\boldsymbol{A}'), \tag{4.12}$$

$$\{[S(\boldsymbol{A} \mid \boldsymbol{X}, \boldsymbol{Y})]^-\} = \{\boldsymbol{A}^- + \boldsymbol{Z}\}, \tag{4.13}$$

where $\boldsymbol{A}^-$ is an arbitrary generalized inverse of $\boldsymbol{A}$ and $\boldsymbol{Z}$ is an arbitrary solution to the equation $\boldsymbol{Y}\boldsymbol{Z}\boldsymbol{X} = \mathbf{0}$. A unique shorted matrix based on either Definition 3 or 4 may exist even when (4.6) does not hold. These are the 'irregular' shorted matrices. The irregularity forces the shorted matrix to lose on one or more of the properties (4.11)–(4.13). Mitra (1986) proves the following converse proposition with respect to (4.13). Here we take into account a lacuna in the statement of Theorem 3.6 in (Mitra, 1986), as pointed out by Tian (1992).

THEOREM 4.1. *Let A, X and Y be defined as in Definition 3 and suppose that there exists a matrix C_0 such that*

$$\{C_0^-\} = \{A^- + Z\}, \tag{4.14}$$

where A^- is an arbitrary generalized inverse of A and the matrix Z is an arbitrary solution to the equation $YZX = 0$. Then either

(a) *precisely one of the matrices X or Y is a null matrix and (4.6) does not hold, or*

(b) *the regular shorted matrix $S(A \mid X, Y)$ exists and satisfies either Definition 3 or 4.*

Furthermore $C_0 = S(A \mid X, Y)$.

We conclude our review of shorted matrices and shorted operators with a short discussion of the paper by Goller (1986). In this extensive study, shorted operators (over an arbitrary field) are characterized through rank decomposition matrices: An $m \times n$ matrix S is a 'rank decomposition matrix' with respect to an $m \times n$ matrix A whenever S and A satisfy the rank-subtractivity equation

$$\operatorname{rank}(A - S) = \operatorname{rank}(A) - \operatorname{rank}(S). \tag{4.15}$$

Let $D(A)$ denote the set of all such $m \times n$ matrices A; then

$$D(A) = \{S\colon\ S \leqslant_{\text{rs}} A\}. \tag{4.16}$$

Goller (1986) provided the general solution to (4.15) and gave a necessary and sufficient condition for the existence of an $m \times n$ matrix S which satisfies (4.15) and which has a particular column space, $\mathcal{S}$ say, and a particular row space, $\mathcal{T}$ say; then using these results in connection with the solution of a rank minimization problem, he obtained the shorted matrix $S(A \mid \mathcal{S}, \mathcal{T})$ of Mitra and Puri (1982), cf. our Definitions 2 and 3. Furthermore, Goller (1986) pointed out that the major drawback of the shorted matrix as defined in our Definition 2 is that it does not always exist; he also drew attention to the key difficulty of choosing $\mathcal{S}$ and $\mathcal{T}$ appropriately. To avoid these problems, an alternative approach is presented using the concept of an anisotropic matrix: a symmetric matrix T is 'anisotropic' on the subspace $\mathcal{S}$ if $x'Tx = 0$ and $x \in \mathcal{S}$ together imply that $x = 0$. Let the symmetric matrix T be anisotropic on $\mathcal{S} \cap \mathcal{M}(A)$; then the two equations

$$\mathcal{M}(S) = \mathcal{S} \cap \mathcal{M}(A) \quad \text{and} \quad S'T(A - S) = 0 \tag{4.17}$$

have a unique solution S, which is the 'shorted matrix'. Indeed if A is symmetric nonnegative definite and $T = A^+$, the Moore–Penrose inverse of A, then this shorted matrix S coincides with the (classical) shorted matrix of our Definition 1 with $A = V$ and $\mathcal{S} = \mathcal{M}(X)$ (Goller, 1986, Theorem 7.2).

Note. Our list of references includes all published work that we are aware of on the subject of shorted matrices and shorted operators. For further references on Schur complements see (Ouellette, 1981 and Styan, 1985). The symbol $\oplus$ before a reference indicates that it was not cited in this paper.

REFERENCES

Albert, A. (1972). *Regression and the Moore–Penrose Pseudoinverse*. Academic Press, New York.

Albert, A. (1973). The Gauss-Markov theorem for regression models with possibly singular covariances. *SIAM J. Appl. Math.* **24**, 182–187.

Anderson, W. N., Jr. (1971). Shorted operators. *SIAM J. Appl. Math.* **20**, 520–525.

Anderson, W. N., Jr. and Duffin, R. J. (1969). Series and parallel addition of matrices. *J. Math. Anal. Appl.* **26**, 576–594.

⊕ Anderson, W. N., Jr., Morley, T. D. and Trapp, G. E. (1986). Cascade addition and subtraction of matrices. *SIAM J. Algebraic Discrete Methods* **7**, 609–626.

⊕ Anderson, W. N., Jr., Morley, T. D. and Trapp, G. E. (1990a). Sets of positive operators with suprema. *SIAM J. Matrix Anal. Appl.* **11**, 207–212.

⊕ Anderson, W. N., Jr., Morley, T. D. and Trapp, G. E. (1990b). Positive solutions to $X = A - BX^{-1}B^*$. *Linear Algebra Appl.* **134**, 53–62.

Anderson, W. N., Jr. and Trapp, G. E. (1975). Shorted operators, II. *SIAM J. Appl. Math.* **28**, 60–71.

⊕ Anderson, W. N., Jr. and Trapp, G. E. (1979). Matrix operations induced by electrical network connections — a survey. In: *Constructive Approaches to Mathematical Models*, pp. 53–73, C. V. Coffman and G. J. Fix (Eds), Academic Press, New York.

Ando, T. (1976). Lebesgue-type decomposition of positive operators. *Acta Sci. Math. (Szeged)* **38**, 253–260.

Ando, T. (1979). Generalized Schur complements. *Linear Algebra Appl.* **27**, 173–186.

Baksalary, J. K. and Hauke, J. (1984). Inheriting independence and chi-squaredness under certain matrix orderings. *Statist. Probab. Lett.* **2**, 35–38.

Baksalary, J. K. and Kala, R. (1983). Partial orderings between matrices one of which is of rank one. *Bull. Polish Acad. Sci. Math.* **31**, 5–7.

Baksalary, J. K., Kala, R. and Kłaczyński, K. (1983). The matrix inequality $M \geqslant B^*MB$. *Linear Algebra Appl.* **54**, 77–86.

Baksalary, J. K. and Markiewicz, A. (1995). Further results on invariance of the eigenvalues of matrix products involving generalized inverses. *Linear Algebra Appl. Fifth Special Issue on Linear Algebra and Statistics*, in press.

Baksalary, J. K., Pukelsheim, F. and Styan, G. P. H. (1989). Some properties of matrix partial orderings. *Linear Algebra Appl.* **119**, 57–85.

Baksalary, J. K. and Puntanen, S. (1990). Characterizations of the best linear unbiased estimator in the general Gauss-Markov model with the use of matrix partial orderings. *Linear Algebra Appl.* **127**, 363–370.

Baksalary, J. K., Puntanen, S. and Styan, G. P. H. (1990). A property of the dispersion matrix of the best linear unbiased estimator in the general Gauss-Markov model. *Sankhyā Ser. A* **52**, 279–296.

⊕ Butler, C. A. and Morley, T. D. (1988a). A note on the shorted operator. *SIAM J. Matrix Anal. Appl.* **9**, 147–155.

⊕ Butler, C. A. and Morley, T. D. (1988b). Six generalized Schur complements. *Linear Algebra Appl.* **106**, 259–269.

Carlson, D. (1986). What are Schur complements, anyway? *Linear Algebra Appl.* **74**, 257–275 [extended summary: **59** (1984): 188–193].

Carlson, D. (1987). Generalized inverse invariance, partial orders, and rank-minimization problems for matrices. In: *Current Trends in Matrix Theory*, pp. 81–87, F. Uhlig and R. Grone (Eds), North-Holland, New York.

Chipman, J. S. (1968). Specification problems in regression analysis. In: *Proceedings of the Symposium on Theory and Applications of Generalized Inverses of Matrices*, pp. 114–176, T. L. Boullion and P. L. Odell (Eds), Texas Tech. Press, Lubbock, Texas.

Chipman, J. S. (1976). Estimation and aggregation in econometrics: an application of the theory of generalized inverses. In: *Generalized Inverses and Applications*, pp. 549–769, M. Z. Nashed (Ed.), Academic Press, New York.

Cline, R. E. and Funderlic, R. E. (1979). The rank of a difference of matrices and associated generalized inverses. *Linear Algebra Appl.* **24**, 185–215.

⊕ Duffin, R. J. and Morley, T. D. (1978). Almost definite operators and electro-mechanical systems. *SIAM J. Appl. Math.* **35**, 21–30.

⊕ Fiedler, M. (1981). Remarks on the Schur complement. *Linear Algebra Appl.* **39**, 189–195.

Gaffke, N. (1985a). Singular information matrices, directional derivatives, and subgradients in optimal design theory. In: *Linear Statistical Inference*, pp. 61–77, T. Caliński and W. Klonecki (Eds), Springer, Berlin.

Gaffke, N. (1985b). Directional derivatives of optimality criteria at singular matrices in convex design theory. *Statistics* **16**, 373–388.

Goller, H. (1986). Shorted operators and rank decomposition matrices. *Linear Algebra Appl.* **81**, 207–236.

Hall, F. J. and Meyer, C. D., Jr. (1975). Generalized inverses of the fundamental bordered matrix used in linear estimation. *Sankhyā Ser. A* **37**, 428–438. [Corrigendum (1978), **40**, p. 399.]

Hartwig, R. E. (1975). 1-2 inverses and the invariance of BA^+C. *Linear Algebra Appl.* **11**, 271–275.

Hartwig, R. E. (1980). How to partially order regular elements. *Math. Japon.* **25**, 1–13.

⊕ Hartwig, R. E. and Shoaf, I. M. (1980). Invariance, group inverses, and parallel sums. *Rev. Roumaine Math. Pures Appl.* **25**, 33–42.

Hartwig, R. E. and Styan, G. P. H. (1986). On some characterizations of the 'star' partial ordering for matrices and rank subtractivity. *Linear Algebra Appl.* **82**, 145–161.

Khatri, C. G. (1968). Some results for the singular normal multivariate regression models. *Sankhyā Ser. A* **30**, 267–280.

Krafft, O. (1983). A matrix optimization problem. *Linear Algebra Appl.* **51**, 137–142.

Krafft, O. and Mitra, S. K. (1982). Least favorable parameters in the Gauss-Markoff model. *Sankhyā Ser. A* **44**, 286–290.

Krein, M. G. (1947). The theory of self-adjoint extensions of semi-bounded Hermitian transformations and its applications, I, II. *Matematicheskii Sbornik, Novaya Seriya* **20**(62), 431–495; **21**(63), 365–404. (In Russian with English summaries.).

Marsaglia, G. and Styan, G. P. H. (1972). When does rank $(A+B)$ = rank (A) + rank(B)? *Canad. Math. Bull.* **15**, 451–452.

Marsaglia, G. and Styan, G. P. H. (1974). Equalities and inequalities for ranks of matrices. *Linear and Multilinear Algebra* **2**, 269–292.

Mathew, T. (1983). Linear estimation with an incorrect dispersion matrix in linear models with a common linear part. *J. Amer. Statist. Assoc.* **78**, 468–471.

Mathew, T. and Mitra, S. K. (1984). Shorted operators and the identification problem — the real case. *IEEE Trans. Circuits and Systems* **CAS-31**, 299–300.

Mitra, S. K. (1972). Fixed rank solutions of linear matrix equations. *Sankhyā Ser. A* **34**, 387–392.

Mitra, S. K. (1975). Optimal inverse of a matrix. *Sankhyā Ser. A* **37**, 550–563.

⊕ Mitra, S. K. (1980). Generalized inverse of matrices and applications to linear models. In: *Handbook of Statistics, Vol. 1*, pp. 471–512, P. R. Krishnaiah (Ed.), North-Holland, Amsterdam.

Mitra, S. K. (1982a). Simultaneous diagonalization of rectangular matrices. *Linear Algebra Appl.* **47**, 139–150.

Mitra, S. K. (1982b). Properties of the fundamental bordered matrix used in linear estimation. In: *Statistics and Probability: Essays in Honor of C. R. Rao*, pp. 505–509, G. Kallianpur *et al.* (Eds), North-Holland, Amsterdam.

Mitra, S. K. (1982c). Shorted operators and the identification problem. *IEEE Trans. Circuits and Systems* **CAS-29**, 581–583.

Mitra, S. K. (1986). The minus partial order and the shorted matrix. *Linear Algebra Appl.* **83**, 1–27.

⊕ Mitra, S. K. (1987). On group inverses and the sharp order. *Linear Algebra Appl.* **92**, 17–37.

⊕ Mitra, S. K. (1988). Infimum of a pair of matrices. *Linear Algebra Appl.* **105**, 163–182.

⊕ Mitra, S. K. (1989). Block independence in generalized inverse: a coordinate free look. In: *Statistical Data Analysis and Inference*, pp. 429–443, Y. Dodge (Ed.), North-Holland, Amsterdam.

Mitra, S. K. (1990a). Shorted matrices in star and related orderings. *Circuits Systems Signal Process* **9**, 197–212.

⊕ Mitra, S. K. (1990b). A pair of simultaneous linear matrix equations $A_1XB_1 = C_1$, $A_2XB_2 = C_2$ and a matrix programming problem. *Linear Algebra Appl.* **131**, 107–123.

Mitra, S. K. (1993). My encounters with the null matrix. Unpublished manuscript.

Mitra, S. K. (1994). Separation theorems. *Linear Algebra Appl.* **208/209**, 239–256.

Mitra, S. K. and Moore, B. J. (1973). Gauss-Markov estimation with an incorrect dispersion matrix. *Sankhyā Ser. A* **35**, 139–152.

Mitra, S. K. and Odell, P. L. (1986). On parallel summability of matrices. *Linear Algebra Appl.* **74**, 239–255.

Mitra, S. K. and Puntanen, S. (1991). The shorted operator statistically interpreted. *Calcutta Statist. Assoc. Bull.* **40**, 97–102.

Mitra, S. K. and Puri, M. L. (1973). On parallel sum and difference of matrices. *J. Math. Anal. Appl.* **44**, 92–97.

Mitra, S. K. and Puri, M. L. (1979). Shorted operators and generalized inverses of matrices. *Linear Algebra Appl.* **25**, 45–56.

Mitra, S. K. and Puri, M. L. (1982). Shorted matrices — an extended concept and some applications. *Linear Algebra Appl.* **42**, 57–79.

Mitra, S. K. and Puri, M. L. (1983). The fundamental bordered matrix of linear estimation and the Duffin–Morley general linear electromechanical systems. *Applicable Anal.* **14**, 241–258.

Mitra, S. K. and Rao, C. R. (1968). Simultaneous reduction of a pair of quadratic forms. *Sankhyā Ser. A* **30**, 313–322.

⊕ Mitra, S. K. and Trapp, G. E. (1975). On hybrid addition of matrices. *Linear Algebra Appl.* **10**, 19–35.

⊕ Morley, T. D. (1979). Several applications of the shorted operator. In: *Constructive Approaches to Mathematical Models*, pp. 439–446, C. V. Coffman and G. J. Fix (Eds), Academic Press, New York.

Nambooripad, K. S. S. (1980). The natural partial order on a regular semigroup. *Proc. Edinburgh Math. Soc.* **23**, 249–260.

⊕ Näther, W. and Reinsch, V. (1981). D_s-optimality and Whittle's equivalence theorem. *Math. Operationsforsch. Statist. Ser. Statist.* **12**, 307–316.

Ouellette, D. V. (1981). Schur complements and statistics. *Linear Algebra Appl.* **36**, 187–295.

Pukelsheim, F. (1993). *Optimal Design of Experiments.* Wiley, New York.

Puntanen, S., Šemrl, P. and Styan, G. P. H. (1995). On the invariance of certain functions of matrix products with respect to the choice of generalized inverse, in preparation.

Rao, C. R. (1971). Unified theory of linear estimation. *Sankhyā Ser. A* **33**, 371–394. [Corrigenda: (1972) **34** 194 ,477.]

Rao, C. R. (1972). A note on the IPM method in the unified theory of linear estimation. *Sankhyā Ser. A* **34**, 285–288.

Rao, C. R. (1973). *Linear Statistical Inference and Its Applications.* 2nd edn. Wiley, New York.

Rao, C. R. and Mitra, S. K. (1971). *Generalized Inverse of Matrices and Its Applications.* Wiley, New York.

Rao, C. R., Mitra, S. K. and Bhimasankaram, P. (1972). Determination of a matrix by its subclasses of generalized inverses. *Sankhyā Ser. A* **34**, 5–8.

Stone, M. and Wang, J. L. (1989). A Bayes, sweep-operational construction for Rao's unified theory of linear estimation. *Statistics* **20**, 3–11.

Styan, G. P. H. (1985). Schur complements and linear statistical models. In: *Proceedings of the First International Tampere Seminar on Linear Statistical Models and Their Applications,* pp. 37–75, T. Pukkila and S. Puntanen (Eds), Dept. of Mathematical Sciences, University of Tampere.

Tian, Yongge. (1992). The maximal and minimal ranks of $A - B_1X_1C_1 - B_2X_2C_2$ and their applications. Preprint, 125 pp. Zhengshou Technical School for Surveying and Mapping, Zhengshou, China.

⊕ Trapp, G. E. (1984). Hermitian semidefinite matrix means and related matrix inequalities — an introduction. *Linear and Multilinear Algebra* **16**, 113–123.

Werner, H. J. (1987). C. R. Rao's IPM method: a geometric approach. In: *New Perspectives in Theoretical and Applied Statistics*, pp. 367–382, M. L. Puri *et al.* (Eds), Wiley, New York.

Zyskind, G. and Martin, F. B. (1969). On best linear estimation and a general Gauss-Markov theorem in linear models with arbitrary nonnegative covariance structure. *SIAM J. Appl. Math.* **17**, 1190–1202.

Multivar. Statist., pp. 313–324
E.-M. Tiit *et al.* (Eds)

ARRAYS OF MULTIVARIATE STATISTICS AND THEIR REPRESENTATION

E. KÄÄRIK and E.-M. TIIT
Department of Mathematical Statistics, University of Tartu,
Liivi 2, EE2400 Tartu, Estonia

ABSTRACT

In this paper the problem of parsimonious representation of the arrays is introduced and the lexicographical ordering of the elements of matrices of higher moments of random vectors by their indices is regarded as a parsimonious representation of the given array.

1. INTRODUCTION

To represent one- and two-dimensional arrays (files, massives) of numbers the vectors and matrices are popular and handy tools. Serious problems arise, when the dimensionality (number of indices) of the array is more than two.

There exist different representations: via tensors (McCullagh, 1989), sets (Karma, 1992), block-matrices (Kollo, 1991; Parring, 1979, etc.), vectors (Traat, 1986), but all of them have some shortages, the lack of perspicuity and laboriousness being the main problems for all of them. At the same time each of these representations has some preferencies in using the arrays in different procedures.

If used in multivariate statistics, then in several problems — in estimation, building asymptotic expansions etc, the arrays, depending on four, six or even eight indices are needed. In multivariate statistics some additional problems arise due to some kind of symmetry of the arrays, causing the identity of some groups of elements. So the representations avoiding the superfluous elements are needed, as well (Kollo and von Rosen, 1993).

It seems that there does not exist any universal representation of arrays for solving all problems at all. Hence it is reasonable:

1° to find transformation rules (and their inverses) to transform any useable representation to any other representation;

2° to find a parsimonious representation of multivariate matrices, avoiding repeated elements caused by some kind of symmetry of the array;

3° to find the transformation rules and their inverses from any other representation to this parsimonious representation;

4° to find the operating rules to handle with the matrices, represented in the parsimonious representation.

In the paper some of the problems listed will be solved:

- the transformation rules for most common representations will be derived;
- the compressed representation for symmetrical multi-dimensional arrays will be introduced (Tiit and Tammet, 1994), and the corresponding transformation rules will be given.

Several examples, concerning the higher moments, will be given.

2. INDEX-VECTORS AND INDEX-SETS

Let us start with the definitions of the basic concepts.

DEFINITION 1. The vector

$$I = (i_1, \ldots, i_k) \tag{1}$$

is said to be an *index-vector*, if all its components are integers, $i_j \leqslant p_j$, $j = 1, \ldots, k$, where $(p_1, \ldots, p_k)$ is a fixed vector.

The natural number k is the *cardinality* of I.
If the condition

$$i_j < i_{j+1}$$

or

$$i_j \leqslant i_{j+1}$$

is satisfied for $j = 1, \ldots, k-1$, then the index-vector I is said to be *ordered.* In the case of opposite inequalities the index-vector is D-ordered (decreasingly).

If all parameters p_j are equal, $p_j = p$, $j = 1, \ldots, k$, p is said to be the *order* of index-vector. The set of index-vectors of order p and cardinality k will be denoted by $\mathcal{I}(p, k)$.

The ordered index-vector $I_0 = (1, \ldots, p)$ is said to be the *full index-vector.* The necessary and sufficient condition for a index-vector to be full is that its order and cardinality are equal, $p = k$.

DEFINITION 2. The set

$$I = \{i_1, \ldots, i_k\} \tag{2}$$

is said to be an *index-set*, if all its elements are the components of an index-vector I.

Notice that in the index-set the equal elements, e.g., $i_g = i_l$ are allowed. Hence for every index-vector I a uniquely defined index-set I corresponds, and as a rule we use the same notation for both of them. In general, the opposite statement is not true — to an index-set many different index-vectors correspond, but only one of them is fully ordered.

DEFINITION 3. Let I_0 be a full index-vector and I an ordered index-vector of order p. Then the index-vector $I^c = (j_1, \ldots, j_k)$, defined as an ordered index-vector by its index-set

$$I^c = I_0 \setminus I$$

is said to be a *complementary index-vector* to I.

The set of all index-vectors (1) of order p and cardinality k will be denoted by $\Upsilon(p, k)$.

3. ARRAYS

3.1. *Basic concepts and conditions*

DEFINITION. Let us have a set of elements $a_{i_1 \cdots i_k}$, depending on an index-vector $I = (i_1, \ldots, i_k)$, where $1 \leqslant i_j \leqslant p_j$, $j = 1, \ldots, k$. Then we say that we have a *k-dimensional array A*.

In some cases we assume that the additional, so-called *symmetricity conditions* are fulfilled as well:

DEFINITION 5. The array A with elements $a_{i_1 \cdots i_k}$ is said to be *symmetric*, if the following conditions are fulfilled:

1) All indices have the same set of values, $p_j = p$, $j = 1, \ldots, k$, i.e, $I \subset \mathcal{I}(p, k)$.
2) The elements $a_{i_1 \cdots i_k}$ are exchangeable by the indices.

Notice that the concept of *exchangeability* (Kotz and Tiit, 1992) is the same as *permutation symmetry* or *full symmetry* (Tiit, 1984).

Almost all useable in practice statistical functions (empirical or theoretical) of k random arguments, calculated by p-variate random vector (e.g., moments, central moments, cumulants and other statistics) satisfy the conditions 1)–2).

Let us examine the following representations of the array, used in statistics.

3.2. *The T-representation or tensor-representation*

Here the array is regarded as k-variate rectangle, where every element is identified by its k indices in fixed order $(i_1, \ldots, i_k)$, $i_j \in \{1, \ldots, p_j\}$, $j = 1, \ldots, k$.

We regard the T-representation as the basic one in all transformations. The shortages of the T-representation are well-known: in general, the matrix-operations are not useable for arrays in this form; no perspicuous picture of elements exists, in case of symmetric arrays there are many repeated, i.e., superfluous elements in this representation.

3.3. *The V-representation or the vector-representation*

Here the elements of the array are regarded as the components of a vector

$$V = (v_1, \ldots, v_P)',$$

where the dimension of the vector P is calculated by the following formula:

$$P = \prod_{j=1}^{k} p_j,$$

or

$$P = p^k$$

in the symmetrical case.

For arrays, represented in V-shape, the matrix and vector operations are available. The shortage of the V-representation is that the results of all matrix operations depend on the order of indices. The lack of perspicouosity and existence of repeated elements are the problems of this representation as well.

Example 1. Symmetric array A with elements (a_{i_1}, a_{i_2}), (index-vector $I = (i_1, i_2)$, order $p = 3$, cardinality $k = 2$). One possible of V-representation is

$$V = (a_{11}, a_{12}, a_{12}, a_{21}, a_{22}, a_{23}, a_{31}, a_{32}, a_{33})',$$

the dimension of the vector is $P = 3^2 = 9$.

Let us define the V-*transformation* (the transformation $T \Rightarrow V$) as an ordering rule, defined by the indices $\{i_1, \ldots, i_p\} \Rightarrow h$. Here h is the index of component v_h of the vector V, $h = 1, \ldots, P$. The rule is the following:

$$h = \sum_{j=1}^{k-1} (i_j - 1) \cdot \prod_{u=j+1}^{k} p_u + i_k. \tag{3}$$

In the special case of symmetric array the rule has more compact shape:

$$h = \sum_{j=1}^{k} (i_j - 1) \cdot p^{k-j} + 1. \tag{3'}$$

For the representation V the unique inverse transformation V^{-1} exists, defined by the following algorithm.

ALGORITHM 1. V^{-1}-transformation.

Step 1. Let $c = 1$.

1.1. Find the value j $(0 \leqslant j < p_j - 1)$, satisfying the following condition:

$$j \cdot \prod_{f=c+1}^{k} p_f < h \leqslant (j+1) \cdot \prod_{f=c+1}^{k} p_f.$$

1.2. Take $i_c := 1 + j$ and $h := h - j \cdot \prod_{c+1}^{k} p_f$.

Step 2. Let c steps be made.

If $c = k$, then end.

If $c < k$, then $c := c + 1$ and repeat step 1 from 1.1.

3.4. *The B-representation or the block-matrix representation*

In B-representation the array is represented as block-matrix

$$B = (b_{(u_1 \cdots u_q), \quad (w_1 \cdots w_r)}),$$

$q+r=k$, where two indices (e.g., u_q and w_r) identify the elements in blocks, the remaining ones — blocks or blocks of blocks. Here the indices u_q are used for notation of rows, w_r for notation of columns.

For defining the B-representation the partition of the index-set $I_0 = \{1, \ldots, k\}$ into two (non-overlapping) subsets $U = \{u_1, \ldots, u_q\}$ of row-indices and $W = \{w_1, \ldots, w_r\}$ of column-indices, $U \cup W = I_0$, $q+r=k$ must be defined. As in this representation all initial indices are used, so the transformation rule $B \Rightarrow T$ is entirely defined by the definition of index-vectors U and $U^c = W$.

For instance, let us use the block-matrix $B = (b_{(u_1 \cdots u_q),(w_1 \cdots w_r)})$, defined by Parring (1979) as the matrix of higher moments of a random vector by the initial vector $a = (a_1, \ldots, a_p)'$ in the following way:

$$E(a \otimes a' \otimes a \otimes a' \otimes \cdots \otimes a*), \tag{4}$$

where the last term $a*$ equals to a, if k is odd and to a', if k is even.

The transformation B is defined for the index-vectors, $(i_1, \ldots, i_k) \Rightarrow ((u_1, \ldots, u_q), (w_1, \ldots w_r))$ in the the following way:

$$q = [(p+1)/2], \quad r = [p/2], \tag{5}$$

$$\begin{aligned} u_j &= i_{2j-1}, \qquad j = 1, \ldots, q; \\ w_j &= i_{2j}, \qquad j = 1, \ldots, r. \end{aligned} \tag{6}$$

As it follows from the definition (4), the odd indices $i_1, i_3, \ldots$ define the row-indices of the block-matrix B, and the even indices – the column indices of B. The existence of the inverse transformation B^{-1} is obvious.

The preference of the B-representation is a possibility to use matrix operations, derived for block matrices. The problems are mainly the same as in case of V-representation, i.e., the dependence on order and number of indices. The most developed results exist for the case $k = 4$.

3.5. *The M-representation or the matrix-representation*

The M-representation is connected with the block-matrix representation. In some operations the elements of the block-matrix B are regarded at the same time as the elements (b_{fg}) of the usual(bivariate) matrix B.

Let us define at first the transformation M' as $B \Rightarrow M$ or

$$((u_1, \ldots, u_q), (w_1, \ldots, w_r)) \Longrightarrow (f, g).$$

Whereas the transformation $T \Rightarrow B$ is already defined we get the transformation $T \Rightarrow M$ as the product: $M = BM'$.

The transformation M is bivariate, containing two rules: $f = f(u_1, \ldots, u_q)$ and $g = g(w_1, \ldots, w_r)$. Both of them can be realized by the formulae (3) where the following values are used:

$k := q$, $i_j := u_j$; $p_j := p_{u_j}$, $j = 1, \ldots, q$, for finding the row indices, and
$k := r$, $i_j := w_j$; $p_j = p_{w_j}$, $j = 1, \ldots, r$, for finding the column indices.

4. COMPRESSING TRANSFORMATIONS

4.1. *Equivalence classes of index-vectors*

Let us examine an index-vector I from $\mathcal{I}(p, k)$, where the parameter p and the dimension k are fixed. Let us define the equivalence classes of index-vectors by permutations of indices in the following way:

DEFINITION 6. Let $I = (i_1, \ldots, i_k)$ and $J = (j_1, \ldots, j_k)$ be two index-vectors from the class $\mathcal{I}(p, k)$. If there exists a permutation P so that the equality $I = PJ$ holds, then we can say that the index-vectors I and J are *equivalent*.

Shortly, the index-vectors, corresponding to the same index-set, are equivalent. As *a representative* of a class of equivalent index-vectors we will take the *D-ordered* index-vector

$$I' = (i'_1, \ldots, i'_k)$$

from this class, where

$$i'_j \leqslant i'_{j-1}, \qquad j = 2, \ldots, k. \tag{7}$$

The index-vectors that belong to different classes are said to be *essentially different*. The set of equivalence classes (essentially different index-vectors) we denote by $\Psi(p, k)$.

Example 2. Let the dimension of an index-vector be $k = 4$ and order $p = 5$. Let us consider index-vectors $I = (5, 4, 4, 1)$, $I' = (4, 1, 5, 4)$, $J = (4, 3, 2, 1)$, then the index-vectors I and I' are equivalent, I is the representative of a class of equivalent index-vectors. Index-vectors I and J are essentially different, they belong to different classes.

4.2. *The ordering of index-vectors*

Let us define the *position* f of a component i_f of an index-vector I as its successive number in the index-vector $(i_1, \ldots, i_k)$. Every ordered index-vector $(i_1, \ldots, i_k)$ can be regarded as an k-positional natural number in the positional number system having the basis p. Notice that for getting the usual form of natural number we should take the transformation $i_j = i_j - 1,\ j = 1, \ldots, p$.

Let us order all index-vectors of the set $\Psi(p,k)$ lexicographically; as a result we receive the sequence $\mathcal{L}(p,k)$ with terms I_h (here I_h is the h-th element of the sequence). The sequence $\mathcal{L}(p,k) = \{I_h\}$ does not include all possible k-figure p-numbers, but only these which satisfy the conditions (7).

Example 3. Let us regard the set of the fourth moments of a five-variate random vector $\mathbf{X} = (X_1, X_2, X_3, X_4, X_5)$. The usual way to present the set of moments is to use the moment's matrix (of order 25×25). In this matrix, which is usually written with the help of 5×5 blocks, there are 625 entries. But most of the elements of the matrix are superfluous, as almost all moments are represented in this matrix repeatedly. The situation is similar to that of correlation matrix, where all correlation coefficients are represented twice: in the upper and in the lower triangle.

The main practical problem is to eliminate the repeated copies of the moments from the matrix. The order of moments k is dimensionality of index-vector and the dimensionality of random vector p is order of index-vector. It is possible to get all different fourth moments of the random vector, identified by the indices, as a finite sequence $\mathcal{L}(p,k)$. In Table 1 all different fourth moments of a five-variate random vector are represented as 4-dimensional index-vectors, identifying the equivalent indices (the sequence $\mathcal{L}(5,4) = \{I_h\}$).

Let us assign to every index-vector I_h from $\mathcal{L}(p,k)$ its rank $h = h(I)$. Our task is to find the ordering rule ϕ, describing the correspondence

$$I_h \Longrightarrow h(I) \tag{8}$$

and ϕ^{-1}:

$$h(I) \Longrightarrow I_h. \tag{8'}$$

In the following we will define several concepts, connected with the sequence $\mathcal{L}(p,k)$.

Table 1.

$h(I)$	I	$h(I)$	I	$h(I)$	I	$h(I)$	I	$h(I)$	I
1	1111	15	3333	29	4431	43	5331	57	5521
2	2111	16	4111	30	4432	44	5332	58	5522
3	2211	17	4211	31	4433	45	5333	59	5531
4	2221	18	4221	32	4441	46	5411	60	5532
5	2222	19	4222	33	4442	47	5421	61	5533
6	3111	20	4311	34	4443	48	5422	62	5541
7	3211	21	4321	35	4444	49	5431	63	5542
8	3221	22	4322	36	5111	50	5432	64	5543
9	3222	23	4331	37	5211	51	5433	65	5544
10	3311	24	4332	38	5221	52	5443	66	5551
11	3321	25	4333	39	5222	53	5444	67	5552
12	3322	26	4411	40	5311	54	5511	68	5553
13	3331	27	4421	41	5321	55	5521	69	5554
14	3332	28	4422	42	5321	56	5522	70	5555

DEFINITION 7. The subsequence of the sequence $\mathcal{L}(p,k)$, consisting of k-variate index-vectors $(i_1,\ldots,i_k)$, satisfying the condition

$$i_1 \leqslant j$$

is said to be a *j-segment* of the sequence $\mathcal{L}(p,k)$ and will be denoted by $H(j,k)$, $j=1,\ldots,p$.

Let $\kappa_H(j,k)$ denote the *cardinality* of the segment $H(j,k)$.

Example 4. Let us see the sequence $\mathcal{L}(5,4)$. Then the 2-segment $H(2,4)$ is the subsequence $(1,1,1,1),(2,1,1,1),(2,2,1,1),(2,2,2,1),(2,2,2,2)$, and $\kappa_H(j,k)=5$ (see Table 1).

DEFINITION 8. The subsequence of the sequence $\mathcal{L}(p,k)$, consisting of k-variate index-vectors $(i_1,\ldots,i_k)$, satisfying the condition

$$i_1 = j$$

is said to be a *j-section* of the sequence $\mathcal{L}(p,k)$ and will be denoted by $G(j,k)$, $j=1,\ldots,p$.

Let $\kappa_G(j,k)$ denote the cardinality of the section $G(j,k)$.

Example 5. The 2-section $G(2,4)$ of the sequence $\mathcal{L}(5,4)$ is the subsequence $(2,1,1,1),(2,2,1,1),(2,2,2,1),(2,2,2,2)$ and $\kappa_G(2,4)=4$ (see Table 1).

LEMMA 1. *The cardinalities of segments and sections are related by the following equation*

$$\kappa_H(j,k)=\sum_{f=1}^{j}\kappa_G(f,k). \tag{9}$$

The correctness of the formula follows immediately from the definitions.

Example 6. Let us see Table 1, then

$$\kappa_H(3,4)=\sum_{f=1}^{3}\kappa_G(f,4)=1+4+10=15.$$

To find the ordering rule ϕ we will start with the simpliest possible index-vectors I. Let us construct the sequence $\mathcal{L}(p,k)$ in case $k=1$. If $k=1$ then $I=i_1$ and we have for the sections of the sequence $\mathcal{L}(p,1)$:

$$G(j,1)=\{j\},\qquad \kappa_G(j,1)=1,\qquad j=1,\ldots,p.$$

and for the segments of $\mathcal{L}(p,1)$:

$$H(j,1)=\{1,\ldots,j\},\qquad \kappa_H(j,1)=j.$$

Thus the following lemma holds.

LEMMA 2. *The ordering rule ϕ for the sequence $\mathcal{L}(p,1)$ is the following*:

$$h(I) = i_1, \qquad i_1 = 1, \ldots, p. \tag{10}$$

From Lemma 2 it follows, that the cardinality of the sequence $\mathcal{L}(p,1)$ is p.

Let us consider the case $k = 2$. In this case the members of the sequence $\mathcal{L}(p,2)$ are the vectors (i_1, i_2). Let us regard a section $G(j,2)$. By the conditions (7) and the lexicographical ordering rule of the sequence $\mathcal{L}(p,2)$ it can be concluded that the index-vectors (i_1, i_2), belonging to the section $G(j,2)$, have the following construction: $i_1 = j$, i_2 has all successive values of the index i_1', from the segment $H(j,1)$ of the sequence $\mathcal{L}(p,1)$.

From this construction immediately follows:

LEMMA 3. *For the sequence $\mathcal{L}(p,2)$ the expression of the cardinality of a section is*:

$$\kappa_G(j,2) = \kappa_H(j,1), \qquad j = 1, \ldots, p. \tag{11}$$

From this lemma another important statement follows, too.

COROLLARY 1. *If the vector $I = (i_1, i_2)$ belongs to the section $G(i_1, 2)$, then for the rank $h(I)$ the following inequalities*:

$$\kappa_H(i_1 - 1, 2) < h(I) \leqslant \kappa_H(i_1, 2), \qquad i_1 = 1, \ldots, p, \tag{12}$$

are valid.

To get the precise value of the rank $h(I)$ we will examine all index-vectors (j, i_2) from the section $G(j,2)$. It is easy to see that there exists *one-to-one correspondence* between the section $G(j,2)$ and the sequence $\mathcal{L}(j,1)$, formed from the second components $\{i_2\}$ of the initial index-vectors. Using the equation (10), we will get from the inequality (12) the following formula:

$$h(I) = \kappa_H(i_1 - 1, 2) + i_2.$$

The last equation can be transformed with the help of the formulae (9) and (11):

$$\kappa_H(j,2) = \sum_{f=1}^{j} \kappa_H(f,1) = (j \cdot (j+1))/2 = A_2^j,$$

where we use the notation $A_g^f = ((f+g-1) \cdot \ldots \cdot f)/(g!) = C_{f+g-1}^g$.

Example 7. $A_4^5 = ((4+5-1) \cdot 7 \cdot 6 \cdot 5)/4! = 70.$

Considering the notation A_g^f the following equation is well-known:

$$A_g^f = \sum_{j=1}^{f} A_{g-1}^j. \tag{13}$$

Using the equality $A_1^g = g$, it will be convenient to rewrite the formula of the transformation rule ϕ in the following, more symmetric (in the sense of indices), shape:

$$h(I) = 1 + A_2^{i_1 - 1} + A_1^{i_2 - 1}. \tag{14}$$

COROLLARY 2. *The cardinality* $\kappa(\mathcal{L}(p,2)) = 1 + A_2^{p-1} + A_1^{p-1} = A_2^p$.

The formula follows from equation (14). Now let us prove Theorem 1, determining the rule ϕ for the arbitrary dimensionality k.

THEOREM 1. *Let Ψ be the set of essentially different index-vectors $(i_1, \ldots, i_k)$, where the dimensionality k and the parameter p are fixed. The rule ϕ for the lexicographical ordering of index-vectors in the sequence $\mathcal{L}(p, k)$ is the following:*

$$h(I) = 1 + \sum_{g=1}^{k} A_{k+1-g}^{i_g - 1}. \tag{15}$$

Proof. The formula (15) is an immediate generalization of the formula (14) by the principle of mathematical induction.

In the first step of the proof we fix the *position* of the index-vector $I = (i_1, \ldots, i_k)$ in the sequence $\mathcal{L}(p, k)$, using the concepts of segments and sections of the sequence.

By Definitions 7 and 8 the index-vector $(i_1, \ldots, i_k)$ belongs to the section $G(i_1, k)$ and hence to the segment $H(i_1, k)$ of the $\mathcal{L}(p, k)$ as well. Let us take the section $G(i_1, k)$ and regard all $(k-1)$-variate subvectors of the index-vectors $(i_1, \ldots, i_k)$ belonging to the section $G(i_1, k)$ and having different values on positions $2, \ldots, k$. This sequence coincides with the sequence $\mathcal{L}(p, k-1)$. The $(k-1)$-subvector $(i_2, \cdots, i_k)$ of the initial index-vector I belongs to the segment $H(i_2, k-1)$ of the sequence $\mathcal{L}(p, k-1)$.

Repeating this discussion, we get on the f-th stage a sequence $\mathcal{L}(p, k+1-f)$ of $(k+1-f)$-variate subvectors of index-vectors from the set Ψ, and fix the segment $H(i_f, k+1-f)$, where the $(k+1-f)$ subvector of the initial index-vector I belongs, $f = 1, \ldots, k$. From here we receive the formula for $h(I)$:

$$h(I) = \sum_{f=1}^{k-1} \kappa_H(i_f - 1, k+1-f) + i_k. \tag{16}$$

The next step is to calculate the cardinalities $\kappa_H(i_1 - 1, k+1-f)$. From the equations (9) and (11) and the construction of the sequence $\mathcal{L}(p, k)$ the following equation can be concluded:

$$\kappa_H(f, g+1) = \sum_{j=1}^{f} \kappa_H(j, g). \tag{11'}$$

Using the equations (13) and (14), we receive the following formula:

$$\kappa_H(f, g) = A_g^{g+f-1}. \tag{17}$$

Substitution of the formula (17) into the equation (16) immediately gives the formula (15). Theorem 1 is proved.

The rule ϕ is a generalization of the well-known operator *vech*, transforming the upper triangle of a symmetric matrix into the vector.

From Theorem 1 immediately Corollary 3 follows.

COROLLARY 3. *The cardinality of the sequence* $\mathcal{L}(p,k)$ *is* A_k^p.

To prove the corollary it is sufficient to use the formula (15), using the values $i_j = p,\ j = 1,\ldots,k$.

Example 8. Let us see the set of the fourth moments of a five-variate index-vector, given in Example 3. By the corollary it is easy to calculate the number of different moments $A_4^5 = (8 \cdot 7 \cdot 6 \cdot 5)/4! = 70$. This number is concordant with the data given in Table 1 and Example 7.

To identify the moment $E(X_1 \cdot X_2^2 \cdot X_5)$ we take the index-vector (1,2,2,5), find its representative (5,2,2,1) and calculate the value of $h(I) = A_4^4 + A_3^1 + A_2^1 + A_1^0 + 1 = 35 + 1 + 1 + 0 + 1 = 38$. The result is the rank $h(I)$ of the index-vector $I = (5,2,2,1)$, as we can see in the Table 1.

THEOREM 2. *The transformation* ϕ *has a unique inverse transformation* ϕ^{-1}.

Proof. Follows immediately from the following Algorithm 2.

ALGORITHM 2. The expanding of compressed array.

I. The construction of index-vector I_h of the representative of an equivalence class by its index h.

Step 1. Let $c = 1$.

1.1. Find the value j ($j = 1,\ldots,p$) satisfying the following condition:

$$A_{k-c+1}^{j-1} < h \leqslant A_{k-c+1}^{j}.$$

1.2. Let $i_c := j$ and $h := h - A_{k-c+1}^{j-1}$.

Step 2. Let c steps be made.

If $c = k$, then end.

If $c < k$, then $c := c + 1$, repeat step 1 from 1.1.

II. The construction of all equivalent index-vectors.

The number of all equivalent index-vectors $(i_1,\ldots,i_k)$, $i_j \leqslant p,\ j = 1,\ldots,k$, is

$$N = \frac{k!}{\prod_{i=1}^{p} r_i!},$$

where k is the cardinality of index-vector and r_i is the number of repetition indices i $(1 < i \leqslant p)$.

The problem of the construction of N equivalent index-vectors is to find all possible different permutations with repetitions.

Example 9. Let us see again the data from Example 3. Let $h(I) = 50$ be given. As $A_4^4 = 35$ and $A_4^5 = 70$, so $i_1 = 5$. The next step is to compare the difference $50 - 35 = 15$ with the values of A_3^j $(j = 1, \ldots, 5)$. As $A_3^3 = 10$ and $A_3^4 = 20$, hence $i_2 = 4$. The difference 5 is between the values of $A_2^2 = 3$ and $A_2^3 = 6$ and the difference 5-3=2 equals to $A_1^2 = A_1^1 + 1$: Hence the index-vector I_{50} has the form (5,4,3,2).

The number of all equivalent index-vectors is $N = 4! = 24$ $(r_i = 1, i = 1, \ldots, 5)$ and all possible equivalent index-vectors are:
(2,3,4,5),(2,3,5,4),(2,4,3,5),(2,4,5,3),(2,5,3,4),(2,5,4,3),(3,2,4,5),(3,2,5,4),
(3,4,2,5),(3,4,5,2),(3,5,2,4),(3,5,4,2),(4,2,3,5),(4,2,5,3),(4,3,2,5),(4,3,5,2),
(4,5,2,3),(4,5,3,2),(5,2,3,4),(5,2,4,3),(5,3,2,4),(5,3,4,2),(5,4,2,3),(5,4,3,2).

Example demonstrates that the rule ϕ and its inverse ϕ^{-1} are both easily programmable.

REFERENCES

Karma, O. (1992). On arrays and array dervatives. *Acta et Comment UT* **942**, 132–150.

Kollo, T. (1991). *Matrix Derivative in Multivariate Statistics.* University of Tartu.

Kollo, T. and von Rosen, D. (1993). Moments and cumulants of symmetric matrices with applications to the wishart distribution. *U.U.D.M. Report* **10**, Dept. of Math., Uppsala University.

Kotz, S. and Tiit, E.-M. (1992). Bounds in multivariate dependence. *Acta et Comment UT* **942**, 35–45.

McCullagh, P. (1989). *Cumulants and Tensor Notation.* Champmann and Hall, London.

Parring, A.-M. (1979). Calculation of asymtotic parameters of sampling functions. *Acta et Comment UT* **492**, 86–90.

Tiit, E.-M. and Tammet, S. (1994). Using index-vectors and partitions in multivariate analysis. *Acta et Comment UT* **968**, 55–75.

Tiit, E.-M. (1984). Construction of discrete multivariate distributions having given moments. *Proceedings of Computing Centre of UT* **51**, 142–165.

Traat, I. (1986). Matrix calculus for multivariate distributions. *Acta et Comment UT* **733**, 64–85.

Multivar. Statist., pp. 325–334
E.–M. Tiit *et al.* (Eds)

ON THE CALCULATIONS WITH THE INDEXIZED SETS OF DATA

O. KARMA
Institute of Applied Mathematics, University of Tartu,
Liivi 2, EE2400 Tartu, Estonia

ABSTRACT

We consider the sets of data depending on indices, whereas the meaning of every index is determined by the index set from which this index takes its values, and not by the location of this index. An attempt has been made to built up the calculus which is close to the matrix calculus for such sets of data. As an application some formulas of multivariate statistics are presented in this new formalism.

1. INTRODUCTION

In multivariate statistics it is necessary to work with the data which depend on indices. For this purpose the matrix formalism is widely exploited. Unfortunately, when we want to use the matrix formalism and our data are depending on more than two indices (e.g., the sets of the higher moments of random vectors, matrix derivatives etc.), we will face serious formal problems which are caused by the necessity to arrange the data as vectors or matrices. Particularly, the formulas may depend on the way of ordering and grouping the indices, and may look more complicated than they are in the one-dimensional case (see, e.g. (Kollo, 1992)). In some cases, it may be even reasonable to use the index notation instead of the more compact matrix notation.

At the same time, the arrangement of the data in the form of the matrix is not necessary for the most operations in multivariate statistics. Instead of the matrices, the indexized sets of data may be used for which the order of the indices is not important. In this case the meaning of every index is determined by the index set from which this index takes its values, and not by the location of this index. We use the term *array* for such data sets in this paper.

In this paper an attempt has been made to present the calculus of arrays in the form close to the matrix calculus. Unfortunately, there are many inaccurate formulations in the previous analogous attempt made by author (see (Karma, 1992)).

Many of the formulas turn out to be formally simpler in the array notation than in the matrix notation and look similar to the one-dimensional case. This gives us an opportunity to use for the multidimensional case the same discussions that we use for the one-dimensional case.

Of course, there are difficulties in the array formalism also. Thus, it may be reasonable to carry out only some intermediate calculations in the array formalism.

Let us explain the difference between matrices and arrays on a simple example.

Let some data $a(i,j)$ represent the results of the measurements M_j, $j = 1,\ldots,n$, on the individuals P_i, $i = 1,\ldots,m$. To be concrete, let $m = n = 2$, and $a(1,1) = 11$, $a(1,2) = 12$, $a(2,1) = 21$, $a(2,2) = 22$.

We may present these data as a matrix, for example arranging them so that $a(i,j)$ stands in the i-th row and j-th column. Let us denote this matrix by A, so that $A = A_{mn} = (a_{ij})$ with $a_{ij} = a(i,j)$, and

$$A = \begin{pmatrix} 11 & 12 \\ 21 & 22 \end{pmatrix}$$

for our concrete data. Every concrete datum is now referred to by indicating the *ordered* pair of the indices, in the first place the row number (the individuals number) and in the second place the column number (the measurements number), e.g., $a_{12} = 12 \neq a_{21} = 21$.

If we want to present these data as an array, we can act as follows. We fix two index sets, say $I = \{1,\ldots,m\}$ for the 'names' of the individuals and $J = \{1,\ldots,n\}$ for the 'names' of the measurements. And we consider this data set as the set of data with two indices, one of which belongs to the index set I and the other to the index set J. Every concrete datum is now referred by indicating, in the arbitrary order, the value and the index set of every index, e.g., $a_{1\in I,2\in J} = a_{2\in J,1\in I} = 12 \neq a_{1\in J,2\in I} = a_{2\in I,1\in J} = 21$. If we agree to denote the index with the values in I by i and the index with the values in J by j, we can also write $a_{i=1,j=2} = a_{j=2,i=1} = 12$. We note that the arrays can be presented in the form which is close to matrices, e.g.,

$$\begin{array}{c|cc} I\setminus J & 1 & 2 \\ \hline 1 & 11 & 12 \\ 2 & 21 & 22 \end{array} = \begin{array}{c|cccc} I & 1 & 1 & 2 & 2 \\ J & 1 & 2 & 1 & 2 \\ \hline & 11 & 12 & 21 & 22 \end{array} = \begin{array}{c|cc} J\setminus I & 1 & 2 \\ \hline 1 & 11 & 21 \\ 2 & 12 & 22 \end{array}.$$

2. FORMALISM

In this paper we shall not use the words 'index', 'index set', and 'array' in the traditional sense but as special terms.

For us an *elementary index set* consists of its label, and of the nonempty finite set of the possible values of the *index*, e.g., $(J;\{1,2,\ldots,9\})$, $(V;\{a,b\})$. No requirements are imposed on the set of the possible values of the index, except that these values must be different pairwise. (Of course, the elementary index set may have its own structure, but we do not use it in our considerations.) We refer to the elementary index set by its label, usually the capital letter, and write $i \in I$ if i is some element from the set of values of the index set I. If two elementary index sets I and J have the same set of possible values for the index, we say that they are set-equal and write $I(=)J$.

A *multi-index set* is a finite (maybe empty) ordered list of the elementary index sets, e.g., $IJ \dots P$. For the multi-index set $IJ \dots P$ the set of possible values for the *multi-index* is the set of all the possible ordered lists $ij \dots p$ with $i \in I, \dots, p \in P$, i.e., the set $\{ij \dots p\colon\ i \in I, \dots,\ p \in P\}$. If the set equalities $I_r(=)J_r$, $r = 1, \dots, m$, hold for two multi-index sets $I = I_1 \dots I_m$ and $J = J_1 \dots J_m$, we say that these multi-index sets are set-equal and write $I(=)J$. If $I = I_1 \dots I_m$ and $J = J_1 \dots J_n$ are two nonempty multi-index sets then the multi-index set $I_1 \dots I_m J_1 \dots J_n$ is denoted by IJ.

An *index set* is a finite (maybe empty) list of the elementary index sets, where the order of listing is ignored. This means that if two lists of the elementary index sets differ only in the order of the listing of their elementary index sets, then they are considered as the same index set. For example, if I and J are the elementary index sets then the index sets IJ and JI are considered as the same index set. For the index set $IJ \dots P$ the set of possible values for the *index* is the set of all the possible unordered lists of the type $\{i \in I,\ j \in J, \dots,\ p \in P\}$. If $I = I_1 \dots I_m$ and $J = J_1 \dots J_m$ are two index sets then by IJ the index set $I_1 \dots I_m J_1 \dots J_n$ is denoted.

Let us denote the empty index set by $\varnothing$. It is useful to agree for formal purposes that $I, I\varnothing$, and $\varnothing I$ denote the same index set.

Usually, we use capital letters and their finite sequenses to denote multi-index sets and index sets. As a rule, the corresponding notation with small letters is used to denote the values of the multi-index and the index. If the index set is listed as $IJ \dots P$, then we shall write $ij \dots p \in IJ \dots P$ instead of writing $i \in I,\ j \in J, \dots,\ p \in P$.

If I is an elementary index set, then by I^m, $m \geqslant 1$, the index set $I \dots I$ (m times I) is denoted. The set of the possible values of the index for I^m is the set of all the unordered lists of the type $\{i_1 \in I, \dots,\ i_m \in I\}$. Thus, if there are $|I|$ possible values for the index in I, then there are $|I|(|I|+1)/2$ possible values for the index in I^2.

An *ordered array* $\underline{a}_I$ is the set of data a_i, depending on the multi-index $i \in I$. In other words, an ordered array is a function on the set of the possible values of some multi-index. We underline the symbol of the ordered array.

An *array* a_I is the set of data a_i depending on the index $i \in I$. In other words, the array a_I is the function on the set of the possible values for the index of the index set I. For the array $a_I = \{(i, a_i)\colon\ i \in I\}$ with the elements a_i, $i \in I$ we shall also use the notations $\{a_i\}_{i \in I}$, $\{a_i\}_I$, and $\{a_i\}$.

We note that both the vectors/matrices and the arrays can be considered as the equivalence classes of the ordered arrays. For this, correspondingly, either the notation or the position of the elementary index sets is ignored, and the equivalence is defined as follows:

$$\underline{a}_I \overset{m}{\equiv} \underline{b}_J \iff I(=)J, \qquad a_{i\in I} = b_{i\in J} \ \forall i \in I(=)J,$$
$$\underline{a}_I \overset{a}{\equiv} \underline{b}_J \iff I = J \ \text{ as the index sets, } a_i = b_j \text{ if } i = j \text{ as the indices.}$$

The arrays correspond here only to these $\overset{a}{\equiv}$-equivalence classes, where the elements satisfy symmetry condition towards all the repeating elementary index sets. For example, if $I = JJK$ with J the elementary index set, then the symmetry condition reads

$$\underline{a}_{j_1 j_2 k} = \underline{a}_{j_2 j_1 k} \qquad \forall j_1 j_2 k \in JJK.$$

3. ARITHMETIC OPERATIONS

We consider only the arrays where the elements are numbers (or variables with the values of numbers). In this case, the arithmetic operations for the arrays can be defined analogously to the operations with the matrices.

The following agreement is useful to simplify the formulas:

if two index sets are denoted by two different letters, then their lists of the elementary index sets do not have any common elementary index set.

We shall always propose that this agreement is fulfilled, *unless it is explicitly stated otherwise.*

We denote by 0_I the array with the index set I, and with all the elements equal to 0. The arrays $1_I, 2_I, \dots$ are defined analogously. The arrays with empty index set can be identified with the numbers.

For the number α and the array a_I we define $\alpha a_I = a_I \alpha = \{\alpha a_i\}_{i\in I}$, i.e., $\alpha a_I = r_I$ with $r_i = \alpha a_i \ \forall i \in I$.

For the arrays a_I and b_I with the same index set I, addition and substraction operations are defined as follows

$$a_I + b_I = \{a_i + b_i\}_{i\in I}, \tag{3.1}$$

$$a_I - b_I = a_I + (-1) b_I . \tag{3.2}$$

Obviously, $\alpha(\beta a_I) = (\alpha\beta) a_I$, $\alpha a_I + \beta a_I = (\alpha + \beta) a_I$, and the addition operation is commutative and associative. Particularly, we do not need brackets in the expression $a_I + b_I + c_I + \cdots + g_I$.

We define the multiplication operation (product) for two arrays a and b as follows. Let J be the maximal common list of the elementary index sets of the arrays a and b, i.e., let $a = a_{IJ}$ and $b = b_{JK}$, where the index sets I and J, as well as the index sets J and K , may have common elementary index sets, but the index sets I and K must have no common elementary index sets. Let $J = J_1 \dots J_r$ with $J_1, \dots, J_r$ the elementary index sets. We define

$$a_{IJ}b_{JK} = \left\{ \sum_{j_1 \in J_1, \dots j_r \in J_r} a_{ij_1 \dots j_r} b_{j_1 \dots j_r k} \right\}_{ik \in IK} \tag{3.3}$$

If $J = \varnothing$, then

$$a_{IJ}b_{JK} = \{a_i b_k\}_{ik \in IK} .$$

For convenience, we denote the sum

$$\sum_{j_1 \in J_1} \cdots \sum_{j_r \in J_r} \alpha_{j_1 \dots j_r} \quad \text{by} \quad \sum\nolimits'_{j \in J} \alpha_j$$

if $J = J_1 \dots J_r$, and by α if $J = \varnothing$. Note that, for example, in the case of $J = J_1 J_1$ the product $a_{IJ}b_{JK}$ is not equal to $\{\sum_{j \in J} a_{ij} b_{jk}\}_{IK}$ (see the definition of the index set J_1^2).

It is easy to see that the multiplication operation is commutative and distributive towards the addition. Associativity holds if there is no elementary index set common for more than two arrays. For example, $(a_{IJ}b_{JK})c_{KJL} = a_{IJ}(b_{JK}c_{KJL})$ and $(a_{IJK}b_I)c_J = a_{IJK}(b_I c_J)$ but not necessarily $(a_{IJ}b_{JK})c_{JL} = a_{IJ}(b_{JK}c_{JL})$.

We note that if the index sets of two arrays a_I and b_K have no common elementary index sets, it is natural to call this multiplication operation the Kronecker multiplication and denote it by $a_I \otimes b_K$.

It is natural to generalize the concept of the Kronecker power of a vector for the array a_I with an elementary index set I as follows:

$$(a_I)^{\otimes 0} = 1 , \qquad (a_I)^{\otimes m} = \{a_{i_1} \cdots a_{i_m}\}_{i_1 \dots i_m \in I^m} , \quad m \geqslant 1 .$$

It is not so obvious to see how to define the Kronecker power for the array with more than one elementary index set, and how to define the Kronecker product of two arrays which have common elementary index sets. We propose here one possible definition. The idea of this definition is that if, for some value of the index, the result is not defined uniquely, then it is defined as an arithmetic mean of all the possibilities for this value of the index.

We shall use the following auxiliary notation. For the unordered list $l = \{i_1, \dots, i_{m+n}\}$ with $m, n \geqslant 0$, we denote the set of all the possible partitions of the list l into the ordered pair of two unordered lists π_1 and π_2 so that there are m elements in π_1 and n elements in π_2 by $\Pi^{m,n} = \Pi^{m,n}\{i_1, \dots, i_{m+n}\}$. For example,

$$\Pi^{2,1}\{i_1, i_2, i_3\} = \{(\{i_1, i_2\}, \{i_3\}), (\{i_1, i_3\}, \{i_2\}), (\{i_2, i_3\}, \{i_1\})\}.$$

The set $\Pi^{m_1,\ldots,m_r}\{i_1,\ldots,i_{m_1+\cdots+m_r}\}$ is defined analogously. As usually,

$$\binom{m_1+\cdots+m_r}{m_1,\ldots,m_r}=\frac{(m_1+\cdots+m_r)!}{m_1!\cdots m_r!},\qquad \binom{m+n}{m}=\binom{m+n}{m,\,n}.$$

Let I and J be the elementary index sets. We define

$$a_I\otimes b_I=\{r_{i_1i_2}\}_{i_1i_2\in I^2}\quad\text{with}\quad r_{i_1i_2}=(a_{i_1}b_{i_2}+a_{i_2}b_{i_1})/2\ \forall i_1,i_2\in I$$

and, in the more general cases, for $m,n,p,r\geqslant 0$,

$$a_{I^mS}\otimes b_{I^nT}=\left\{\left[\sum_{(\pi_1,\pi_2)}a_{\pi_1s}b_{\pi_2t}\right]\Big/\binom{m+n}{m}\right\}_{I^{m+n}ST},$$

$$\begin{aligned}&a_{I^mJ^pS}\otimes b_{I^nJ^rT}\\&=\left\{\left[\sum_{(q_1,q_2)}\sum_{(\pi_1,\pi_2)}a_{\pi_1q_1s}b_{\pi_2q_2t}\right]\Big/\binom{m+n}{m}\binom{p+r}{p}\right\}_{I^{m+n}J^{p+r}ST}\end{aligned}\tag{3.4}$$

with $(\pi_1,\pi_2)\in\Pi^{m,n}\{i_1,\ldots,i_{m+n}\}$, $(q_1,q_2)\in\Pi^{p,r}\{j_1,\ldots,j_{p+r}\}$. (The further generalization is obvious, but we do not consider it here.) Some particular cases of (3.4) are as follows:

$$a_{IJ}\otimes b_{IJ}=\left\{\left[(a_{i_1j_1}b_{i_2j_2}+a_{i_1j_2}b_{i_2j_1})+(a_{i_2j_1}b_{i_1j_2}+a_{i_2j_2}b_{i_1j_1})\right]/4\right\},$$

$$a_{IJ}\otimes a_{IJ}=\left\{\left[a_{i_1j_1}a_{i_2j_2}+a_{i_1j_2}a_{i_2j_1}\right]/2\right\}_{i_1i_2\in I^2,\,j_1j_2\in J^2},$$

$$(c_I\otimes d_J)\otimes(c_I\otimes d_J)=\left\{c_{i_1}c_{i_2}d_{j_1}d_{j_2}\right\}_{i_1i_2\in I^2,\,j_1j_2\in J^2}.$$

For $m=n=p=r=0$ we get from (3.4)

$$a_S\otimes b_T=\{a_sb_t\}_{ST}=a_Sb_T$$

and thus we can write, for example,

$$\begin{aligned}(a_{SK}\otimes b_{TL})(c_{UK}\otimes d_{VL})&=(a_{SK}b_{TL})(c_{UK}d_{VL})\\&=(a_{SK}c_{UK})(b_{TL}d_{VL})=(a_{SK}c_{UK})\otimes(b_{TL}d_{VL}).\end{aligned}$$

The Kronecker multiplication operation defined in (3.4) is obviously commutative and distributive towards the addition. Its associativity can be verifed by the direct calculations which give

$$\begin{aligned}(a_{I^mS}\otimes b_{I^nT})\otimes c_{I^kU}&=\left\{\left[\sum_{(\pi_1,\pi_2,\pi_3)}a_{\pi_1s}b_{\pi_2t}c_{\pi_3u}\right]\Big/\binom{m+n+k}{m,n,k}\right\}\\&=a_{I^mS}\otimes(b_{I^nT}\otimes c_{I^kU}),\\(a_{I^mJ^pS}\otimes b_{I^nJ^rT})\otimes c_{I^kJ^qU}&=a_{I^mJ^pS}\otimes(b_{I^nJ^rT}\otimes c_{I^kJ^qU}).\end{aligned}\tag{3.5}$$

Thus we can omit the brackets which determine the order of the Kronecker multiplication operations.

We give the following definition of the Kronecker power for the array a_S, where the index set S has the form $S = I^m J^p$, with I and J the elementary index sets and $m, p \geqslant 0$:

$$(a_S)^{\otimes 0} = 1, \qquad (a_S)^{\otimes k} = (a_S)^{\otimes(k-1)} \otimes a_S = \underbrace{a_S \otimes \cdots \otimes a_S}_{k \text{ times}}, \quad k \geqslant 1. \tag{3.6}$$

Particularly, if I and J are the elementary index sets, we have

$$\begin{aligned}(a_I)^{\otimes k} &= \left\{ \left[\sum_{(\pi_1,\dots,\pi_k)\in\Pi^{1,\dots,1}} a_{\pi_1} \cdots a_{\pi_k} \right] \Big/ \binom{k}{1,\dots,1} \right\}_{I^k} \\ &= \left\{ a_{i_1} \cdots a_{i_k} \right\}_{i_1 \dots i_k \in I \dots I},\end{aligned} \tag{3.7}$$

$$\begin{aligned}(a_{IJ})^{\otimes k} &= \left\{ \left[\sum_{(q_1,\dots,q_k)} \sum_{(\pi_1,\dots,\pi_k)} a_{\pi_1 q_1} \cdots a_{\pi_k q_k} \right] \Big/ (k!\,k!) \right\}_{I^k J^k} \\ &= \left\{ \left[\sum_{(q_1,\dots,q_k)\in\Pi^{1,\dots,1}} a_{i_1 q_1} \cdots a_{i_k q_k} \right] \Big/ k \right\}_{I^k J^k} \\ &= \left\{ \left[\sum_{(\pi_1,\dots,\pi_k)\in\Pi^{1,\dots,1}} a_{\pi_1 j_1} \cdots a_{\pi_k j_k} \right] \Big/ k! \right\}_{I^k J^k}.\end{aligned} \tag{3.8}$$

By associativity and commutativity of the Kronecker multiplication operation, we have for $S = I^m J^p$ and $T = I^n J^r$, with I and J the elementary index sets and $m, p, n, r \geqslant 0$:

$$(a_S)^{\otimes(k+l)} = (a_S)^{\otimes k} \otimes (a_S)^{\otimes l}, \tag{3.9}$$

$$(a_S \otimes b_T)^{\otimes k} = (a_S)^{\otimes k} \otimes (b_T)^{\otimes k}. \tag{3.10}$$

When I and J are the elementary index sets, the direct calculations give us

$$(a_{IJ} x_J)^{\otimes k} = \left\{ \left(\sum_j a_{i_1 j} x_j \right) \cdots \left(\sum_j a_{i_k j} x_j \right) \right\}_{I^k},$$

$$\begin{aligned}(a_{IJ})^{\otimes k} (x_J)^{\otimes k} &= \Bigg\{ \sum_{j_1,\dots,j_k \in J} \left[\sum_{(q_1,\dots,q_k)\in\Pi^{1,\dots,1}} a_{i_1 q_1} \cdots a_{i_k q_k} \right] \\ &\qquad \times \left[x_{j_1} \dots x_{j_k} \right] / k! \Bigg\}_{I^k} \\ &= \left\{ \sum_{(q_1,\dots,q_k)\in\Pi^{1,\dots,1}} \left[\sum_{j_1,\dots,j_k \in J} (a_{i_1 q_1} x_{q_1}) \cdots (a_{i_k q_k} x_{q_k}) \right] \Big/ k! \right\}_{I^k} \\ &= \left\{ \sum_{j_1,\dots,j_k \in J} (a_{i_1 j_1} x_{j_1}) \cdots (a_{i_k j_k} x_{j_k}) \right\}_{I^k}.\end{aligned}$$

Thus we have

$$(a_{IJ}x_J)^{\otimes k} = (a_{IJ})^{\otimes k}(x_J)^{\otimes k}. \tag{3.11}$$

Analogously, with I, J, and L the elementary index sets, the direct calculations give that

$$(a_{IJ}x_{JL})^{\otimes k} = (a_{IJ})^{\otimes k}(x_{JL})^{\otimes k}.$$

Let $S = I^m J^p$, with I and J the elementary index sets, and m, $n \geqslant 0$. The following formula may be proved, for example, by the mathematical induction:

$$(a_S + b_S)^{\otimes k} = \sum_{l=0}^{k} \binom{k}{l} a_S^{\otimes l} \otimes b_S^{\otimes(k-l)}, \qquad k \geqslant 0. \tag{3.12}$$

4. RANDOM ARRAYS

Analogously to the case of the vectors and the matrices, the *random array* is defined as an array where the elements are random variables. Here we consider only some connections between the moments of the random array which has elementary index set. We see that they have formally the same form as in the case of the random variables.

We denote the mean of the random variable X_i by EX_i. If, for the random array X_I, for every $i \in I$ there exists (i.e., is finite) the mean EX_i then we call the array $\{EX_i\}_{i \in I}$ the *mean of the random array*, and denote it by EX_I.

Let I be the elementary index set. If, for the random array X_I and the array a_I, for any set of indices $\{i_1, \dots, i_k\}$ with $i_1, \dots, i_k \in I$, $k \geqslant 1$, exists the mean $E[(X_{i_1} - a_{i_1}) \dots (X_{i_k} - a_{i_k})]$ then the array

$$\{E[(X_{i_1} - a_{i_1}) \cdots (X_{i_k} - a_{i_k})]\}_{i_1 \dots i_k \in I^k}$$

is called the *k-th moment of the random array X_I with respect to the array a_I*, and denoted by $m_a^{(k)}(X)$ or $m_a^{(k)}$. We also define $m_a^{(0)}(X) = 1$.

Thus, for the random array X_I with the elementary index set I, we have

$$\begin{aligned} m_a^{(k)}(X_I) &= \{E[(X_{i_1} - a_{i_1}) \cdots (X_{i_k} - a_{i_k})]\} \\ &= E\{(X_{i_1} - a_{i_1}) \cdots (X_{i_k} - a_{i_k})\} = E[(X_I - a_I)^{\otimes k}]. \end{aligned} \tag{4.1}$$

The most important particular cases of $m_a^{(k)}(X)$ are the cases of $a_I = 0_I$, and $a_I = EX_I$. The array $m_0^{(k)}(X)$ is called the *k-th moment of X* and denoted by $m^{(k)}(X)$ or $m^{(k)}$. The array $m_{EX}^{(k)}$ is called the *k-th central moment of X* and denoted by $\mu^{(k)}(X)$ or $\mu^{(k)}$. Particularly,

$$m_a^{(k)}(X) = m^{(k)}(X_I - a_I).$$

From the properties of the random variables we will get straightforwardly the following equalities for the random arrays X and Y, and array b (for the simplicity of the formulations we always assume that all the necessary means exist):

(1) $Eb_I = b_I$,

(2) $E(cX_I) = cEX_I$,

(3) $E(X_I + Y_I) = EX_I + EY_I$,

(4) $E(b_{IJ}X_I) = \left\{E\sum_{i\in I} b_{ij}X_i\right\}_J = \left\{\sum_{i\in I} b_{ij}EX_i\right\}_J = b_{IJ}EX_I$,

$E(b_IX_I) = E\sum_{i\in I} b_iX_i = b_IEX_I$,

$E(b_J \otimes X_I) = E(b_JX_I) = \{E(b_jx_i)\}_{IJ} = b_J \otimes EX_I$,

$E(b_I \otimes X_I) = \{E(b_{i_1}X_{i_2} + b_{i_2}X_{i_1})/2\}_{i_1i_2\in I^2} = b_I \otimes EX_I$,

(5) if the random arrays X_I and Y_J are *independent*, i.e., if the elements X_i and Y_j of these arrays are independent random variables for every pair (X_i, Y_j), then

$$E(X_IY_I) = E\sum_{i\in I} x_iy_i = \sum_{i\in I} Ex_iEy_i = (EX_I)(EY_I),$$

$$E(X_I \otimes Y_J) = E(X_IY_J) = EX_I \otimes EY_J,$$

$$E(X_I \otimes Y_I) = \{E(X_{i_1}Y_{i_2} + X_{i_2}Y_{i_1})/2\}_{i_1i_2\in I} = EX_I \otimes EY_I.$$

Let the index sets I and J be the elementary index sets. For the random arrays X and Y, the arrays a and b, and the constant c the following equalities follow from the equalities (1)–(5) and (4.1):

(1) $m_a^{(k)}(b_I) = E[(b_I - a_I)^{\otimes k}] = (b_I - a_I)^{\otimes k}$,

(2) $m_{ca}^{(k)}(cX_I) = E[(cX_I - ca_I)^{\otimes k}] = E[c^k(X_I - a_I)^{\otimes k}] = c^k m_a^{(k)}(X_I)$,

(3) $m_{a+b}^{(k)}(X_I + Y_I) = E[(X_I - a_I + Y_I - b_I)^{\otimes k}]$

$$= E\Big[\sum_{l=0}^{k} \tbinom{k}{l}(X_I - a_I)^{\otimes k} \otimes (Y_I - b_I)^{\otimes(k-l)}\Big]$$

$$= \sum_{l=0}^{k} \tbinom{k}{l} E[(X_I - a_I)^{\otimes l} \otimes (Y_I - b_I)^{\otimes(k-l)}],$$

where in the particular case of the independent random arrays X_I and Y_I

$$m_{a+b}^{(k)}(X_I + Y_I) = \sum_{l=0}^{k} \tbinom{k}{l} m_a^{(l)}(X_I) \otimes m_b^{\otimes(k-l)}(Y_I),$$

(4) $m_{ba}^{(k)}(b_{IJ}X_I) = E[(b_{IJ}X_I - b_{IJ}a_I)^{\otimes k}] = E[b_{IJ}^{\otimes k}(X_I - a_I)^{\otimes k}]b_{IJ}^{\otimes k}m_a^{(k)}(X_I)$.

Also, it holds that

$$m_a^{(k)}(X_I) = E[(X_I - b_I + b_I - a_I)^{\otimes k}] = \textstyle\sum_{l=0}^{k} \tbinom{k}{l}(b_I - a_I)^{\otimes l} \otimes m_b^{(k-l)}(X_I).$$

Particularly, for $a_I = EX_I = m^{(1)}$ and $b_I = 0_I$, we have

$$\mu^{(k)} = \sum_{l=0}^{k}(-1)^l \binom{k}{l} \left(m^{(1)}\right)^{\otimes l} \otimes m^{(k-l)},$$

which gives us for the first three central moments

$$\begin{aligned}
\mu^{(1)} &= m^{(1)} - m^{(1)} = 0_I, \\
\mu^{(2)} &= m^{(2)} - 2m^{(1)} \otimes m^{(1)} + \left(m^{(1)}\right)^{\otimes 2} = m^{(2)} - \left(m^{(1)}\right)^{\otimes 2}, \\
\mu^{(3)} &= m^{(3)} - 3m^{(1)} \otimes m^{(2)} + 3\left(m^{(1)}\right)^{\otimes 2} \otimes m^{(1)} - \left(m^{(1)}\right)^{\otimes 3} \\
&= m^{(3)} - 3m^{(2)} \otimes m^{(1)} + 2\left(m^{(1)}\right)^{\otimes 3}.
\end{aligned}$$

All these equalities for the random arrays are in the same form as the corresponding equalities for the random variables. The Kronecker multiplication of the arrays plays the role of multiplication of the numbers.

REFERENCES

Karma, O. (1992). On arrays and array-derivatives. *Acta et Comm. Univ. Tartuensis* **942**, 132–150.

Kollo, T. (1992). On the correspondence between matrix derivative and some basic notions of multivariate statistics. *Acta et Comm. Univ. Tartuensis* **942**, 28–34.

Multivar. Statist., pp. 335–342
E.-M. Tiit *et al.* (Eds)

ON STATISTICAL APPLICATIONS OF MATRIX MAJORIZATION

J. HAUKE[1] and A. MARKIEWICZ[2]
[1,2]Department of Mathematical and Statistical Methods,
The Agricultural University of Poznań, Wojska Polskiego 28, 60-637 Poznań

[1]Institute of Socio-Economic Geography and Spatial Planning,
Adam Mickiewicz University, Fredry 10, 61-701 Poznań, Poland

ABSTRACT

The paper presents a discussion of the criteria of comparison of experiments and the general sufficient condition of universal optimality. The theory of group weak majorization introduced by Giovagnoli and Wynn (1985) is applied. The examples of matrices defining appriopriate groups for the purpose of the comparison of experiments as well as determining the universal optimality are given.

1. INTRODUCTION

We are interested in applications of group majorization for comparison and finding of the optimal experimental designs under linear models. It is worth noticing that several other authors, e.g., Bailey (1981), and Dawid (1988), have applied group theory to linear models, and that group majorization was used for fixed effects models by Giovagnoli *et al.* (1987) and Pukelsheim (1987). We also emphasize that the problems of the comparison of experiments and optimal designs are recently often discussed in different aspects (see, e.g. (Jones *et al.*, 1992; Peres and Ting Hui-Ching, 1993; Pukelsheim, 1993; Yurramendi, 1993)).

Throughout this paper, we write $\mathbf{L} \in \mathcal{M}_{m,n}$ if $\mathbf{L}$ is an $m \times n$ real matrix, $\mathbf{L} \in \mathcal{M}_n$ if $\mathbf{L} \in \mathcal{M}_{n,n}$, $\mathbf{L} \in \mathcal{M}_n^o$ if $\mathbf{L} \in \mathcal{M}_n$ and is orthogonal, $\mathbf{L} \in \mathcal{P}_n$ if $\mathbf{L} \in \mathcal{M}_n^o$ and is a permutation matrix, $\mathbf{L} \in \mathcal{M}_n^s$ if $\mathbf{L} \in \mathcal{M}_n$ and is symmetric, $\mathbf{L} \in \mathcal{M}_n^{\geqslant}$ if $\mathbf{L} \in \mathcal{M}_n^s$ and is nonnegative definite, and $\mathbf{L} \in \mathcal{M}_n^{>}$ if $\mathbf{L} \in \mathcal{M}_n^{\geqslant}$ and is nonsingular. Given $\mathbf{L} \in \mathcal{M}_{m,n}$, the symbols $\mathbf{L}'$ and $\mathbf{L}^-$ denote the transpose and a generalized inverse of $\mathbf{L}$, respectively, while $\mathbf{Q}_L = \mathbf{I}_m - \mathbf{L}(\mathbf{L}'\mathbf{L})^-\mathbf{L}'$ denotes the orthogonal projector on the orthocomplement of the range of $\mathbf{L}$, where $\mathbf{I}_m \in \mathcal{M}_m$ is an identity matrix. Moreover, given $\mathbf{L}, \mathbf{M} \in \mathcal{M}_n^s$, the expression $\mathbf{L} \preceq_L \mathbf{M}$ will mean that $\mathbf{M} - \mathbf{L} \in \mathcal{M}_n^{\geqslant}$, i.e., that $\mathbf{L}$ is below $\mathbf{M}$ with respect to the Löwner partial ordering (Marshall and Olkin, 1979, p. 462). The direct sum of the two given sets of matrices $\mathcal{A} \subseteq \mathcal{M}_a$ and $\mathcal{B} \subseteq \mathcal{M}_b$ is a subset of $\mathcal{M}_{a+b}$ consisting of block diagonal matrices

$$\mathcal{A} \oplus \mathcal{B} = \{\mathrm{diag}(\mathbf{A}, \mathbf{B})\colon\ \mathbf{A} \in \mathcal{A},\ \mathbf{B} \in \mathcal{B}\}.$$

Let us present some preliminary results in group majorizations: $\mathcal{G}$-majorization and $\mathcal{GW}$-majorization. For a general concept of them with applications to matrix orderings and suggestions how to verify them, see (Giovagnoli and Wynn, 1985).

Let $\mathcal{O}(ab)$ denote a group of $ab \times ab$ orthogonal matrices acting on $\mathbb{R}^{ab}$, and suppose that $\mathcal{G}$ is a closed subgroup of $\mathcal{O}(ab)$ (with respect to multiplication). For $\mathbf{x} \in \mathbb{R}^{ab}$, let $\mathcal{C}(\mathbf{x})$ denote the convex hull of the $\mathcal{G}$-orbit $\{g(\mathbf{x})\colon\ g \in \mathcal{G}\}$. The group $\mathcal{G}$ determines a preordering called $\mathcal{G}$-majorization, on the vector space of real $a \times b$ matrices. It is denoted by $\leqslant_G$ and, using the vec operator, can be defined as follows.

DEFINITION 1. For any matrices $\mathbf{A}, \mathbf{B} \in \mathcal{M}_{a,b}$,

$$\mathbf{A} \prec_G \mathbf{B} \quad \text{if} \quad \mathrm{vec}(\mathbf{A}') \in \mathcal{C}[\mathrm{vec}(\mathbf{B}')].$$

Let us note that for any $\mathbf{A}$ and $\mathbf{B} \in \mathcal{M}_k^s$

$$\mathbf{A} \prec_G \mathbf{B} \quad \text{if and only if} \quad \mathbf{A} = \sum_{i \in \mathcal{I}} \alpha_i \mathbf{\Pi}_i \mathbf{B} \mathbf{\Pi}_i',$$

where $\mathcal{I}$ is a finite index set and $\alpha_i \geqslant 0$, $\sum_{i \in \mathcal{I}} \alpha_i = 1$, and $\mathbf{\Pi}_i \in \mathcal{M}_k^o$ for $i \in \mathcal{I}$.

Further, let $<\cdot$ denote another ordering on $\mathcal{M}_k^s$ which is assumed to satisfy the condition

$$\mathbf{A} <\cdot\, \mathbf{B} \Longrightarrow \sum_{i \in \mathcal{I}} \alpha_i \mathbf{\Pi}_i \mathbf{A} \mathbf{\Pi}_i' <\cdot \sum_{i \in \mathcal{I}} \alpha_i \mathbf{\Pi}_i \mathbf{B} \mathbf{\Pi}_i'$$

for all finite sets of $\alpha_i \geqslant 0$, $\sum_{i \in \mathcal{I}} \alpha_i = 1$, and $\mathbf{\Pi}_i \in \mathcal{M}_k^o$ for $i \in \mathcal{I}$. The ordering $\prec_G$ can be combined with $<\cdot$ to produce a new ordering called upper weak $\mathcal{GW}$-majorization $\prec^{GW}$ (Giovagnoli and Wynn, 1985, p. 120).

DEFINITION 2. For any $\mathbf{A}$ and $\mathbf{B} \in \mathcal{M}_k^s$

$$\mathbf{A} \prec^{GW} \mathbf{B} \quad \text{if} \quad \mathbf{C} <\cdot\, \mathbf{A} \quad \text{and} \quad \mathbf{C} \prec_G \mathbf{B} \quad \text{for some} \quad \mathbf{C} \in \mathcal{M}_k^s.$$

The notion of the (generalized) Schur complement (see, e.g. (Styan, 1983)) plays an important role in the theory presented below.

DEFINITION 3. Let $\mathbf{A} \in \mathcal{M}_n^{\geqslant}$ be partitioned as follows

$$\mathbf{A} = \begin{bmatrix} \mathbf{E} & \mathbf{F} \\ \mathbf{F}' & \mathbf{G} \end{bmatrix}.$$

The function $\psi(\mathbf{A}, \mathbf{G}) = (\mathbf{A}/\mathbf{G}) = \mathbf{E} - \mathbf{F}\mathbf{G}^{-}\mathbf{F}'$ is called the Schur complement of $\mathbf{G}$ in $\mathbf{A}$.

2. RESULTS

Let us consider the general linear model

$$\mathbf{y} = \mathbf{X}_1\gamma + \mathbf{X}_2\delta + \mathbf{Z}\xi + \varepsilon, \tag{1}$$

where $\mathbf{y}$ is an $n \times 1$ vector of observation, $\mathbf{X}_1, \mathbf{X}_2, \mathbf{Z}$ are known matrices of size $n \times p$, $n \times q$, and $n \times r$, respectively, γ is a $p \times 1$ vector of parameters which are of main interest, δ is a $q \times 1$ vector containing all the other (nuisance) parameters in the model (apart from γ). Further, ξ is an $r \times 1$ vector of random effects with the expectation zero and the dispersion matrix $\mathrm{Cov}(\xi) = \mathrm{diag}(\sigma_1^2\mathbf{I}_{r_1}, \dots, \sigma_s^2\mathbf{I}_{r_s})$, $\sigma_i^2 \geqslant 0$, $i = 1, 2, \dots, s$, and $\sum_{i=1}^{s} r_i = r$, and ε is a vector of i.i.d. random variables with variance, say, $\sigma_0^2 = 1$. The partition of $\mathrm{Cov}(\xi)$ corresponds to the partitioning of ξ into s subvectors, $\xi = (\xi_1', \dots, \xi_s')'$, and $\mathbf{Z}$ into s submatrices, $\mathbf{Z} = (\mathbf{Z}_1, \dots, \mathbf{Z}_s)$. As a result, the dispersion matrix of $\mathbf{y}$ takes the form $\mathbf{V} = \sum_{i=0}^{s} \sigma_i^2 \mathbf{V}_i$, where $\mathbf{V}_i = \mathbf{Z}_i\mathbf{Z}_i'$ for $i = 1, 2, \dots, s$ and $\mathbf{V}_0 = \mathbf{I}_n$.

We can investigate the following three information matrices for the estimation of γ: the information matrix $\mathbf{C}_0$ for the model (1) without ξ, i.e., $\sigma_i^2 = 0$, $i = 1, 2, \dots, s$, the information matrix $\mathbf{C}_\infty$ for model (1) with fixed effects ξ, and the information matrix $\mathbf{C}_\sigma$ for model (1) and $\sigma = (\sigma_1^2, \dots, \sigma_s^2)'$ known. The respective cases of model (1) will be denoted by M_0, M_∞, and M_σ.

Using the notion of the Schur complement, the information matrix $\mathbf{C}_0$ for the estimation of γ in model (1) without ξ, i.e., $\sigma_i^2 = 0$, $i = 1, 2, \dots, s$, can be written

$$\mathbf{C}_0 = (\mathbf{T}_0 / \mathbf{X}_2'\mathbf{X}_2), \tag{2}$$

where

$$\mathbf{T}_0 = \begin{pmatrix} \mathbf{X}_1'\mathbf{X}_1 & \mathbf{X}_1'\mathbf{X}_2 \\ \mathbf{X}_2'\mathbf{X}_1 & \mathbf{X}_2'\mathbf{X}_2 \end{pmatrix};$$

similarly, the information matrix $\mathbf{C}_\infty$ for the estimation of γ in model (1) with fixed effects ξ is

$$\mathbf{C}_\infty = (\mathbf{T}/\mathbf{S}), \tag{3}$$

where

$$\mathbf{T} = \begin{pmatrix} \mathbf{X}_1'\mathbf{X}_1 & \mathbf{X}_1'\mathbf{X}_2 & \mathbf{X}_1'\mathbf{Z} \\ \mathbf{X}_2'\mathbf{X}_1 & \mathbf{X}_2'\mathbf{X}_2 & \mathbf{X}_2'\mathbf{Z} \\ \mathbf{Z}'\mathbf{X}_1 & \mathbf{Z}'\mathbf{X}_2 & \mathbf{Z}'\mathbf{Z} \end{pmatrix} \tag{4}$$

and

$$\mathbf{S} = \begin{pmatrix} \mathbf{X}_2'\mathbf{X}_2 & \mathbf{X}_2'\mathbf{Z} \\ \mathbf{Z}'\mathbf{X}_2 & \mathbf{Z}'\mathbf{Z} \end{pmatrix}.$$

The information matrix $\mathbf{C}_\sigma$ for model (1) and $\sigma = (\sigma_1^2, \dots, \sigma_s^2)'$ known, is

$$\mathbf{C}_\sigma = (\mathbf{T}_\sigma / \mathbf{S}_\sigma), \tag{5}$$

where

$$\mathbf{T}_\sigma = \begin{pmatrix} \mathbf{X}_1'\mathbf{X}_1 & \mathbf{X}_1'\mathbf{X}_2 & \mathbf{X}_1'\mathbf{U} \\ \mathbf{X}_2'\mathbf{X}_1 & \mathbf{X}_2'\mathbf{X}_2 & \mathbf{X}_2'\mathbf{U} \\ \mathbf{U}'\mathbf{X}_1 & \mathbf{U}'\mathbf{X}_2 & \mathbf{U}'\mathbf{U}+\mathbf{I}_r \end{pmatrix} \tag{6}$$

and

$$\mathbf{S}_\sigma = \begin{pmatrix} \mathbf{X}_2'\mathbf{X}_2 & \mathbf{X}_2'\mathbf{U} \\ \mathbf{U}'\mathbf{X}_2 & \mathbf{U}'\mathbf{U}+\mathbf{I}_r \end{pmatrix}$$

with $\mathbf{U} = (\sigma_1\mathbf{Z}_1, \ldots, \sigma_s\mathbf{Z}_s)$.

The following theorem, given by Markiewicz (1994), is a generalization of theorem of Hauke and Markiewicz (1994).

THEOREM. *The information matrices $\mathbf{C}_\infty$ and $\mathbf{C}_\sigma$ are matrix-concave functions of $\mathbf{T}$ and if*

$$\mathbf{T}_* = \operatorname{diag}(\Pi_{X_1}, \Pi_{X_2}, \Pi_Z)\mathbf{T}\operatorname{diag}(\Pi'_{X_1}, \Pi'_{X_2}, \Pi'_Z),$$

where $\Pi_{X_1} \in \mathcal{M}_p^o$, $\Pi_{X_2} \in \mathcal{M}_q^o$, while $\Pi_Z = \operatorname{diag}(\Pi_{Z_1}, \ldots, \Pi_{Z_s})$ with $\Pi_{Z_i} \in \mathcal{M}_{r_i}^o$, $i = 1, 2, \ldots, s$, then

$$\mathbf{C}_\infty(\mathbf{T}_*) = \Pi_{X_1}\mathbf{C}_\infty(\mathbf{T})\Pi'_{X_1}$$

and

$$\mathbf{C}_\sigma(\mathbf{T}_*) = \Pi_{X_1}\mathbf{C}_\sigma(\mathbf{T})\Pi'_{X_1}.$$

Moreover, the information matrices $\mathbf{C}_\infty$ and $\mathbf{C}_\sigma$ are matrix-concave functions of the matrix

$$\mathbf{W} = \begin{pmatrix} \mathbf{X}_1'\mathbf{Q}_{X_2}\mathbf{X}_1 & \mathbf{X}_1'\mathbf{Q}_{X_2}\mathbf{Z} \\ \mathbf{Z}'\mathbf{Q}_{X_2}\mathbf{X}_1 & \mathbf{Z}'\mathbf{Q}_{X_2}\mathbf{Z} \end{pmatrix}, \tag{7}$$

and

$$\mathbf{C}_\infty(\mathbf{W}_*) = \Pi_{X_1}\mathbf{C}_\infty(\mathbf{W})\Pi'_{X_1},$$

and

$$\mathbf{C}_\sigma(\mathbf{W}_*) = \Pi_{X_1}\mathbf{C}_\sigma(\mathbf{W})\Pi'_{X_1}$$

for all matrices $\mathbf{W}_*$, such that

$$\mathbf{W}_* = \operatorname{diag}(\Pi_{X_1}, \Pi_Z)\mathbf{W}\operatorname{diag}(\Pi'_{X_1}, \Pi'_Z),$$

where $\Pi_{X_1} \in \mathcal{M}_p^o$ and $\Pi_Z = \operatorname{diag}(\Pi_{Z_1}, \ldots, \Pi_{Z_s})$ with $\Pi_{Z_i} \in \mathcal{M}_{r_i}^o$, $i = 1, 2, \ldots, s$.

We will now apply the results to the general linear model for the comparison of experimental designs and for the study of universal optimality.

Let $\mathbf{\Phi}$ be the class of all functionals φ on a given set of nonnegative definite matrices, satisfying the following conditions:

(a) φ is concave,

(b) φ is isotonic (increasing) with respect to the Löwner partial ordering,

(c) φ is invariant under each permutation of rows and (the same on) of columns.

Following Kiefer (1975) and Pukelsheim (1993, p. 354) we can compare an experimental design via comparison of the information matrices.

DEFINITION 4. An experimental design $d_2 \in \mathcal{D}$ is said to be at least as good as an experimental design $d_1 \in \mathcal{D}$ with respect to Kiefer ordering, in symbols $d_1 \leqslant_\varphi d_2$, if

$$\varphi(\mathbf{C}_{d_2}) - \varphi(\mathbf{C}_{d_1}) \geqslant 0$$

for every $\varphi \in \Phi$.

Consequently, a design d^* will be termed *universally optimal* in the class $\mathcal{D}$ of designs under consideration if d^* maximizes $\varphi(\mathbf{C}_d)$ for every functional $\varphi \in \Phi$ (Kiefer, 1975; Bondar, 1983; Giovagnoli *et al.*, 1987; Pukelsheim, 1987,1993; Shah and Sinha, 1989; David, 1994). The general sufficient condition of universal optimality can be expressed via $\mathcal{GW}$-majorization as follows (Giovagnoli and Wynn, 1985; Giovagnoli *et al.*, 1987; Pukelsheim, 1993, Chapter 14).

PROPOSITION. *Let $\mathcal{G}$ be a subgroup of $\mathcal{P}_k$, $\lessdot$ be the Löwner partial ordering, and $\mathcal{C} = \{\mathbf{C}_d,\ d \in \mathcal{D}\}$ be the class of information matrices. If $\mathbf{C}_{d_1}$ and $\mathbf{C}_{d_1} \in \mathcal{C}$ then*

$$\mathbf{C}_{d_1} \prec^{\mathrm{GW}} \mathbf{C}_{d_2} \quad \textit{implies} \quad d_1 \leqslant_\varphi d_2.$$

Moreover, if $\mathcal{C}$ contains $\mathbf{C}_{d^*}$ which is $\mathcal{GW}$-minimal in $\mathcal{C}$, i.e.,

$$\mathbf{C}_{d^*} \prec^{\mathrm{GW}} \mathbf{C}_d \quad \textit{for all} \quad \mathbf{C}_d \in \mathcal{C},$$

then d^* is universally optimal in $\mathcal{D}$.

If $\mathcal{G} = \mathcal{P}_k$ and $\mathbf{C}_d \mathbf{1}_k = \mathbf{0}$ for all $\mathbf{C}_d \in \mathcal{C}$ then the Proposition implies as a special case Proposition 1 of Kiefer (1975) (Giovagnoli and Wynn, 1985, p. 133). Some examples of $\mathcal{GW}$-minimal elements are presented in (Giovagnoli and Wynn, 1985) and the general way of finding such elements is given in their Proposition 1.

Now, let us consider experimental designs $d \in \mathcal{D}$ in models M_0, M_∞, and M_σ with information matrices and matrices $\mathbf{T}$ and $\mathbf{W}$, defined in (4) and (7), denoted by additional letter subscripts indicating the design. The matrix

$$\Pi = \mathrm{diag}(\Pi_{\mathrm{X}_1}, \Pi_{\mathrm{X}_2}, \Pi_{\mathrm{Z}})$$

from Theorem consists (generally) of suitable permutation matrices corresponding to parameter relabelling. The relabelling group $\mathcal{G}_t$ is a subgroup of $\mathcal{P}_p \oplus \mathcal{P}_q \oplus \mathcal{P}_r$. In the following corollaries we consider two groups: $\mathcal{G}_t$ and $\mathcal{G}_w \subseteq \mathcal{P}_p \oplus \mathcal{P}_r$, and the Löwner partial ordering as $\lessdot$. The related $\mathcal{G}_t$-majorization and $\mathcal{G}_w\mathcal{W}$-majorization will be denoted by $\prec_{\mathrm{G}_t}$ and $\prec^{\mathrm{G}_w\mathrm{W}}$, respectively.

COROLLARY 1. *For $d_1, d_2 \in \mathcal{D}$*

$$\mathbf{W}_{d_2} \prec^{\mathrm{G}_w\mathrm{W}} \mathbf{W}_{d_1} \quad \textit{implies} \quad d_1 \leqslant_\varphi d_2.$$

Moreover, in the class of designs $\mathcal{D}$ in which for some $d^* \in \mathcal{D}$

$$\mathbf{W}_{d^*} \prec^{\mathrm{G}_w\mathrm{W}} \mathbf{W}_d \quad \textit{for all} \quad \mathbf{W}_d \in \mathcal{W} = \{\mathbf{W}_d, d \in \mathcal{D}\}$$

the design d^* is universally optimal under models M_0, M_∞, and M_σ.

COROLLARY 2. *For* $d_1, d_2 \in \mathcal{D}$

$$\mathbf{T}_{d_2} \prec^{G_t} \mathbf{T}_{d_1} \quad \textit{implies} \quad d_1 \leqslant_{\varphi} d_2.$$

Moreover, in the class of designs $\mathcal{D}$ *in which for some* $d^* \in \mathcal{D}$

$$\mathbf{T}_{d^*} \prec^{G_t} \mathbf{T}_d \quad \textit{for all} \quad \mathbf{T}_d \in \mathcal{T} = \{\mathbf{T}_d,\ d \in \mathcal{D}\}$$

the design d^* *is universally optimal under models* M_0, M_∞, *and* M_σ.

Let us notice that Corollary 1 generalizes Strategy I from (Kunert, 1983), its second part is comparable with Proposition 3 in (Giovagnoli and Wynn, 1985), while the second part of Corollary 2 is comparable with Theorem 5 in (Giovagnoli *et al.*, 1987).

Example 1. *Row-column designs.* Let us consider the setting of a row-column design, involving t treatments, in which ab experimental units are arranged in the form of an $a \times b$ array of entries from the set $\{1, 2, \ldots, t\}$, with a/t integers. Further, let $\mathbf{N}_1 \in \mathcal{M}_{t,a}$ and $\mathbf{N}_2 \in \mathcal{M}_{t,b}$ be the incidence matrices of treatments vs rows and treatments vs columns, respectively. We will consider two models: with all effects fixed and with column effects random, respectively. The respective matrices $\mathbf{T}_d$ and $\mathbf{W}_d$, as in (4) and (7), can be written as

$$\mathbf{T}_d = \begin{bmatrix} \mathbf{r}^\delta & \mathbf{N}_1 & \mathbf{N}_2 \\ \mathbf{N}_1' & bt\mathbf{I}_a & t\mathbf{J}_{ab} \\ \mathbf{N}_2' & t\mathbf{J}_{ba} & at\mathbf{I}_b \end{bmatrix} \tag{8}$$

and

$$\mathbf{W}_d = \begin{bmatrix} \mathbf{r}^\delta - (1/bt)\mathbf{N}_1\mathbf{N}_1' & \mathbf{N}_2 - (1/b)\mathbf{N}_1\mathbf{J}_{ab} \\ \mathbf{N}_2' - (1/b)\mathbf{J}_{ba}\mathbf{N}_1' & at\mathbf{I}_b - (t/b)\mathbf{J}_{bb} \end{bmatrix}, \tag{9}$$

where $\mathbf{r} = \mathbf{N}_1\mathbf{1}_a = \mathbf{N}_2\mathbf{1}_b$ is a vector of replications and $\mathbf{r}^\delta = \mathrm{diag}(\mathrm{r}_1, \mathrm{r}_2, \ldots, \mathrm{r}_t)$.

Let us observe that in the case of the comparison of designs the relabelling group can be chosen as $\mathcal{G}_w = \mathcal{P}_t \oplus \mathcal{P}_b$, but in the case of the finding of the optimal design we can take the smaller group $\mathcal{G}_w^* = \mathcal{P}_t \oplus \{\mathbf{I}_b\}$. Then (Giovagnoli *et al.*, 1987)

$$\mathbf{W}_{d^*} = \begin{bmatrix} \mathbf{C}_* & 0 \\ 0 & at\mathbf{I}_b - (t/b)\mathbf{J}_{bb} \end{bmatrix},$$

which is the $\mathbf{W}$-matrix of some regular Generalized Youden Design d^*, is $\mathcal{G}_w^*$-invariant and $\prec^{G_w \mathrm{W}}$-minimal in the set $\mathcal{W} = \{\mathbf{W}_d,\ d \in \mathcal{D}\}$. It follows from Corollary 1 that d^* is universally optimal under the models M_∞ (Kiefer, 1975) and M_σ (Mukhopadhyay, 1981; Shah and Sinha, 1989, Chapter 5). Moreover, d^* is universally optimal under the model M_0, i.e., it is a BBD (BIBD) when considered as a block design with rows as blocks.

The appriopriate choice of the relabelling group can be illustrated by a comparison of two row-column designs with the same matrix $\mathbf{N}_1$ and matrix

$$\mathbf{N}_2 = \begin{bmatrix} 1 & 1 & 3 \\ 2 & 3 & 1 \\ 3 & 2 & 2 \end{bmatrix}$$

and

$$\mathbf{N}_2 = \begin{bmatrix} 1 & 2 & 2 \\ 2 & 2 & 2 \\ 3 & 2 & 2 \end{bmatrix},$$

respectively. Such designs can be compared taking group $\mathcal{G}_w$, but it is impossible taking group $\mathcal{G}_w^*$.

Example 2. *Row-column designs for comparing treatments with a control.* Let us consider the setting of a row-column design for comparing t treatments with a control treatment, say 0, in which ab experimental units are arranged in the form of an $a \times b$ array of entries from the set $\{0, 1, 2, \ldots, t\}$, with $r_0/b = k_0$ and $(a - k_0)/t$ integers, where r_0 is a fixed number of replications of the 0 treatment. Further, let $\mathbf{N}_1 \in \mathcal{M}_{t+1,a}$ and $\mathbf{N}_2 \in \mathcal{M}_{t+1,b}$ be the incidence matrices of treatments vs rows and treatments vs columns, respectively. In this case, matrices $\mathbf{T}_d$ and $\mathbf{W}_d$ are of the form (8) and (9), respectively, while $\mathcal{G}_w = \{1\} \oplus \mathcal{P}_t \oplus \mathcal{P}_b$ (Giovagnoli *et al.*, 1987; Shah and Sinha, 1989, p. 131). Now, the $\mathcal{G}_w$-symmetric and $\prec^{\mathrm{G}_w\mathrm{W}}$-minimal element in the set $\mathcal{W}$ is

$$\mathbf{W}_{\tilde{d}} = \begin{bmatrix} \widetilde{\mathbf{C}} & 0 \\ 0 & a(t+1)\mathbf{I}_b - [(t+1)/b]\mathbf{J}_{bb} \end{bmatrix},$$

where $\widetilde{\mathbf{C}}$ is a $\mathbf{C}$-matrix of BTBD (BTIBD); (Bechhofer and Tamhane, 1981). The design $\tilde{d}$ is BTBD (BTIBD) with rows as blocks, and has test treatments equally replicated, and the total number of replications for each treatment, test or control, divided equally among b columns. According to Corollary 2, $\tilde{d}$ is universally optimal under models M_∞ (Jacroux, 1986; Hedayat *et al.*, 1988; Giovagnoli and Verdinelli, 1988, p. 484) and M_σ.

REFERENCES

Bailey, R. A. (1981). Distributive block structures and their automorphismums. In: *Combinatorial Mathematics VIII, Lect. Notes Math.* **884**, pp. 115–124, K. L. McAveney (Ed.), Springer, Berlin.

Bechhofer, R. E. and Tamhane, A. C. (1981). Incomplete block designs for comparing treatments with a control: General theory. *Technometrics* **23**, 45–57.

Bondar, J.V. (1983). Universal optimality of experimental designs: definitions and a criterion. *Canad. J. Statist.* **11**, 325–331.

David, O. (1994). Balanced block designs under interactive linear models. *J. Statist. Plann. Inference* **39**, 33–41.

Dawid, A. P. (1988). Symmetry models and hypotheses for structured data layouts. *J. Roy. Statist. Soc. Ser. B* **50**, 1–34.

Giovagnoli, A., Pukelsheim, F. and Wynn, H. P. (1987). Group invariant ordering and experimental designs. *J. Statist. Plann. Inference* **17**, 159–171.

Giovagnoli, A. and Wynn, H. P. (1985). G-majorization with applications to matrix orderings. *Linear Algebra Appl.* **67**, 111–135.

Giovagnoli, A. and Verdinelli, I. (1988). Comment to: Optimum designs for comparing test treatments with controls by A. S. Hedayat, M. Jacroux and D. Majumdar. *Statist. Sci.* **3**, 482–484.

Hauke, J. and Markiewicz, A. (1994). Properties of comparison criteria of normal experiments. In: *Proc. International Conference on Linear Statistical Inference, LINSTAT–93.*

Hedayat, A. S., Jacroux, M. and Majumdar, D. (1988). Optimum designs for comparing test treatments with controls (with discussion). *Statist. Sci.* **3**, 462-476.

Jacroux, M. (1986). On the usage of refined linear models for determining N-way classification designs which are optimal for comparing test treatments with a standard treatment. *Ann. Inst. Statist. Math.* **38**, 569–581.

Jones, B., Kunert, J. and Wynn, H. P. (1992). Information matrices for mixed effects models with applications to the optimality of repeated measurements designs. *J. Statist. Plann. Inference* **33**, 261–274.

Kiefer, J. (1975). Construction and optimality of generalized Youden designs. In: *A Survey of Statistical Designs and Linear Models.* J. N. Srivastava (Ed.), North-Holland, Amsterdam.

Kunert, J. (1983). Optimal design and refinement of the linear model with applications to repeated measurements designs. *Ann. Statist.* **11**, 247-257.

Markiewicz, A. (1994). Properties of information matrices for linear models and universal optimality of experimental designs. Technical Report, Department of Mathematical and Statistical Methods, Agricultural University of Poznań.

Marshall, A. W. and Olkin, I. (1979). *Inequalities: Theory of Majorization and Its Applications.* Academic, New York.

Mukhopadhyay, S. (1981). On the optimality of block designs under mixed effects model. *Cal. Statist. Assoc. Bull.* **30**, 171–185.

Peres, C. A. and Ting Hui-Ching. (1993). Optimum allocation for point estimation of the intraclass correlation coefficient. In: *Bulletin of the International Statistical Institute, 49th Session. Contributed Papers*, pp. 329–330,

Pukelsheim, F. (1987). Majorization ordering for linear regression designs. In: *Proceedings of the Second International Tampere Conference in Statistics*, pp. 261–274, T. Pukkila and S. Puntanen (Eds).

Pukelsheim, F. (1993). *Optimal Designs of Experiments.* Wiley, New York.

Shah, K. R. and Sinha, B. K. (1989). *Theory of Optimal Designs, Lect. Notes Statist.* **54**. Springer, Berlin-Heidelberg.

Styan, G. P. H. (1983). Schur complements and linear statistical models. In: *Proceedings of the First Tampere Seminar in Linear Models*, pp. 35–75, T. Pukkila and S. Puntanen (Eds).

Yurramendi, Y. (1993). Structural comparison of hierarchical classifications. In: *4th Conference of the Federation of Classification Societies, Collection of Abstracts*, pp. 298–299, Paris.

Bei Fragen zur Produktsicherheit wenden Sie sich bitte an:
If you have any questions regarding product safety,
please contact:

Walter de Gruyter GmbH
Genthiner Straße 13
10785 Berlin
productsafety@degruyterbrill.com